LONDON MATHEMATICAL SOCIETY LECTURE NOTE SERIES

Managing Editor: Professor Endre Süli, Mathematical Institute, University of Oxford, Woodstock Road, Oxford OX2 6GG, United Kingdom

The titles below are available from booksellers, or from Cambridge University Press at www.cambridge.org/mathematics

383 Motivic integration and its interactions with model theory and non-Archimedean geometry I, R. CLUCKERS, J. NICAISE & J. SEBAG (eds)
384 Motivic integration and its interactions with model theory and non-Archimedean geometry II, R. CLUCKERS, J. NICAISE & J. SEBAG (eds)
385 Entropy of hidden Markov processes and connections to dynamical systems, B. MARCUS, K. PETERSEN & T. WEISSMAN (eds)
386 Independence-friendly logic, A.L. MANN, G. SANDU & M. SEVENSTER
387 Groups St Andrews 2009 in Bath I, C.M. CAMPBELL et al (eds)
388 Groups St Andrews 2009 in Bath II, C.M. CAMPBELL et al (eds)
389 Random fields on the sphere, D. MARINUCCI & G. PECCATI
390 Localization in periodic potentials, D.E. PELINOVSKY
391 Fusion systems in algebra and topology, M. ASCHBACHER, R. KESSAR & B. OLIVER
392 Surveys in combinatorics 2011, R. CHAPMAN (ed)
393 Non-abelian fundamental groups and Iwasawa theory, J. COATES et al (eds)
394 Variational problems in differential geometry, R. BIELAWSKI, K. HOUSTON & M. SPEIGHT (eds)
395 How groups grow, A. MANN
396 Arithmetic differential operators over the p-adic integers, C.C. RALPH & S.R. SIMANCA
397 Hyperbolic geometry and applications in quantum chaos and cosmology, J. BOLTE & F. STEINER (eds)
398 Mathematical models in contact mechanics, M. SOFONEA & A. MATEI
399 Circuit double cover of graphs, C.-Q. ZHANG
400 Dense sphere packings: a blueprint for formal proofs, T. HALES
401 A double Hall algebra approach to affine quantum Schur–Weyl theory, B. DENG, J. DU & Q. FU
402 Mathematical aspects of fluid mechanics, J.C. ROBINSON, J.L. RODRIGO & W. SADOWSKI (eds)
403 Foundations of computational mathematics, Budapest 2011, F. CUCKER, T. KRICK, A. PINKUS & A. SZANTO (eds)
404 Operator methods for boundary value problems, S. HASSI, H.S.V. DE SNOO & F.H. SZAFRANIEC (eds)
405 Torsors, étale homotopy and applications to rational points, A.N. SKOROBOGATOV (ed)
406 Appalachian set theory, J. CUMMINGS & E. SCHIMMERLING (eds)
407 The maximal subgroups of the low-dimensional finite classical groups, J.N. BRAY, D.F. HOLT & C.M. RONEY-DOUGAL
408 Complexity science: the Warwick master's course, R. BALL, V. KOLOKOLTSOV & R.S. MACKAY (eds)
409 Surveys in combinatorics 2013, S.R. BLACKBURN, S. GERKE & M. WILDON (eds)
410 Representation theory and harmonic analysis of wreath products of finite groups, T. CECCHERINI-SILBERSTEIN, F. SCARABOTTI & F. TOLLI
411 Moduli spaces, L. BRAMBILA-PAZ, O. GARCÍA-PRADA, P. NEWSTEAD & R.P. THOMAS (eds)
412 Automorphisms and equivalence relations in topological dynamics, D.B. ELLIS & R. ELLIS
413 Optimal transportation, Y. OLLIVIER, H. PAJOT & C. VILLANI (eds)
414 Automorphic forms and Galois representations I, F. DIAMOND, P.L. KASSAEI & M. KIM (eds)
415 Automorphic forms and Galois representations II, F. DIAMOND, P.L. KASSAEI & M. KIM (eds)
416 Reversibility in dynamics and group theory, A.G. O'FARRELL & I. SHORT
417 Recent advances in algebraic geometry, C.D. HACON, M. MUSTAŢĂ & M. POPA (eds)
418 The Bloch–Kato conjecture for the Riemann zeta function, J. COATES, A. RAGHURAM, A. SAIKIA & R. SUJATHA (eds)
419 The Cauchy problem for non-Lipschitz semi-linear parabolic partial differential equations, J.C. MEYER & D.J. NEEDHAM
420 Arithmetic and geometry, L. DIEULEFAIT et al (eds)
421 O-minimality and Diophantine geometry, G.O. JONES & A.J. WILKIE (eds)
422 Groups St Andrews 2013, C.M. CAMPBELL et al (eds)
423 Inequalities for graph eigenvalues, Z. STANIĆ
424 Surveys in combinatorics 2015, A. CZUMAJ et al (eds)
425 Geometry, topology and dynamics in negative curvature, C.S. ARAVINDA, F.T. FARRELL & J.-F. LAFONT (eds)
426 Lectures on the theory of water waves, T. BRIDGES, M. GROVES & D. NICHOLLS (eds)
427 Recent advances in Hodge theory, M. KERR & G. PEARLSTEIN (eds)
428 Geometry in a Fréchet context, C.T.J. DODSON, G. GALANIS & E. VASSILIOU
429 Sheaves and functions modulo p, L. TAELMAN
430 Recent progress in the theory of the Euler and Navier–Stokes equations, J.C. ROBINSON, J.L. RODRIGO, W. SADOWSKI & A. VIDAL-LÓPEZ (eds)
431 Harmonic and subharmonic function theory on the real hyperbolic ball, M. STOLL
432 Topics in graph automorphisms and reconstruction (2nd Edition), J. LAURI & R. SCAPELLATO

433	Regular and irregular holonomic D-modules,	M. KASHIWARA & P. SCHAPIRA
434	Analytic semigroups and semilinear initial boundary value problems (2nd Edition),	K. TAIRA
435	Graded rings and graded Grothendieck groups,	R. HAZRAT
436	Groups, graphs and random walks,	T. CECCHERINI-SILBERSTEIN, M. SALVATORI & E. SAVA-HUSS (eds)
437	Dynamics and analytic number theory,	D. BADZIAHIN, A. GORODNIK & N. PEYERIMHOFF (eds)
438	Random walks and heat kernels on graphs,	M.T. BARLOW
439	Evolution equations,	K. AMMARI & S. GERBI (eds)
440	Surveys in combinatorics 2017,	A. CLAESSON et al (eds)
441	Polynomials and the mod 2 Steenrod algebra I,	G. WALKER & R.M.W. WOOD
442	Polynomials and the mod 2 Steenrod algebra II,	G. WALKER & R.M.W. WOOD
443	Asymptotic analysis in general relativity,	T. DAUDÉ, D. HÄFNER & J.-P. NICOLAS (eds)
444	Geometric and cohomological group theory,	P.H. KROPHOLLER, I.J. LEARY, C. MARTÍNEZ-PÉREZ & B.E.A. NUCINKIS (eds)
445	Introduction to hidden semi-Markov models,	J. VAN DER HOEK & R.J. ELLIOTT
446	Advances in two-dimensional homotopy and combinatorial group theory,	W. METZLER & S. ROSEBROCK (eds)
447	New directions in locally compact groups,	P.-E. CAPRACE & N. MONOD (eds)
448	Synthetic differential topology,	M.C. BUNGE, F. GAGO & A.M. SAN LUIS
449	Permutation groups and cartesian decompositions,	C.E. PRAEGER & C. SCHNEIDER
450	Partial differential equations arising from physics and geometry,	M. BEN AYED et al (eds)
451	Topological methods in group theory,	N. BROADDUS, M. DAVIS, J.-F. LAFONT & I. ORTIZ (eds)
452	Partial differential equations in fluid mechanics,	C.L. FEFFERMAN, J.C. ROBINSON & J.L. RODRIGO (eds)
453	Stochastic stability of differential equations in abstract spaces,	K. LIU
454	Beyond hyperbolicity,	M. HAGEN, R. WEBB & H. WILTON (eds)
455	Groups St Andrews 2017 in Birmingham,	C.M. CAMPBELL et al (eds)
456	Surveys in combinatorics 2019,	A. LO, R. MYCROFT, G. PERARNAU & A. TREGLOWN (eds)
457	Shimura varieties,	T. HAINES & M. HARRIS (eds)
458	Integrable systems and algebraic geometry I,	R. DONAGI & T. SHASKA (eds)
459	Integrable systems and algebraic geometry II,	R. DONAGI & T. SHASKA (eds)
460	Wigner-type theorems for Hilbert Grassmannians,	M. PANKOV
461	Analysis and geometry on graphs and manifolds,	M. KELLER, D. LENZ & R.K. WOJCIECHOWSKI
462	Zeta and L-functions of varieties and motives,	B. KAHN
463	Differential geometry in the large,	O. DEARRICOTT et al (eds)
464	Lectures on orthogonal polynomials and special functions,	H.S. COHL & M.E.H. ISMAIL (eds)
465	Constrained Willmore surfaces,	Á.C. QUINTINO
466	Invariance of modules under automorphisms of their envelopes and covers,	A.K. SRIVASTAVA, A. TUGANBAEV & P.A. GUIL ASENSIO
467	The genesis of the Langlands program,	J. MUELLER & F. SHAHIDI
468	(Co)end calculus,	F. LOREGIAN
469	Computational cryptography,	J.W. BOS & M. STAM (eds)
470	Surveys in combinatorics 2021,	K.K. DABROWSKI et al (eds)
471	Matrix analysis and entrywise positivity preservers,	A. KHARE
472	Facets of algebraic geometry I,	P. ALUFFI et al (eds)
473	Facets of algebraic geometry II,	P. ALUFFI et al (eds)
474	Equivariant topology and derived algebra,	S. BALCHIN, D. BARNES, M. KĘDZIOREK & M. SZYMIK (eds)
475	Effective results and methods for Diophantine equations over finitely generated domains,	J.-H. EVERTSE & K. GYŐRY
476	An indefinite excursion in operator theory,	A. GHEONDEA
477	Elliptic regularity theory by approximation methods,	E.A. PIMENTEL
478	Recent developments in algebraic geometry,	H. ABBAN, G. BROWN, A. KASPRZYK & S. MORI (eds)
479	Bounded cohomology and simplicial volume,	C. CAMPAGNOLO, F. FOURNIER-FACIO, N. HEUER & M. MORASCHINI (eds)
480	Stacks Project Expository Collection (SPEC),	P. BELMANS, W. HO & A.J. DE JONG (eds)
481	Surveys in combinatorics 2022,	A. NIXON & S. PRENDIVILLE (eds)
482	The logical approach to automatic sequences,	J. SHALLIT
483	Rectifiability: a survey,	P. MATTILA
484	Discrete quantum walks on graphs and digraphs,	C. GODSIL & H. ZHAN
485	The Calabi problem for Fano threefolds,	C. ARAUJO et al
486	Modern trends in algebra and representation theory,	D. JORDAN, N. MAZZA & S. SCHROLL (eds)
487	Algebraic combinatorics and the Monster group,	A.A. IVANOV (ed)
488	Maurer–Cartan methods in deformation theory,	V. DOTSENKO, S. SHADRIN & B. VALLETTE
489	Higher dimensional algebraic geometry,	C. HACON & C. XU (eds)
490	C^∞-algebraic geometry with corners,	K. FRANCIS-STAITE & D. JOYCE
491	Groups and graphs, designs and dynamics,	R.A. BAILEY, P.J. CAMERON & Y. WU (eds)
492	Homotopy theory of enriched Mackey functors,	N. JOHNSON & D. YAU

London Mathematical Society Lecture Note Series: 493

Surveys in Combinatorics 2024

Edited by

FELIX FISCHER
Queen Mary University of London

ROBERT JOHNSON
Queen Mary University of London

Shaftesbury Road, Cambridge CB2 8EA, United Kingdom

One Liberty Plaza, 20th Floor, New York, NY 10006, USA

477 Williamstown Road, Port Melbourne, VIC 3207, Australia

314–321, 3rd Floor, Plot 3, Splendor Forum, Jasola District Centre, New Delhi – 110025, India

103 Penang Road, #05–06/07, Visioncrest Commercial, Singapore 238467

Cambridge University Press is part of Cambridge University Press & Assessment, a department of the University of Cambridge.

We share the University's mission to contribute to society through the pursuit of education, learning and research at the highest international levels of excellence.

www.cambridge.org
Information on this title: www.cambridge.org/9781009490535
DOI: 10.1017/9781009490559

© Cambridge University Press & Assessment 2024

This publication is in copyright. Subject to statutory exception and to the provisions of relevant collective licensing agreements, no reproduction of any part may take place without the written permission of Cambridge University Press & Assessment.

When citing this work, please include a reference to the DOI 10.1017/9781009490559

First published 2024

A catalogue record for this publication is available from the British Library

ISBN   978-1-009-49053-5   Paperback

Cambridge University Press & Assessment has no responsibility for the persistence or accuracy of URLs for external or third-party internet websites referred to in this publication and does not guarantee that any content on such websites is, or will remain, accurate or appropriate.

Contents

	Preface *Felix Fischer and Robert Johnson*	*page* vii
1	**Intersection Theory of Matroids: Variations on a Theme** *Federico Ardila-Mantilla*	1
2	**Erdős Covering Systems** *Paul Balister*	31
3	**The Cluster Expansion in Combinatorics** *Matthew Jenssen*	55
4	**Sublinear Expanders and Their Applications** *Shoham Letzter*	89
5	**Transversals in Latin Squares** *Richard Montgomery*	131
6	**Finite Field Models in Arithmetic Combinatorics – Twenty Years On** *Sarah Peluse*	159
7	**The Slice Rank Polynomial Method – A Survey a Few Years Later** *Lisa Sauermann*	201
8	**An Introduction to Transshipments Over Time** *Martin Skutella*	239
9	**Oriented Trees and Paths in Digraphs** *Maya Stein*	271

Preface

The thirtieth British Combinatorial Conference takes place at Queen Mary University of London, from Monday 1st to Friday 5th of July 2024. The British Combinatorial Committee invited nine distinguished combinatorialists to give survey lectures in areas of their expertise, and this volume contains the survey articles on which these lectures are based. We hope this collection of surveys will be a useful resource to bring researchers and graduate students up to date with the latest developments in some key areas of extremal, probabilistic and additive combinatorics, as well as graph theory and combinatorial optimisation.

In compiling this volume, we are indebted to the authors for preparing their articles so accurately and professionally, and to the referees for their rapid responses and keen eye for detail. We would also like to thank Holly Paveling and Anna Scriven at Cambridge University Press. Finally, we thank the editors of previous Surveys and the British Combinatorial Committee for helpful guidance.

Intersection Theory of Matroids:
Variations on a Theme

Federico Ardila–Mantilla

Abstract

Chow rings of toric varieties, which originate in intersection theory, feature a rich combinatorial structure of independent interest. We survey four different ways of computing in these rings, due to Billera, Brion, Fulton Sturmfels, and Allermann Rau. We illustrate the beauty and power of these methods by giving four proofs of Huh and Huh Katz's formula $\mu^k(\mathsf{M}) = \deg_{\mathsf{M}}(\alpha^{r-k}\beta^k)$ for the coefficients of the reduced characteristic polynomial of a matroid M as the mixed intersection numbers of the hyperplane and reciprocal hyperplane classes α and β in the Chow ring of M. Each of these proofs sheds light on a different aspect of matroid combinatorics, and provides a framework for further developments in the intersection theory of matroids.

Our presentation is combinatorial, and does not assume previous knowledge of toric varieties, Chow rings, or intersection theory. This survey was prepared for the Clay Lecture to be delivered at the 2024 British Combinatorics Conference.

1 Introduction

Our starting point is the *chromatic polynomial* $\chi_G(t)$ of a graph $G = (V, E)$. For a positive integer q,

$$\chi_G(q) := \text{ number of proper vertex-colorings of G with } q \text{ colors,}$$

where a coloring is *proper* if no two neighboring vertices have the same color. For example, the chromatic polynomial of the graph below is $\chi_G(q) = q(q-1)^2(q-2)$.

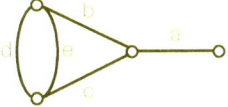

Figure 1: A graph G with $\chi_G(q) = q^4 - 4q^3 + 5q^2 - 2q$. and $\mu^0 = 1, \mu^1 = 3, \mu^2 = 2$.

More generally, the *characteristic polynomial* $\chi_{\mathsf{M}}(t)$ of a matroid $\mathsf{M} = (E, r)$ is

$$\chi_{\mathsf{M}}(q) := \sum_{A \subseteq E} (-1)^{|A|} q^{r-r(A)}. \tag{1.1}$$

It is one of the most important invariants of a matroid; it is introduced in detail in Section 3 and [4, Sections 6, 7]. The characteristic polynomial generalizes the chromatic polynomial in the sense that if $\mathsf{M}(G)$ is the cycle matroid of a graph G that has c connected components, then $\chi_G(q) = q^c \chi_{\mathsf{M}(G)}(q)$. This polynomial is a multiple of $q - 1$, and we define the *reduced characteristic polynomial* of M to be

$$\overline{\chi}_{\mathsf{M}}(q) := \frac{\chi_{\mathsf{M}}(q)}{q-1} = \mu^0 q^r - \mu^1 q^{r-1} + \cdots + (-1)^r \mu^r q^0$$

where $r+1$ is the rank of M. In the example above, $\overline{\chi}_{\mathsf{M}}(q) = q^2 - 3q + 2$.

It is not too difficult to prove recursively that the numbers $\mu^0, \mu^1, \ldots, \mu^r$ are non-negative. A combinatorialist then asks: Do they count something? An algebraic combinatorialist then asks: Do they have an algebraic, geometric, or topological interpretation? Such questions often give rise to a deeper understanding of the objects under study. In this case, and in numerous others, they lead to proofs of long-standing conjectures for which no purely combinatorial proof is known.

1.1 Theme

Our recurring theme will be Huh [31] and Huh-Katz [34]'s remarkable interpretation of $\mu^0, \ldots, \mu^r$. Beautiful in its own right, their Theorem 1.1 also lies at the heart of the celebrated proof of the conjecture that this sequence is log-concave [1].

Let M be a matroid of rank $r+1$ on a set E with $n+1$ elements. The *Chow ring* $A(\mathsf{M})$ is the $\mathbb{Z}$-algebra generated by variables x_F for each non-empty proper flat, with relations

$$x_F x_G = 0 \quad \text{for any flats } F, G \text{ such that } F \subsetneq G \text{ and } F \not\supseteq G,$$
$$\sum_{F \ni i} x_F = \sum_{F \ni j} x_F \quad \text{for any elements } i, j \in E.$$

One can show that the Chow ring is graded $A(\mathsf{M}) = A^0(\mathsf{M}) \oplus \cdots \oplus A^r(\mathsf{M})$, and that there is a canonical isomorphism $\deg_\mathsf{M} : A^r(\mathsf{M}) \xrightarrow{\sim} \mathbb{R}$ called the *degree map* [1].

Consider the following two elements of $A^1(\mathsf{M})$, which we call the *hyperplane* and *reciprocal hyperplane* classes:

$$\alpha = \alpha_i = \sum_{i \in F} x_F, \qquad \beta = \beta_i = \sum_{i \notin F} x_F.$$

One readily verifies that they do not depend on i.

> **Theorem 1.1** *Let M be a matroid of rank $r+1$. Let α, β be the hyperplane and reciprocal hyperplane classes in the Chow ring $A(\mathsf{M})$. Then*
> $$\deg_\mathsf{M}(\alpha^{r-k} \beta^k) = \mu^k(\mathsf{M}) \qquad \text{for } 0 \leq k \leq r.$$

The Chow ring $A(\mathsf{M})$ has remarkable Hodge-theoretic properties [1] surveyed in [33, 5, 13, 25]. In particular, $A(\mathsf{M})$ satisfies the *Hodge-Riemann relations*, which give

$$\deg_\mathsf{M}(\ell_1 \ell_2 \ell_3 \cdots \ell_d)^2 \geq \deg_\mathsf{M}(\ell_1 \ell_1 \ell_3 \cdots \ell_d) \deg_\mathsf{M}(\ell_2 \ell_2 \ell_3 \cdots \ell_d),$$

for any $\ell_1, \ell_2, \ldots, \ell_d$ in a certain cone $\mathcal{K}(\mathsf{M}) \subseteq A^1(\mathsf{M})$ whose closure contains α and β. In light of Theorem 1.1, this proves the following inequalities conjectured by Rota [46], Heron [30], and Welsh [50] in the 1970s:

$$(\mu^k)^2 \geq \mu^{k+1} \mu^{k-1} \qquad \text{for } 1 \leq k \leq r-1.$$

This survey focuses on the combinatorial aspects of this program:

Question 1.2 *How does one discover and prove combinatorially interesting formulas in Chow rings like Theorem 1.1?*

This question fits within the framework of intersection theory of toric varieties, in ways that can be understood combinatorially. The Chow ring $A(X_\Sigma)$ of a toric variety X_Σ corresponding to a rational polyhedral fan Σ is a beautifully rich object that can be understood from several different points of view. We will present four, due to Billera, Brion, Fulton–Sturmfels, and Allermann–Rau [16, 18, 28, 3]. Each of these points of view gives us a different ways to compute in a Chow ring, and teaches us different things about the objects at hand. This machinery is relevant to Theorem 1.1 because the Chow ring of a matroid M equals the Chow ring of the toric variety X_{Σ_M} and is closely related to the permutahedral toric variety X_{Σ_E}, where Σ_E and Σ_M are the *matroid fan* of M and the *braid fan* of E, discussed in detail in Sections 2.0 and 3.0.

Our presentation will be combinatorial, and will not assume previous knowledge of toric varieties, Chow rings, or intersection theory. A familiarity with the basics of enumerative matroid theory will be helpful; see for example [4, 17, 44].

This survey is organized as follows. In Section 2 we discuss the general intersection theory of simplicial rational fans Σ and toric varieties X_Σ, giving four different combinatorial points of view on the Chow ring $A(\Sigma) = A(X_\Sigma)$. We pay special attention to the Chow ring of the braid fan Σ_E for a finite set E. In Section 3 we discuss some basic aspects of the intersection theory of matroids. The general theory gives us four different ways to think about the Chow ring of a matroid M. We illustrate each one of these approaches by using it to give a different proof of Theorem 1.1.

2 Intersection Theory of Toric Varieties: A Case Study

Intersection theory studies how subvarieties of an algebraic variety X intersect. For example, Bezout's theorem tells us that two generic plane curves of degrees m and n intersect at mn points. We want a robust theory that will keep track of multiplicities correctly, and where the answer to such intersection questions does not change under rational equivalence. The Chow ring $A(X)$ provides an algebraic framework to carry out such computations. Because this ring encodes the answers to very subtle questions, it is generally an unwieldy object.

The situation is much better behaved when $X = X_\Sigma$ is the toric variety associated to a simplicial rational fan Σ. In this case, the Chow ring $A(X_\Sigma)$ can be described entirely in terms of the fan Σ in several ways. This leads to algebraic, geometric, and combinatorial methods for computing in $A(X_\Sigma)$, and to combinatorial results of independent interest. Those methods and results are the subjects of this survey.

Let $N_\mathbb{Z} \cong \mathbb{Z}^n$ be a lattice and $N = \mathbb{R} \otimes N_\mathbb{Z} \cong \mathbb{R}^n$ the corresponding real vector space. A *rational cone* $\{\lambda_1 \mathbf{v}_1 + \cdots + \lambda_k \mathbf{v}_k : \lambda_1, \ldots, \lambda_k \geq 0\}$ is a cone in N generated by finitely many lattice vectors $\mathbf{v}_1, \ldots, \mathbf{v}_k \in N_\mathbb{Z}$; it is *strongly convex* if it contains no lines. A *rational fan* Σ in N is a set of strongly convex rational cones that are glued along common faces; that is, any face of a cone in Σ is a cone in Σ, and the intersection of any two cones in Σ is a cone in Σ. We say a fan Σ is *simplicial* if every d-dimensional cone is generated by d vectors, *unimodular* if those d vectors always form a basis for $N_\mathbb{Z}$, and *complete* if the union of the cones in Σ is all of N. We say Σ is *pure* if all maximal cones have the same dimension, and write $\Sigma(d)$ for the set of d-dimensional cones. A rational fan Σ in N determines a toric variety $X = X_\Sigma$;

for details see [20, 27].

The goal of this section is to explain the following theorem. After explaining each of its parts, we use it to compute explicitly the Chow ring of the two-dimensional braid fan.

Theorem 2.1 *Let Σ be a complete simplicial rational fan in $N = \mathbb{R} \otimes N_{\mathbb{Z}}$. The following rings are isomorphic:*

1. *The quotient $A(\Sigma) = S(\Sigma)/(I(\Sigma) + J(\Sigma))$ where*

$$\begin{aligned}
S(\Sigma) &= \mathbb{Q}[x_\rho : \rho \text{ is a ray of } \Sigma]/(I(\Sigma) + J(\Sigma)), \\
I(\Sigma) &= \langle x_{\rho_1} \cdots x_{\rho_k} : \rho_1, \ldots, \rho_k \text{ do not generate a cone of } \Sigma \rangle, \\
J(\Sigma) &= \langle \sum_{\rho \text{ ray of } \Sigma} \ell(\mathbf{e}_\rho) x_\rho : \ell \text{ is a linear function on } N \rangle.
\end{aligned}$$

2. *The ring $\mathrm{PP}(\Sigma)/\langle N^\vee \rangle$ of piecewise polynomial functions on Σ modulo the ideal generated by the space $N^\vee$ of (global) linear functions on N.*

3. *The ring $\mathrm{MW}(\Sigma)$ of Minkowski weights on Σ under stable intersection.*

4. *The ring $\mathrm{MW}(\Sigma)$ of Minkowski weights on Σ under tropical intersection.*

5. *The cohomology ring of the toric variety $X(\Sigma)$.*

6. *The Chow ring of the toric variety $X(\Sigma)$.*

When Σ is unimodular, these isomorphisms also hold over $\mathbb{Z}$.

2.0 The Braid Fan

For a finite set E, we let $\{\mathbf{e}_i : i \in E\}$ be the standard basis of $\mathbb{R}^E$, and we write

$$\mathbf{e}_S := \sum_{s \in S} \mathbf{e}_s \qquad \text{for } S \subseteq E.$$

The fans considered in this paper will live in $N_E := \mathbb{R}^E / \mathbb{R} \mathbf{e}_E$. The image of $\mathbf{e}_S \in \mathbb{R}^E$ in this quotient will also be denoted $\mathbf{e}_S \in N_E$. We will often consider $E = [0, n] := \{0, 1, \ldots, n\}$.

Definition 2.2 Let E be a finite set. The *braid fan* Σ_E in $N_E := \mathbb{R}^E/\mathbb{R}\mathbf{e}_E$ has

- rays: $\mathbf{e}_S$ for the nonempty proper subsets $\emptyset \subsetneq S \subsetneq E$
- cones: $\sigma_S = \mathrm{cone}(\mathbf{e}_{S_1}, \ldots, \mathbf{e}_{S_k})$ for the flags $\mathcal{S} = (\emptyset \subsetneq S_1 \subsetneq \cdots \subsetneq S_k \subsetneq E)$

The braid fan is the decomposition of N_E determined by the *braid arrangement* in N_E, which consists of the hyperplanes $t_i = t_j$ for $i, j \in E$. If $|E| = n+1$, the braid fan Σ_E is n-dimensional, and has a facet $\sigma_{\mathcal{S}} = \sigma_\pi = \{\mathbf{t} \in N_E : t_{s_0} \geq t_{s_1} \geq \cdots \geq t_{s_n}\}$ for each complete flag $\mathcal{S} = (\emptyset \subsetneq \{s_0\} \subsetneq \cdots \subsetneq \{s_0, s_1, \ldots, s_{n-1}\} \subsetneq E)$, or equivalently, each bijection $\pi : [0, n] \to E$ given by $\pi(i) = s_i$. Slightly abusing terminology, we will call π a *permutation* of E and write $\pi = s_0 \ldots, s_n$. It follows that the braid fan is complete, simplicial, and unimodular.

Figure 2 shows the braid fan Σ_E for $E = [0, 2] = \{0, 1, 2\}$. It is the complete fan in N_E cut out by the braid arrangement consisting of the lines $t_0 = t_1, t_1 = t_2$, and $t_2 = t_0$ in N_E.

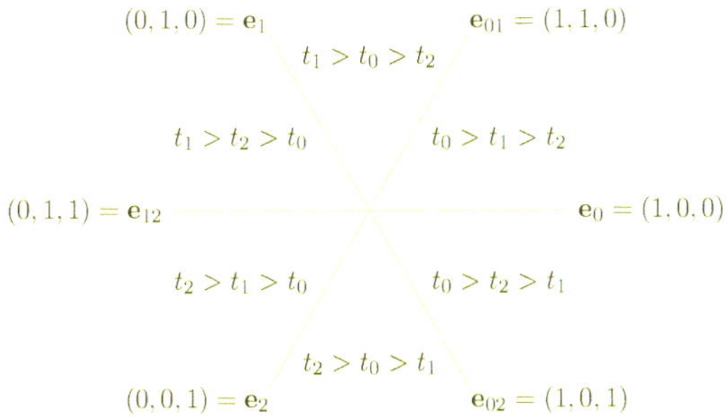

<pre>
            (0,1,0) = e₁   e₀₁ = (1,1,0)
                         t₁ > t₀ > t₂

                  t₁ > t₂ > t₀   t₀ > t₁ > t₂

(0,1,1) = e₁₂ ──────────────────────────── e₀ = (1,0,0)

                  t₂ > t₁ > t₀   t₀ > t₂ > t₁

                         t₂ > t₀ > t₁
            (0,0,1) = e₂   e₀₂ = (1,0,1)
</pre>

Figure 2: The braid fan $\Sigma_{[0,2]}$.

We will return to this picture many times in what follows; the reader may wish to keep it within reach. We will call its toric variety and Chow ring the *permutahedral variety* and the *permutahedral Chow ring*.

2.1 The Chow Ring as a Quotient of a Polynomial Ring

For the remainder of Section 2, Σ will be a simplicial rational fan in $N = \mathbb{R} \otimes N_{\mathbb{Z}}$.

The Chow Ring The *Chow ring* of Σ is the graded algebra

$$A(\Sigma) := S(\Sigma)/(I(\Sigma) + J(\Sigma)),$$

where

$$\begin{aligned}
S(\Sigma) &= \mathbb{Z}[x_\rho : \rho \text{ is a ray of } \Sigma]/(I(\Sigma) + J(\Sigma)), \\
I(\Sigma) &= \langle x_{\rho_1} \cdots x_{\rho_k} : \rho_1, \ldots, \rho_k \text{ do not generate a cone of } \Sigma \rangle, \\
J(\Sigma) &= \langle \sum_{\rho \text{ ray of } \Sigma} \ell(e_\rho) x_\rho : \ell \text{ is a linear function on } N \rangle.
\end{aligned}$$

The ideal $I(\Sigma)$ is called the *Stanley-Reisner ideal* of Σ and $S(\Sigma)/I(\Sigma)$ is called its *Stanley-Reisner ring*. In $J(\Sigma)$, it is sufficient to let ℓ range over a basis of the space $N^\vee$ of linear functions on N.

Example 2.3 (The Chow ring $A(\Sigma_{[0,2]})$.) Let us compute the Chow ring of the braid fan Σ_E for $E = [0, 2]$. We have

$$\begin{aligned}
S(\Sigma_E) &= \mathbb{R}[x_0, x_1, x_2, x_{01}, x_{02}, x_{12}] \\
I(\Sigma_E) &= \langle x_i x_j : i \neq j \rangle + \langle x_i x_{jk} : i, j, k \text{ distinct} \rangle + \langle x_{ij} x_{jk} : i, j, k \text{ distinct} \rangle \\
J(\Sigma_E) &= \langle (x_0 + x_{02}) - (x_1 + x_{12}), (x_0 + x_{01}) - (x_2 + x_{12}) \rangle
\end{aligned}$$

where we use $t_0 - t_1$ and $t_0 - t_2$ as a basis for $N^\vee$ in the description of $J(\Sigma_E)$.

We claim that $A = A(\Sigma_E)$ has degree 2 and
$$A^0 = \mathbb{Z}\{1\} \cong \mathbb{Z}^1, \qquad A^1 = \mathbb{Z}\{x_0, x_1, x_2, x_{12}\} \cong \mathbb{Z}^4, \qquad A^2 = \mathbb{Z}\{x_0 x_{01}\} \cong \mathbb{Z}.$$

The description of A^0 is clear. The description of A^1 follows from the two linear relations in $J(\Sigma_E)$ that express x_{01} and x_{02} in terms of the four chosen generators. To compute A^2, notice that
$$x_0^2 = x_0(x_2 + x_{12} - x_{01}) = -x_0 x_{01}, \qquad x_{01}^2 = x_{01}(x_2 + x_{12} - x_0) = -x_0 x_{01},$$
and similarly for the squares of the other terms $x_i x_{ij}$. This implies that
$$-x_0^2 = -x_1^2 = -x_2^2 = -x_{01}^2 = -x_{02}^2 = -x_{12}^2 = x_i x_{ij} \text{ for all } i \neq j. \tag{2.1}$$

Thus A^2 is indeed generated by $x_0 x_{01}$, and we have an isomorphism
$$\deg : A^2 \simeq \mathbb{Z}, \qquad \deg(x_i x_{ij}) = 1 \text{ for all facets } \sigma_{i \subset ij} \text{ of } \Sigma_{[0,2]}.$$

Any monomial of degree 3 can be reduced via (2.1) to a square free monomial of degree 3, which is in $I(\Sigma_E)$ and hence vanishes in $A(\Sigma_E)$.

Computing Degrees When Σ is complete, the Chow ring $A(\Sigma)$ is graded of degree n, and there is a canonical *degree map* $\deg : A^n(\Sigma) \simeq \mathbb{Z}$. If Σ is unimodular, this map is characterized by the property that the degree of any facet monomial is 1: $\deg(x_\sigma) = 1$ for any facet σ, where $x_\sigma = \prod_{\rho \text{ ray}} x_\rho$. Any $f \in A^n(\Sigma)$ can be expressed as a linear combination of facet monomials [1, Prop. 5.5], and this expression gives the degree of f.

The Hyperplane and Reciprocal Hyperplane Classes We will pay special attention to two special elements α, β in the degree one piece $A^1(\Sigma_E)$ of the permutahedral Chow ring:
$$\alpha := \alpha_i = \sum_{i \in S} x_S, \qquad \beta := \beta_i = \sum_{i \notin S} x_S, \qquad \text{for } i \in E.$$

We invite the reader to check that these do not depend on the choice of $i \in E$.

Example 2.4 (The degree of $\alpha\beta$ in $A(\Sigma_{[0,2]})$.) For $E = [0,2]$ we have
$$\begin{aligned} \alpha &= \alpha_0 = x_0 + x_{01} + x_{02} & \beta &= \beta_0 = x_1 + x_2 + x_{12} \\ &= \alpha_1 = x_1 + x_{01} + x_{12} & &= \beta_1 = x_0 + x_2 + x_{02} \\ &= \alpha_2 = x_2 + x_{02} + x_{12} & &= \beta_2 = x_0 + x_1 + x_{01}. \end{aligned}$$

Let us compute the intersection degree of α and β. Using the relations in the Chow ring, we can write
$$\alpha\beta = \alpha_0 \beta_0 = (x_0 + x_{01} + x_{02})(x_1 + x_2 + x_{12}) = x_1 x_{01} + x_2 x_{02}.$$

This implies that
$$\deg(\alpha\beta) = 2.$$

Note that a different choice of representatives, such as $\alpha_0 \beta_1$, leads to a more complicated computation.

2.2 The Chow Ring in Terms of Piecewise Polynomials

The Chow Ring A *piecewise polynomial* on Σ is a continuous function on N whose restriction to each cone in Σ agrees with a polynomial function. Let $PP(\Sigma)$ be the ring of piecewise polynomials on Σ, with pointwise addition and multiplication. Let $\langle N^\vee \rangle$ be the ideal of $PP(\Sigma)$ generated by the set $N^\vee$ of (global) linear functions on N. Thanks to work of Billera [16], the Chow ring of Σ can be described as:

$$A(\Sigma) \cong PP(\Sigma)/\langle N^\vee \rangle.$$

The Dictionary Billera [16] constructed an isomorphism from the Stanley-Reisner ring $S(\Sigma)/I(\Sigma)$ to the algebra $PP(\Sigma)$ of continuous piecewise polynomial functions on Σ, by identifying the variable x_ρ with the piecewise linear *Courant function* on Σ determined by the condition

$$x_\rho(\mathbf{e}_{\rho'}) = \begin{cases} 1, & \text{if } \rho \text{ is equal to } \rho', \\ 0, & \text{if } \rho \text{ is not equal to } \rho', \end{cases} \quad \text{for each ray } \rho \text{ of } \Sigma.$$

Conversely, this isomorphism identifies a piecewise linear function $\ell \in PP(\Sigma)$ on Σ with the linear form

$$\ell = \sum_{\rho \text{ ray}} \ell(\mathbf{e}_\rho) x_\rho,$$

and allows us to regard the elements of $A(\Sigma)$ as equivalence classes of piecewise polynomial functions on Σ, modulo the linear functions on Σ.

Example 2.5 (The ring $A(\Sigma_{[0,2]})$) Let us carry out this computation for the braid arrangement $\Sigma_{[0,2]}$, referring to Figure 2. The Courant functions representing the ray variables $x_0, x_1, x_2, x_{01}, x_{02}, x_{12}$ of the previous section are the following, where $t_{ij} := t_i - t_j$:

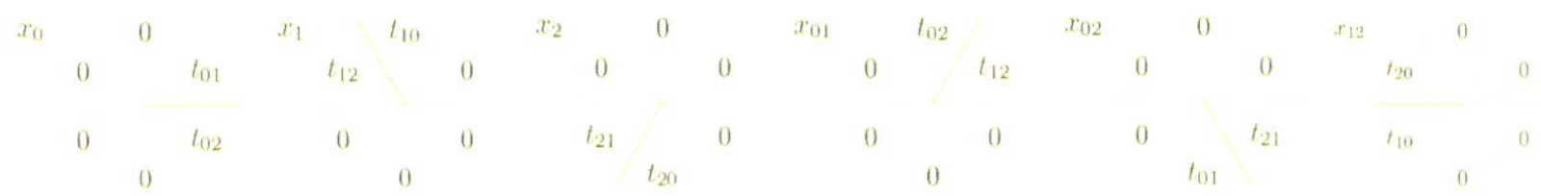

Figure 3: The Courant functions $x_0, x_1, x_2, x_{01}, x_{02}, x_{12}$ on $\Sigma_{[0,2]}$. Each function x_S equals 1 on the marked primitive ray $\mathbf{e}_S$ and 0 on the others.

As we saw in the previous section, A^0 is generated by the constant function 1, A^1 is generated by x_0, x_1, x_2, x_{01}, and A^2 is generated by

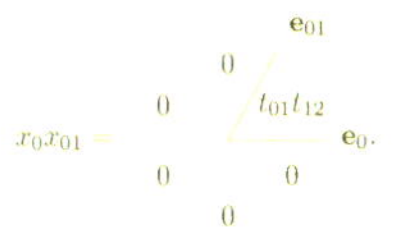

This expression for the generator $x_0 x_{01}$ is supported on the chamber cone$\{\mathbf{e}_0, \mathbf{e}_{01}\} = \{\mathbf{t} \in N_E : t_0 > t_1 > t_2\}$. Its unique non-zero polynomial $t_{01} t_{12}$ is the product of the linear forms $t_0 - t_1$ and $t_1 - t_2$ defining the inequalities of the chamber.

It is instructive to double check that two adjacent chambers (and hence any two chambers) give the same generator of A^2. The neighbor chamber cone$\{e_0, e_{02}\} = \{t \in N_E : t_0 > t_2 > t_1\}$ separated by the wall $t_{12} = 0$, gives generator $x_0 x_{02}$. Their difference is

$$x_0 x_{01} - x_0 x_{02} = \begin{matrix} 0 \\ 0 \quad t_{01} t_{12} \\ \overline{\quad \quad \quad} \\ 0 \quad -t_{02} t_{21} \\ 0 \end{matrix} = t_{12} \cdot \begin{matrix} 0 \\ 0 \quad t_{01} \\ \overline{\quad \quad \quad} = 0 \\ 0 \quad t_{02} \\ 0 \end{matrix}$$

since it is the product of the linear function t_{12} of the wall separating them and a piecewise polynomial function: we have $t_{01} = 0$ on e_{01}, $t_{01} = t_{02}$ on e_0, and $t_{02} = 0$ on e_{02}. This generalizes to any two neighbor chambers in any braid fan, and further, in any simplicial rational fan.

Computing Degrees There is a very elegant way to compute the degree of an element $f \in A^n(\Sigma)$ given by a piecewise polynomial $f = (f_\sigma : \sigma \in \Sigma(n))$. To describe it, we first associate a rational function to each facet σ of Σ. If σ is simplicial and unimodular, it is generated by n inequalities $f_1(x) \geq 0, \ldots, f_n(x) \geq 0$, where $\{f_1, \ldots, f_n\}$ is the basis dual to the rays generating σ. This determines a rational function $e_\sigma := 1/(f_1 \cdots f_n)$ in $\mathrm{Sym}^\pm(N^\vee)$. In general, we can triangulate σ into simplicial unimodular cones $\sigma_1, \ldots, \sigma_n$ and define $e_\sigma := e_{\sigma_1} + \cdots + e_{\sigma_n}$, which turns out to be independent of the triangulation [19]. We then have

$$\deg(f) = \sum_{\sigma \in \Sigma(n)} e_\sigma f_\sigma.$$

It is pleasant and not a priori obvious that this is always a constant, after significant cancellation. It is not so difficult to prove it, though, by verifying that the above formula gives $\deg(x_\sigma) = 1$ for every facet monomial and 0 for every other square-free monomial.

The Hyperplane and Reciprocal Hyperplane Classes The elements α and β of the permutahedral Chow ring can be described by the following piecewise linear functions, for any $i \in E$:

$$\alpha = \alpha_i = \max(t_i - t_j : j \in E), \qquad \beta = \beta_i = \max(t_j - t_i : j \in E).$$

For any $i \neq i'$ the function $\alpha_i - \alpha_{i'} = t_i - t_{i'}$ is linear, and hence in $N^\vee$, so α is well-defined.[1] To verify the formula for α_i, notice that the value of $\max(t_i - t_j : j \in E)$ on e_S is 1 if $i \in S$ and 0 if $i \notin S$. A similar argument works for β.

Example 2.6 (The degree of $\alpha\beta$ in $A(\Sigma_{[0,2]})$.) The special element $\alpha \in A(\Sigma_E)$ is given by the expressions $\alpha_0 = x_0 + x_{01} + x_{02}$, $\alpha_1 = x_1 + x_{01} + x_{12}$, and $\alpha_2 = x_2 + x_{02} + x_{12}$, which give:

---

[1] It is tempting but incorrect to think that t_i is linear so we can write $\alpha = max(-t_j : j \in E)$: in fact t_i is not even a well defined function on the ambient space $N = \mathbb{R}^{[0,2]}/\mathbb{R}\,e_{[0,2]}$.

$$\alpha = \begin{matrix} & & t_{02} & & \\ 0 & & & & 0 \\ & t_{02} & & t_{01} & \\ & & t_{01} & & \end{matrix} = \begin{matrix} & & t_{12} & & \\ 0 & t_{10} & & t_{12} & \\ & t_{10} & & 0 & \\ & & 0 & & \end{matrix} = \begin{matrix} & & 0 & & \\ 0 & t_{20} & & 0 & \\ & t_{20} & & t_{21} & \\ & & t_{21} & & \end{matrix}.$$

These look different, but they are equal modulo global linear functions on N: the first two differ by t_{01} and the latter two differ by t_{12}. Similarly, there are three natural piecewise linear representatives for β, namely $\beta_0, \beta_1, \beta_2$. Let's compute the degree of $\alpha\beta$ in two ways, referring to Figure 2 again. Since

$$\alpha_0 \beta_0 = \begin{matrix} & & t_{02} & & \\ 0 & t_{02} & & t_{10} & 0 \\ & t_{01} & & t_{20} & \\ & & t_{01} & t_{20} & \end{matrix} = \begin{matrix} & & t_{02}t_{10} & & \\ 0 & & & 0 & \\ & 0 & & 0 & \\ & & t_{01}t_{20} & & \end{matrix} \quad \alpha_1\beta_0 = \begin{matrix} & & t_{10}t_{12} & & \\ & t_{10}^2 & & 0 & \\ & t_{10}t_{20} & & & \\ & & 0 & & \end{matrix}$$

we have that $\deg(\alpha\beta)$ equals

$$\frac{t_{02}t_{10}}{t_{02}t_{10}} + \frac{t_{01}t_{20}}{t_{01}t_{20}} = 2 \text{ and } \frac{t_{10}t_{12}}{t_{02}t_{10}} + \frac{t_{10}^2}{t_{12}t_{20}} + \frac{t_{10}t_{20}}{t_{21}t_{10}} = 2,$$

where the first computation is immediate and the second involves a fun cancellation.

2.3 The Chow Ring in Terms of Minkowski Weights

The Chow Ring A k-*dimensional Minkowski weight* on Σ is a real-valued function ω on the set $\Sigma(k)$ of k-dimensional cones that satisfies the *balancing condition*: For every $(k-1)$-dimensional cone τ in Σ,

$$\sum_{\tau \subset \sigma} \omega(\sigma)\mathbf{e}_{\sigma/\tau} = 0 \text{ in the quotient space } N/\operatorname{span}(\tau),$$

where $\mathbf{e}_{\sigma/\tau}$ is the primitive generator of the ray $(\sigma + \operatorname{span}(\tau))/\operatorname{span}(\tau)$. We say that w is *positive* if $w(\sigma)$ is positive for every σ in $\Sigma(k)$. We write $\operatorname{MW}_k(\Sigma)$ for the space of k-dimensional Minkowski weights on Σ, and set $\operatorname{MW}(\Sigma) = \bigoplus_{k \geq 0} \operatorname{MW}_k(\Sigma)$.

The product in $\operatorname{MW}(\Sigma)$ is given by the following *fan displacement rule*. If X_1 and X_2 are Minkowski weights of codimension k and ℓ on Σ, then their product is defined to be the *stable intersection*

$$X_1 \cdot X_2 := \lim_{\epsilon \to 0} X_1 \cdot (X_2 + \epsilon \mathbf{v})$$

for any vector $\mathbf{v} \in N$ such that X_1 and $X_2 + \epsilon \mathbf{v}$ intersect transversally for sufficiently small $\epsilon > 0$. The facets of $X_1 \cdot X_2$ are the $(k+\ell)$-codimensional intersections of a facet of X_1 and a facet of X_2. The weight of a facet τ of $X_1 \cdot X_2$ is

$$w(\tau) = \sum_{\sigma_1, \sigma_2} w(\sigma_1) w(\sigma_2) [\mathbb{Z}^n : L_\mathbb{Z}(\sigma_1) + L_\mathbb{Z}(\sigma_2)],$$

summing over the facets σ_1 and σ_2 of X_1 and X_2 respectively such that $\tau = \sigma_1 \cap \sigma_2$ and $\sigma_1 \cap (\sigma_2 + \epsilon \mathbf{v}) \neq 0$ for small $\epsilon > 0$. It is non-trivial that the construction above is independent of the choice of a (generic) vector $\mathbf{v}$, and that it is also a Minkowski weight, that is, it satisfies the balancing condition [28, 35].

When Σ is complete, Fulton and Sturmfels [28] proved that

$$A(\Sigma) \cong \mathrm{MW}(\Sigma)$$

so understanding the Chow ring of Σ is equivalent to understanding Minkowski weights on Σ and their stable intersections.

The Dictionary For Σ complete, Katz and Payne [37] described the canonical[2] map from $\mathrm{PP}(\Sigma)$ to $\mathrm{MW}(\Sigma)$ that descends to an isomorphism $A^k(\Sigma) \cong \mathrm{MW}_{n-k}(\Sigma)$. We focus on a different description for a special case: when $f \in \mathrm{PP}^1(\Sigma)$ is a piecewise linear function that is convex, that is, $f((\mathbf{x}+\mathbf{y})/2) \leq (f(\mathbf{x})+f(\mathbf{y}))/2$ for all $\mathbf{x}, \mathbf{y} \in \mathrm{N}$. In this case, f can be written as a *tropical polynomial*; that is, the maximum of a finite number of linear functions:

$$f(\mathbf{x}) = \max\{\mathbf{v}_1(\mathbf{x}), \ldots, \mathbf{v}_m(\mathbf{x})\} \qquad \text{for } \mathbf{v}_1, \ldots, \mathbf{v}_m \in \mathrm{N}^\vee.$$

The *corner locus*, where this function is not linear, is the *tropical hypersurface*:

$$\operatorname{trop} f = \{\mathbf{x} \in \mathrm{N} : \max_{1 \leq i \leq m}\{\mathbf{v}_i(\mathbf{x})\} \text{ is achieved at least twice}\}.$$

This is the $(n-1)$-skeleton of the normal fan of the *Newton polytope* $\mathrm{Newt}(f) = \mathrm{conv}(\mathbf{v}_1, \ldots, \mathbf{v}_m)$. It turns into a balanced fan with a natural choice of weights: for each facet F of $\operatorname{trop} f$ the weight $w(F) = \ell(F^\vee)$ equals the lattice length of the corresponding edge of $\mathrm{Newt}(f)$. This balanced fan is the Minkowski weight in $\mathrm{MW}_{n-1}(\Sigma)$ corresponding to f. For details, see [40, 41, 43].

Example 2.7 (The Chow ring $A(\Sigma_{[0,2]})$.) Let $\Sigma = \Sigma[0,2]$. For $k = 0$ the balancing condition is vacuous and a Minkowski weight is a choice of a weight on the origin. For $k = 1$, we need to put a weight on each of the six rays so that the weighted sum of the rays is 0. The four choices of weight below generate all others. For $k = 2$ we need weights on each maximal cone of Σ. Each ray τ is in two cones σ_1 and σ_2 which satisfy $\mathbf{e}_{\sigma_1/\tau} = -\mathbf{e}_{\sigma_2/\tau}$, so the balancing condition says that $w(\sigma_1) = w(\sigma_2)$, and hence all weights are equal. Thus $\mathrm{MW}(\Sigma)$ is spanned by the following Minkowski weights:

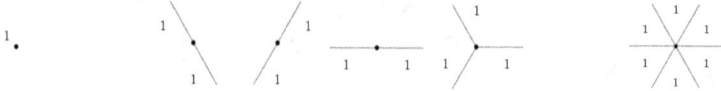

Computing Degrees One can use the *fan displacement rule* to compute the degree of a product: $X_1 \cdot X_2 = \lim_{\epsilon \to 0} X_1 \cdot (X_2 + \epsilon \mathbf{v})$, where $\mathbf{v} \in \mathrm{N}$ is any vector such that X_1 and $X_2 + \epsilon \mathbf{v}$ intersect transversally for sufficiently small $\epsilon > 0$. This requires one to understand how these fans intersect by solving systems of linear equations and inequalities. Sometimes a clever choice of $\mathbf{v}$ – for example one whose coordinates increase very quickly – can simplify the computations.

---

[2] This is canonical in the sense that $\mathrm{PP}(\Sigma) \cong A_T(X_\Sigma)$ and $\mathrm{MW}(\Sigma) \cong A(X_\Sigma)$ are isomorphic to the equivariant and the ordinary Chow cohomology rings of the toric variety X_Σ, respectively, and there is a canonical map $A_T(X_\Sigma) \to A(X_\Sigma)$.

Intersection Theory of Matroids

The Hyperplane and Reciprocal Hyperplane Classes In $\Sigma_E = \Sigma_{[0,n]}$, the Minkowski weights of $\alpha = \alpha_i = \max(t_i - t_j : j \in E)$ and $\beta = \beta_i = \max(t_j - t_i : j \in E)$ are the $(n-1)$-skeleta of the normal fans of $\mathrm{Newt}(\alpha_i) = \mathbf{e}_i - \Delta_E$ and $\mathrm{Newt}(\beta_i) = \Delta_E - \mathbf{e}_i$ where $\Delta_E = \mathrm{conv}(\mathbf{e}_0, \ldots, \mathbf{e}_n)$ is the standard simplex. Explicitly, the facets of α and β are:

$$\alpha: \quad \{\sigma_{i_0 \subset i_0 i_1 \subset \cdots \subset i_0 i_1 \ldots i_{n-2}} : i_0 i_1 \ldots i_{n-1} i_n \text{ permutation of } E\}$$
$$\beta: \quad \{\sigma_{i_0 i_1 \subset \cdots \subset i_0 i_1 \ldots i_{n-1} \subset i_0 i_1 \ldots i_{n-1}} : i_0 i_1 \ldots i_{n-1} i_n \text{ permutation of } E\}$$

with unit weights on all facets. The supports of these fans are

$$|\alpha| = \{\mathbf{t} \in N_E : \min_{i \in E} t_i \text{ is achieved at least twice}\}$$
$$|\beta| = \{\mathbf{t} \in N_E : \max_{i \in E} t_i \text{ is achieved at least twice}\}.$$

Notice that $\min_{i \in E} t_i$ is not a well defined function on $N_E = \mathbb{R}^E / \mathbb{R}\,\mathbf{e}_E$, but whether or not this minimum is achieved at least twice is well defined; similarly for β.

Example 2.8 (The degree of $\alpha\beta$ in $A(\Sigma_{[0,2]})$.) In $A(\Sigma_{[0,2]})$, we can draw the Minkowski weights of α and β using the description obtained above. Alternatively, we can look at the expressions for α and β as piecewise linear functions in Section 2.2 and draw their corner loci, where the functions are not locally linear.

Using these Minkowski weights, we compute the degree of $\alpha\beta$ in $A(\Sigma_{[0,2]})$ in two ways:

In each case the index of intersection is 1, so $\deg(\alpha\beta) = 2$.

2.4 The Chow Ring in Terms of Tropical Intersection

The Chow Ring When Σ is complete, there is an alternative description of the product in the Chow ring that combines piecewise polynomials and Minkowski weights [3, 36, 42, 45]. Let $w \in \mathrm{MW}_k(\Sigma)$ be a Minkowski weight on Σ, and $f \in A^1(\Sigma)$ be a piecewise linear function (modulo global linear functions) on Σ, regarded as a codimension 1 Minkowski weight. The Minkowski weight $f \cdot w \in \mathrm{MW}_{k-1}(\Sigma)$ is given by

$$f \cdot w\,(\tau) := \sum_{\substack{\sigma \in \Sigma^{(k)} \\ \sigma \supset \tau}} f(w(\sigma)\,\mathbf{e}_{\sigma/\tau}) - f\left(\sum_{\substack{\sigma \in \Sigma^{(k)} \\ \sigma \supset \tau}} w(\sigma)\,\mathbf{e}_{\sigma/\tau}\right) \qquad (2.2)$$

for each $(k-1)$-cone τ of Σ. In tropical geometry, the Minkowski weight $f \cdot w$ is known as the *divisor* $\text{div}_w(f)$. Intuitively, it measures the non-linearity of f on w. In particular, if w is linear on f locally around τ, then the divisor equals 0 at τ.

The Dictionary Underlying this description is an isomorphism

$$\text{MW}(\Sigma) \simeq \text{Hom}(A(\Sigma), \mathbb{R})$$

given by the maps

$$\begin{aligned}\text{MW}_k(\Sigma) &\simeq \text{Hom}(A^k(\Sigma), \mathbb{R}) \\ w &\longmapsto (x_\sigma \mapsto w(\sigma)/\text{mult}(\sigma) \text{ for each } k\text{-face } \sigma)\end{aligned}$$

for any simplicial Σ [1, 7]. This isomorphism gives $\text{MW}(\Sigma)$ the structure of a graded $A(\Sigma)$-module. [3] When Σ is complete, we can compute a product $w_1 w_2$ by regarding $w_1 \in \text{MW}_{n-n_1}(\Sigma)$ as the image of $f_1 \in A^{n_1}(\Sigma)$ under the isomorphism of Section 2.3, and letting it act on $w_2 \in \text{MW}_{n-n_2}$ to obtain $w_1 w_2 = f_1 \cdot w_2 \in \text{MW}_{n-n_1-n_2}$.

Computing Degrees Since A is generated in degree 1, we can iterate (2.2) to compute the product of any two Minkowski weights. In particular, this gives a method for computing $\deg(w_1 \cdots w_k)$ for any $w_i \in \text{MW}_{n-n_i}(\Sigma)$ with $n_1 + \cdots + n_k = n$.

The Hyperplane and Reciprocal Hyperplane Classes From the description of α and β as Minkowski weights in $\text{MW}_{n-1}(\Sigma_E) \cong \text{Hom}(A^{n-1}(\Sigma_E), \mathbb{R})$, we get representations of α and β in $\text{Hom}(A^{n-1}(\Sigma_E), \mathbb{R})$ as

$$\alpha(x_\mathcal{F}) = \begin{cases} 1 & \text{if } \mathcal{F} = \{i_0 \subset i_0 i_1 \subset \cdots \subset i_0 i_1 \ldots i_{n-2}\} \\ 0 & \text{otherwise,} \end{cases}$$

$$\beta(x_\mathcal{F}) = \begin{cases} 1 & \text{if } \mathcal{F} = \{i_0 i_1 \subset \cdots \subset i_0 i_1 \ldots i_{n-2} \subset i_0 i_1 \ldots i_{n-2} i_{n-1}\} \\ 0 & \text{otherwise.} \end{cases}$$

We invite the reader to check that these are precisely the results of multiplying $x_\mathcal{F}$ by $\alpha, \beta \in A^1(\Sigma_E)$, as described in Section 2.1.

Example 2.9 (The degree of $\alpha\beta$ in $\Sigma_{[0,2]}$.) Let's regard α as a piecewise linear function and β as a Minkowski weight:

$$\alpha_0 = \begin{array}{c} t_{02} \\ 0 \quad t_{02} \\ \diagup\diagdown \\ 0 \quad t_{01} \\ t_{01} \end{array} = \min_i(t_0 - t_i) \in A^1(\Sigma), \qquad \beta = \begin{array}{c} \mathbf{e}_{12} \end{array} \qquad \begin{array}{c} \mathbf{e}_{01} \\ \diagup \\ \diagdown \\ \mathbf{e}_{02} \end{array} \in \text{MW}_1(\Sigma).$$

Then $\alpha \cdot \beta$ is a 0-dimensional Minkowski weight, whose weight at the origin $\bullet$ is

$$\begin{aligned}(\alpha \cdot \beta)(\bullet) &= \alpha(\mathbf{e}_{12}) + \alpha(\mathbf{e}_{01}) + \alpha(\mathbf{e}_{02}) - \alpha(\mathbf{e}_{12} + \mathbf{e}_{01} + \mathbf{e}_{02}) \\ &= 1 + 1 + 0 - 0 = 2,\end{aligned}$$

so the degree of $\alpha\beta$ is 2.

---

[3] The map $\cdot : A(\Sigma) \times \text{MW}(\Sigma) \to \text{MW}(\Sigma)$ is sometimes called the *cap product* $\cap$.

2.5 Morphisms

A *morphism* from a fan Σ in $N = \mathbb{R} \otimes N_\mathbb{Z}$ to a fan Σ' in $N' = \mathbb{R} \otimes N'_\mathbb{Z}$ is an integral linear map from N to N' such that the image of any cone in Σ is a subset of a cone in Σ'. In the context of toric geometry, a morphism from Σ to Σ' can be identified with a toric morphism from the toric variety of Σ to the toric variety of Σ' [20, Chapter 3].

Let $f : \Sigma \to \Sigma'$ be a morphism of simplicial fans. The pullback of functions defines the *pullback homomorphism* between the Chow rings

$$f^* : A(\Sigma') \longrightarrow A(\Sigma),$$

whose dual is the *pushforward homomorphism* of Minkowski weights

$$f_* : \mathrm{MW}(\Sigma) \longrightarrow \mathrm{MW}(\Sigma').$$

Since f^* is a homomorphism of graded rings, f_* is a homomorphism of graded modules. In other words, the pullback and the pushforward homomorphisms satisfy the *projection formula*

$$\eta \cap f_* w = f_*(f^* \eta \cap w).$$

for any $\eta \in A(\Sigma')$ and $w \in \mathrm{MW}(\Sigma)$.

2.6 Geometry: The Cohomology and Chow Ring of a Toric Variety

The Chow Ring When Σ is complete and simplicial, the ring $A(\Sigma)$ is both the cohomology ring and the Chow ring of the toric variety X_Σ [19, 20, 26][4]:

$$H^\bullet(X_\Sigma, \mathbb{Q}) \cong A(X_\Sigma) \cong A(\Sigma)$$

This isomorphism also holds over $\mathbb{Z}$ when Σ is unimodular [21].

The Dictionary Under this isomorphism, the class of the torus orbit closure of a cone σ in Σ is identified with $\mathrm{mult}(\sigma) x_\sigma$, where x_σ is the monomial $\prod_{\rho \subseteq \sigma} x_\rho$ and $\mathrm{mult}(\sigma)$ is the index of the sublattice $(\sum_{\rho \subseteq \sigma} \mathbb{Z} \mathbf{e}_\rho)$ in the lattice $N_\mathbb{Z} \cap (\sum_{\rho \subseteq \sigma} \mathbb{R} \mathbf{e}_\rho)$. All the fans appearing in this paper will be unimodular, so $\mathrm{mult}(\sigma) = 1$ for every σ in Σ.

Computing Degrees In the Chow ring of a general algebraic variety, computing degrees is a rich and subtle problem for which intersection theory provides a powerful toolkit; see for example [27]. In the special case of toric varieties, the previous sections provide several useful methods.

The Hyperplane and Reciprocal Hyperplane Classes The braid fan Σ_E refines the normal fans Δ_E and $-\Delta_E$ of the standard and inverted simplices $\mathrm{conv}\{\mathbf{e}_i : i \in E\}$ and $\mathrm{conv}\{-\mathbf{e}_i : i \in E\}$. This gives morphisms of toric varieties $\pi_1 : X_{\Sigma_E} \to X_{\Delta_E} \cong \mathbb{P}^E$ and $\pi_2 : X_{\Sigma_E} \to X_{-\Delta_E} \cong \mathbb{P}^E$, where the two copies of $\mathbb{P}^E$ are related to each

---
[4]In [19], Brion identifies $A(\Sigma)$ with the Chow group of X_Σ with real coefficients. For the existence of the ring structure and the pullback, see [49].

other by the Cremona transformation $\mathbb{P}^E \dashrightarrow \mathbb{P}^E$ given by $(z_i)_{i \in E} \mapsto (z'_i)_{i \in E}$ where $z'_i = z_i^{-1}$. The classes

$$\alpha_i = \pi_1^*(z_i = 0), \qquad \beta_i = \pi_2^*(z'_i = 0)$$

in the Chow ring $A(X_{\Sigma_E})$ are the pullbacks of the hyperplane classes $z_i = 0$ and $z'_i = 0$ in the respective copies of $A(\mathbb{P}^E)$.

Example 2.10 (The degree of $\alpha\beta$ in $\Sigma_{[0,2]}$.) Let's compute the degree of $\alpha\beta$ from first principles. Away from the coordinate subspaces, we need to compute the number of intersections of a generic hyperplane α and a generic reciprocal hyperplane β:

$$z_0 = az_1 + bz_2, \qquad \frac{1}{z_0} = \frac{c}{z_1} + \frac{d}{z_2}.$$

Setting $z = z_1/z_2$, this system is equivalent to the equation $adz^2 + (ac+bd)z + bc = 0$, which has two solutions for generic a, b, c, d. Therefore $\deg(\alpha\beta) = 2$.

3 Intersection Theory of Matroids: Four Approaches

3.0 Matroids, Characteristic Polynomials, and Matroid Fans

Let us introduce some basic definitions on matroids, and discuss three combinatorial ways to compute the coefficients $\mu^0, \ldots, \mu^r$ of the reduced characteristic polynomial of M.

A *matroid* $M = (E, r)$ consists of a finite set E and a function $r : 2^E \to \mathbb{Z}$, called the *rank function* such that
(R1) $0 \leq r(A) \leq |A|$ for all $A \subseteq E$,
(R2) $r(A) \leq r(B)$ for all $A \subseteq B \subseteq E$, and
(R3) $r(A) + r(B) \geq r(A \cup B) + r(A \cap B)$ for all $A, B \subseteq E$.

A motivating example is the matroid of a vector configuration $E \subset \mathbb{F}^d$, whose rank function is given by

$$r(A) = \dim(\text{span } A) \qquad \text{for } A \subseteq E.$$

Such a matroid is said to be *linear* over $\mathbb{F}$.

The Lattice of Flats A *flat* of M is a subset $F \subseteq E$ such that $r(F \cup e) > r(F)$ for all $e \notin F$. We say a flat F is *proper* if it does not have rank 0 or r. The *lattice of flats* L_M is the set of flats, partially ordered by inclusion. Its minimum and maximum element are called $\widehat{0}$ and $\widehat{1}$, and its least upper bound and greatest lower bound maps are denoted $\wedge$ and $\vee$. When M is the matroid of a vector configuration E in a vector space V, the flats of M correspond to the subspaces of V spanned by subsets of E, as illustrated in Figure 4.

The *Möbius function* of L_M is the function $\mu : L_M \to \mathbb{Z}$ defined by

$$\sum_{G \leq F} \mu(G) = \begin{cases} 1 & \text{if } F = \widehat{0}, \\ 0 & \text{otherwise.} \end{cases} \qquad (3.1)$$

Intersection Theory of Matroids 15

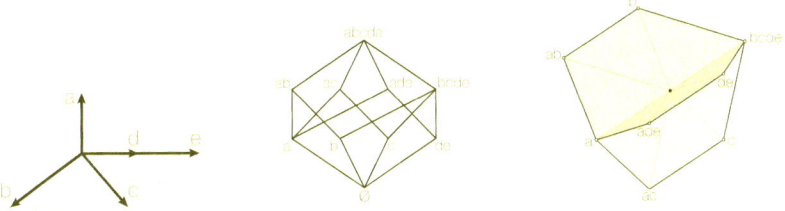

Figure 4: A vector configuration, its lattice of flats, and its matroid fan.

The *Möbius number* of M is $\mu(\mathsf{M}) := \mu(\hat{1})$. The characteristic polynomial, which was defined in terms of the rank function in Section 1, can be expressed in terms of the Möbius function:

$$\chi_\mathsf{M}(q) = \sum_{F \in L_\mathsf{M}} \mu(F) q^{r(\mathsf{M}) - r(F)}. \tag{3.2}$$

Whitney's theorem gives the alternative expression in (1.1).

Matroid Constructions and Three Properties of μ^k Let e be an element of E. The *deletion* $\mathsf{M} \setminus e$ and *contraction* M/e are the matroids on $E - e$ with rank functions

$$r_{\mathsf{M} \setminus e}(A) = r_\mathsf{M}(A) \quad \text{and} \quad r_{\mathsf{M}/e}(A) = r_\mathsf{M}(A \cup e) - r_\mathsf{M}(e) \quad \text{for } A \subseteq E - e.$$

If M is the matroid of a vector configuration $E \subseteq \mathbb{F}^d$ then $\mathsf{M} \setminus e$ and M/e are the matroids of the vector configuration $E - e \subseteq \mathbb{F}^d$ and its image $\overline{E - e}$ in the quotient vector space $\mathbb{F}^d / \mathbb{F} e$. It follows from the definition in Section 1 that the characteristic polynomial satisfies the *deletion-contraction* recurrence $\chi_\mathsf{M}(q) = \chi_{\mathsf{M} \setminus e}(q) - \chi_{\mathsf{M}/e}(q)$, which gives

$$\mu^k(\mathsf{M}) = \mu^{k-1}(\mathsf{M}/e) + \mu^k(\mathsf{M} \setminus e) \quad \text{for } 0 \leq k \leq r. \tag{3.3}$$

The *truncation* Tr M is the rank r matroid obtained from M by omitting the flats of rank r. If M is the matroid of a vector configuration $E \subseteq \mathbb{F}^d$ over a field of characteristic 0, then Tr M is the matroid of the projection of E onto a generic hyperplane H of $\mathbb{F}^d$. It follows from (3.2) that the first $r - 1$ coefficients of $\chi_\mathsf{M}(q)$ and $\chi_{\mathrm{Tr}\,\mathsf{M}}(q)$ match; a simple calculation then gives

$$\mu^k(\mathsf{M}) = (-1)^k \mu(\mathrm{Tr}^{r+1-k}\,\mathsf{M}) \quad \text{for } 0 \leq k \leq r. \tag{3.4}$$

Let's label each edge from F to G in the Hasse diagram of L_M with the element $\min_<(G - F)$. The *Jordan-Hölder sequence* $\pi(\mathbf{m})$ of a maximal chain $\mathbf{m}$ from $\hat{0}$ to $\hat{1}$ is the sequence of labels from the bottom to the top. Its *descent set* records the positions where this sequence decreases: $D(\mathbf{m}) = \{i \in [r] : \pi(\mathbf{m})_i > \pi(\mathbf{m})_{i+1}\}$. Stanley proved [17, Theorem 2.7] that the number of maximal chains $\mathbf{m}$ whose Jordan-Hölder sequence $\pi(\mathbf{m})$ has descent set $D(\mathbf{m}) = S$ equals the Möbius number $(-1)^{|S|+1}\mu((L_\mathsf{M})_S)$ of the rank-selected subposet $(L_\mathsf{M})_S = \{F \in L_\mathsf{M} : r(F) \in S\}$. In particular, if $S = [k]$ then $(L_\mathsf{M})_{[k]}$ is the lattice of flats of $\mathrm{Tr}^{r+1-k}\,\mathsf{M}$. Therefore

$$\mu^k(\mathsf{M}) = \# \text{ of maximal chains } \mathbf{m} \text{ in } L_\mathsf{M} \text{ with descent set } D(\mathbf{m}) = [k]. \tag{3.5}$$

This edge labeling is important in the study of the topology of L_M; see [17].

Matroid Fans Sturmfels [48] and Ardila and Klivans [9] introduced the *matroid fan* or *Bergman fan* of a matroid:

Definition 3.1 [9] Let M be a matroid on ground set E. The *matroid fan* or *Bergman fan* Σ_M in $\mathrm{N}_E = \mathbb{R}^E / \mathbb{R}\,\mathbf{e}_E$ has

- rays: $\mathbf{e}_F$ for the proper flats $\emptyset \subsetneq F \subsetneq E$, and
- cones: $\sigma_\mathcal{F} = \mathrm{cone}(\mathbf{e}_{F_1}, \ldots, \mathbf{e}_{F_k})$ for the flags $\mathcal{F} = (\emptyset \subsetneq F_1 \subsetneq \cdots \subsetneq F_k \subsetneq E)$.

If M has rank $r+1$, the braid fan Σ_M is a pure r-dimensional subfan of the braid fan Σ_E. Notice that Σ_E is the matroid fan for the *Boolean matroid* where $r(A) = |A|$ for all $A \subseteq E$.

Proposition 3.2 *The matroid fan is balanced with unit weights.*

Proof Consider any $(r-1)$-face of the braid fan; we can write it as $\tau = \sigma_\mathcal{F}$ for $\mathcal{F} = (\emptyset \subsetneq F_1 \subsetneq \cdots F_{i-1} \subsetneq F_{i+1} \subsetneq \cdots \subsetneq F_r \subsetneq E)$ where $r(F_j) = j$ for all j. The facets of Σ_E containing τ are those of the form $\sigma = \sigma_{\mathcal{F} \cup F}$ for the rank i flats F with $F_{i-1} \subsetneq F \subsetneq F_{i+1}$. These correspond to the lines $F - F_{i-1}$ of the rank 2 matroid $\mathsf{M}[F_{i-1}, F_{i+1}] = (\mathsf{M}\backslash(E - F_{i+1}))/F_i$, whose union is its ground set $F_{i+1} - F_{i-1}$. Therefore

$$\sum_{\tau \subset \sigma} w(\sigma)\,\mathbf{e}_{\sigma/\tau} = \sum_{F_{i-1} \subsetneq F \subsetneq F_{i+1}} \mathbf{e}_{F - F_{i-1}} = \mathbf{e}_{F_{i+1} - F_{i-1}} = 0 \text{ in } \mathrm{N}/\mathrm{span}\,\tau.$$

as desired. $\square$

It follows that we can regard Σ_M, with unit weights, as a Minkowski weight 1_M on the matroid fan Σ_M or on the permutahedral fan Σ_E.

The *Chow ring* $A(\mathsf{M})$ of M is the Chow ring $A(\Sigma_\mathsf{M})$ of its matroid fan, as defined in Section 2.1. Even though Σ_M is not complete, $A(\mathsf{M})$ also has a degree map [1]:

$$\deg_\mathsf{M} : A^r(\mathsf{M}) \to \mathbb{Z}$$
$$\eta \mapsto \eta \cdot 1_\mathsf{M}.$$

The Theme The inclusion $i : \Sigma_\mathsf{M} \to \Sigma_E$ of the matroid fan in the braid fan is a morphism of fans. As explained in Section 2.5, this gives pullback and pushforward homomorphisms

$$i^* : A(\Sigma_E) \longrightarrow A(\Sigma_\mathsf{M}), \qquad i_* : \mathrm{MW}(\Sigma_\mathsf{M}) \longrightarrow \mathrm{MW}(\Sigma_E),$$

satisfying the *projection formula* $\eta \cdot i_* w = i_*(i^*\eta \cdot w)$.

The classes α_E and β_E of the braid Chow ring $A(\Sigma_E)$ described in Section 2 pull back to the *hyperplane* and *reciprocal hyperplane classes*

$$\alpha_\mathsf{M} := i^*(\alpha_E), \qquad \beta_\mathsf{M} := i^*(\beta_E)$$

of the matroid Chow ring $A(\Sigma_\mathsf{M})$. Also, the top-dimensional constant Minkowski weight 1_M on the matroid fan Σ_M pushes forward to the Minkowski weight $i_*(1_\mathsf{M}) = \Sigma_\mathsf{M}$ on the braid fan Σ_E. The projection formula then gives

$$\deg_\mathsf{M}(\alpha_\mathsf{M}^{r-k}\beta_\mathsf{M}^k) = \deg_E(\Sigma_\mathsf{M} \cdot \alpha_E^{r-k}\beta_E^k).$$

where $\deg_M : A^r(\Sigma_M) \xrightarrow{\sim} \mathbb{R}$ and $\deg_E : A^n(\Sigma_E) \xrightarrow{\sim} \mathbb{R}$ are the degree map of Σ_M and Σ_E, respectively. Now we restate our main theme:

> **Theorem 1.1** *Let* M *be a matroid of rank* $r+1$. *Let* α, β *be the hyperplane and reciprocal hyperplane classes in the Chow ring* $A(M)$. *Then*
> $$\deg_M(\alpha^{r-k}\beta^k) = \mu^k(M) \qquad \text{for } 0 \leq k \leq r.$$

and devote the rest of the paper to four variations on its proof.

3.1 The Chow Ring as a Quotient of a Polynomial Ring

Let M be a loopless matroid of rank $r+1$ on a set E with $n+1$ elements. The Chow ring $A(M)$ is the $\mathbb{Z}$-algebra generated by variables x_F for each non-empty proper flat and relations

$$x_F x_G = 0 \qquad \text{for flats } F, G \text{ such that } F \subsetneq G \text{ and } F \supsetneq G,$$
$$\sum_{F \supset i} x_F = \sum_{F \ni j} x_F \qquad \text{for elements } i, j \in E.$$

The Chow ring is graded $A(M) = A^0(M) \oplus \cdots \oplus A^r(M)$, and the isomorphism $\deg_M : A^r(M) \to \mathbb{Z}$ is characterized by its value on square-free monomials:

$$\deg(x_{F_1} \cdots x_{F_k}) = \begin{cases} 1 & \text{if } F_1, \ldots, F_k \text{ form a flag,} \\ 0 & \text{otherwise.} \end{cases} \qquad \text{for } F_1, \ldots, F_k \text{ distinct.}$$

In this presentation, the *hyperplane* and *reciprocal hyperplane* classes α_M and β_M are given by:

$$\alpha = \alpha_i = \sum_{i \in F} x_F, \qquad \beta = \beta_i = \sum_{i \notin F} x_F.$$

As before, that they do not depend on i.

Our goal is to compute the degree of $\alpha^{r-k}\beta^k$ in the Chow ring $A(M)$. To do so, we seek to express $\alpha^{r-k}\beta^k$ as a sum of square-free monomials, each of which have degree one. One fundamental feature of this computation, which is simultaneously a challenge and an advantage, is that there are many ways to carry it out. We are free to choose any one of the E different expressions for α and β to compute. To have control over the computation, we require some structure amidst that freedom. Let us prescribe a precise way of carrying out these computations, in terms of a fixed linear order $<$ on the ground set E of M.

Definition 3.3 Let $\mathcal{F} = \{\emptyset \subsetneq F_1 \subsetneq \cdots \subsetneq F_k \subsetneq E\}$ be a flag of flats of M.
• The *lexicographic expansion* of $x_{\mathcal{F}} \alpha$ is the expression

$$x_{\mathcal{F}} \alpha = x_{\mathcal{F}} \alpha_e = \sum_{F \supset F_k \cup e} x_{\mathcal{F}} x_F,$$

where $e = \min_<(E - F_k)$ is the $<$-smallest element of E that **is not** in F_k. Note that since $e \in F$ and $e \notin F_k$, the new flat F in each term must be the maximal flat in the new flag $\mathcal{F} \cup F$.

- The *lexicographic expansion of* $x_\mathcal{F}\,\beta$ is the expression
$$x_\mathcal{F}\,\beta = x_\mathcal{F}\,\beta_e = \sum_{F \subset F_1 - e} x_F x_\mathcal{F},$$
where $e = \min_< F_1$ is the $<$-smallest element of E that **is** in F_1. Note that since $e \notin F$ and $e \in F_1$, the new flat F must be the minimal flat in the new flag $F \cup \mathcal{F}$.
- The *lexicographic expansion of* $x_\mathcal{F}\,\beta^t$ is obtained recursively by multiplying each monomial in the lexicographic expansion of $x_\mathcal{F}\,\beta^{t-1}$ by β, again using the lexicographic expansion.
- The *lexicographic expansion of* $x_\mathcal{F}\,\alpha^s\beta^t$ is obtained recursively by multiplying each monomial in the lexicographic expansion of $x_\mathcal{F}\,\alpha^{s-1}\beta^t$ by α, again using the lexicographic expansion.

By construction, these lexicographic expansions are sums of non-zero square-free monomials in $A(\mathsf{M})$. We invite the reader to compute the lexicographic expansions of $\alpha^2, \alpha\beta,$ and β^2 for the matroid in Figure 4. We now describe the outcome of this computation in general.

A flag $\mathcal{F} = \{\emptyset \subsetneq F_1 \subsetneq \cdots \subsetneq F_k \subsetneq E\}$ gives rise to a word
$$\mathbf{m}(\mathcal{F}) = m_1 m_2 \ldots m_{k+1} \qquad \text{where } m_i = \min_<(F_i - F_{i-1})$$
and a *descent set*
$$D(\mathcal{F}) = \{i \in [k] : m_i > m_{i+1}\}.$$

Lemma 3.4 *The lexicographic expansion of $\alpha^s \beta^t$ is*
$$\alpha^s \beta^t = \sum_{\mathcal{F}\,:\,|\mathcal{F}|=s+t,\,D(\mathcal{F})=[t]} x_\mathcal{F}.$$

Proof In the terms $x_{F \cup \mathcal{F}}$ of the lexicographic expansion of $x_\mathcal{F}\,\beta$, the condition $F \subset F_1 - e$ is equivalent to $\min F > \min(F_1 - F) = e$, that is, to an initial descent in the word of $F \cup \mathcal{F}$.

In the terms $x_{\mathcal{F} \cup F}$ of the lexicographic expansion of $x_\mathcal{F}\,\alpha$, the condition $F \supset F_k \cup e$ is equivalent to $\min(F - F_k) = e > \min(E - F)$, that is, to a final ascent in the word of $\mathcal{F} \cup F$.

Since the lexicographic expansion in question is computed recursively in the order $1, \beta, \beta^2, \ldots, \beta^t, \alpha\beta^t, \alpha^2\beta^t, \ldots, \alpha^s\beta^t$, its terms correspond to the flags of length $s+t$ whose words have t initial descents and s final ascents. $\square$

Proof 1 of Theorem 1.1: The flags of length $(r-k)+k$ whose words have k initial descents and $r-k$ final ascents correspond to the maximal chains in the lattice L_M whose Jordan-Hölder sequence has descent set $[k]$. As we discussed in Section 3.0, there are μ^k such flags. $\square$

This first proof of Theorem 1.1 is based on [1]. For an alternative deletion-contraction proof motivated by the intersection theory of moduli spaces of curves, see [22].

3.2 The Chow Ring in Terms of Piecewise Polynomials

For a permutation $\sigma : [0,n] \to E$, let B_σ be the lexicographically smallest basis of M with respect to the order σ. It can be constructed greedily, by sequentially adding each of $\sigma(0), \sigma(1), \ldots, \sigma(n)$ as long as it is independent from the previously added elements.

The following piecewise polynomial functions are representatives for the classes of Σ_M, α, and β in $\mathrm{PP}(\Sigma_E)/N^\vee$. For any fixed element $f \in E$,

$$[\alpha]_\sigma = t_f - t_{\sigma(n)}, \quad [\beta]_\sigma = t_{\sigma(0)} - t_f, \quad [\Sigma_M]_\sigma = (-\mathbf{t})_{E-B_\sigma} := \prod_{i \notin B_\sigma}(t_f - t_i), \quad (3.6)$$

for each permutation σ of E: The first two equalities were shown in Section 2.2; for the third, see [11, Lemma 4.3] and [14, Theorem 7.6]. Therefore, as explained in Section 2.2,

$$\deg_E(\alpha^{r-k}\beta^k \Sigma_M) = \sum_{\sigma \in S_E} \frac{(t_{\sigma(0)} - t_f)^k (t_f - t_{\sigma(n)})^{r-k} \prod_{i \notin B_\sigma}(t_f - t_i)}{(t_{\sigma(0)} - t_{\sigma(1)})(t_{\sigma(1)} - t_{\sigma(2)}) \cdots (t_{\sigma(n-1)} - t_{\sigma(n)})}$$

and we need to prove that this rational function equals the constant $\mu^k(M)$.

Let us write $\mathbf{t}_\sigma := (t_{\sigma(0)} - t_{\sigma(1)})(t_{\sigma(1)} - t_{\sigma(2)}) \cdots (t_{\sigma(n-1)} - t_{\sigma(n)})$ for each bijection $\sigma : [0,n] \to E$, written in "one-line notation" as the word $\sigma(0) \ldots \sigma(n)$. Also recall that $t_{ij} := t_i - t_j$. Finally write $[m_k(M)] := [\alpha^{r-k}\beta^k \Sigma_M]$ regarded as a piecewise polynomial function on Σ_E, so that

$$\deg_E(\alpha^{r-k}\beta^k \Sigma_M) = \sum_{\sigma \in S_E} \frac{[m_k(M)]_\sigma}{\mathbf{t}_\sigma}.$$

Proof 2 of Theorem 1.1: We will prove that this sum equals $\mu^k(M)$ by showing that it satisfies a deletion-contraction recurrence. Let e be an element that is neither a loop nor a coloop.[5] Each permutation of E can be written uniquely in the form $\tau^i = \tau(0) \ldots \tau(i-1) \, e \, \tau(i) \ldots \tau(n-1)$ for a permutation $\tau = \tau(0) \ldots \tau(n-1)$ of $E - e$ and an index $0 \leq i \leq n$. For each permutation τ of $E - e$, there is an element $j \in E - e$ such that

$$B_\tau(M/e) =: B_\tau, \qquad B_\tau(M \backslash e) = B_\tau \cup \tau(j).$$

Then we have

$$B_{\tau^i}(M) = \begin{cases} B_\tau \cup e & \text{if } i \leq j, \\ B_\tau \cup \tau(j) & \text{if } i > j. \end{cases}$$

Now we use this to compute the parts of a piecewise polynomial function representing $[m_k(M)] := [\alpha^{r-k}\beta^k \Sigma_M]$ recursively. We use the equations in (3.6) which are valid for any $f \in E$; we will choose $f = e$.[6] Notice that $[\beta]_{\tau^0} = 0$, $[\alpha]_{\tau^n} = 0$, and $[\Sigma_M]_{\tau^i} = 0$ for $i > j$ since $f = e$ and $e \notin B_{\tau^i}$. Therefore

$$[m_k(M)]_{\tau^i} = \begin{cases} (t_{\tau(0)e})^k (t_{e\tau(n-1)})^{r-k}(-\mathbf{t})_{E-B-e} & \text{for } i = 1, \ldots, j, \\ 0 & \text{for } i = 0, j+1, \ldots, n-1, n. \end{cases}$$

---
[5] A similar analysis, which we omit, will hold when e is a loop or a coloop.
[6] One can prove the recurrence without making this choice, but that requires additional ideas.

Let us sum the contributions from permutations $\tau^0, \ldots, \tau^n$ to $\deg[m_k(\mathbf{M})]$:

$$\sum_{i=0}^{n} \frac{[m_k(\mathbf{M})]_{\tau^i}}{\mathbf{t}_{\tau^i}}$$

$$= \frac{(t_{\tau(0)e})^k (t_{e\tau(n-1)})^{r-k} (-\mathbf{t})_{E-B-e}}{\mathbf{t}_\tau} \left[\sum_{i=1}^{j} \frac{\mathbf{t}_\tau}{\mathbf{t}_{\tau^i}} \right]$$

$$= \frac{(t_{\tau(0)e})^k (t_{e\tau(n-1)})^{r-k} (-\mathbf{t})_{E-B-e}}{\mathbf{t}_\tau} \cdot \left[\frac{1}{t_{\tau(0)e}} + \frac{1}{t_{e\tau(j)}} \right]$$

$$= \frac{(t_{\tau(0)e})^{k-1} (t_{e\tau(n-1)})^{r-k} (-\mathbf{t})_{(E-e)-B}}{\mathbf{t}_\tau} + \frac{(t_{\tau(0)e})^k (t_{e\tau(n-1)})^{r-k} (-\mathbf{t})_{(E-e)-B-\tau(j)}}{\mathbf{t}_\tau}$$

$$= \frac{[m_{k-1}(\mathbf{M}/e)]_\tau}{\mathbf{t}_\tau} + \frac{[m_k(\mathbf{M}\setminus e)]_\tau}{\mathbf{t}_\tau},$$

using the fact that the sum of $\frac{\mathbf{t}_\tau}{\mathbf{t}_{\tau^i}} = \frac{t_{\tau(i-1)\tau(i)}}{(t_{\tau(i-1)e})(t_{e\tau(i)})} = \frac{1}{t_{\tau(i-1)e}} - \frac{1}{t_{\tau(i)e}}$ telescopes.
Summing this equality over all $\tau \in S_{E-e}$, we obtain

$$\begin{aligned}\deg_E(m_k(\mathbf{M})) &= \deg_E(m_{k-1}(\mathbf{M}/e)) + \deg_E(m_k(\mathbf{M}\setminus e)) \\ &= \mu^{k-1}(\mathbf{M}/e) + \mu^k(\mathbf{M}\setminus e) \\ &= \mu^k(\mathbf{M})\end{aligned}$$

as desired. □

For a similar recursive proof of a much more general statement, see [14].

3.3 The Chow Ring in Terms of Minkowski Weights

Let us now compute the degree of $\alpha^{r-k}\beta^k \Sigma_\mathbf{M}$ stable intersection of the respective Minkowski weights. We have already described the matroid fan $\Sigma_\mathbf{M}$. The tropical fans of α^{r-k} and β^k are the subfans of Σ_E with unit weights and support $\Sigma_{E,r-k}$ and $-\Sigma_{E,k}$ where

$$\Sigma_{E,i} = \{\mathbf{x} \in \mathbf{N}_E : \text{the smallest } i+1 \text{ coordinates of } \mathbf{x} \text{ are equal}\}.$$

For each $S \subseteq E$ we consider the cone of $\mathbf{N}_E$ where the S coordinates are minimal:

$$\Sigma_{E,S} = \{\mathbf{x} \in \mathbf{N}_E : x_s \leq x_t \text{ for all } s \in S, t \in E\}.$$

The relative interiors are pairwise disjoint, and $\Sigma_{E,i} = \bigcup_{|I|=i+1} \Sigma_{E,I}$ for $1 \leq i \leq n$.

Proof 3 of Theorem 1.1: Let $\mathbf{a}$ and $\mathbf{b}$ be generic vectors with $\mathbf{a}$ decreasing and $\mathbf{b}$ increasing, so $a_0 > a_1 > \cdots > a_n$ and $b_0 < b_1 < \cdots < b_n$ We need to find the points in the intersection $\Sigma_\mathbf{M} \cap (\mathbf{a} + \Sigma_{E,r-k}) \cap (\mathbf{b} - \Sigma_{E,k})$. Let us compute the points

$$\mathbf{x} \in \sigma_\mathcal{F} \cap (\mathbf{a} + \Sigma_{E,I}) \cap (\mathbf{b} - \Sigma_{E,J}), \qquad \text{for } |\mathcal{F}| = r, |I| = r-k+1, |J| = k+1.$$

Let the flag $\mathcal{F}$ induce the ordered set partition $\mathcal{S} = S_1|\cdots|S_{r+1}$ of E with parts $S_i = F_i - F_{i-1}$ for $1 \leq i \leq r+1$.

Since the codimensions of these cones add up to $(n-r)+(r-k)+k = n = \dim N_E$ and $\mathbf{a}$ and $\mathbf{b}$ are generic, for this intersection to be nonempty, the sets I and J cannot share a pair of elements with each other or with any S_h. Therefore the parts $S_1, \ldots, S_{r+1}$ split into three types:

(I) $r - k$ parts containing exactly one element of I and none of J,
(J) k parts containing no element of I and exactly one of J,
(IJ) one part containing exactly one element of I and one element of J.

We claim that in any part of $\mathcal{S}$, the minimum element is the one belonging to I or J. To show the claim, assume $i \in I$ is in a block $S \ni i$ of $\mathcal{S}$. Since $\mathbf{x} - \mathbf{a} \in \Sigma_{E,I}$, $(x-a)_i \leq (x-a)_s$ for all $s \in S$, so $a_i = \max\{a_s : s \in S\}$. Since $\mathbf{a}$ is decreasing, $i = \min\{s : s \in S\}$. An analogous argument shows that any $j \in J$ is the smallest in its block $S \ni j$. In particular, the minimum element m of the part of type (IJ) is the element belonging to I and J.

Now let us choose the representatives of $\mathbf{a}, \mathbf{b}, \mathbf{x} \in N_E = \mathbb{R}^E / \mathbb{R}\mathbf{e}_E$ that make $a_m = b_m = x_m = 0$. Since $\mathbf{x} - \mathbf{a} \in \Sigma_{E,I}$ and $\mathbf{x} - \mathbf{b} \in -\Sigma_{E,J}$, the smallest coordinates of $\mathbf{x} - \mathbf{a}$ and the largest coordinates of $\mathbf{x} - \mathbf{b}$, both achieved at $m \in I \cap J$, equal 0. It follows that

$$x_s = \begin{cases} x_i = a_i & \text{for } s, i \in S \text{ of type (I)} \\ x_j = b_j & \text{for } s, j \in S \text{ of type (J)} \\ x_m = 0 & \text{for } s, m \in S \text{ of type (IJ)}, \end{cases} \quad \text{and } a_e \leq x_e \leq b_e \text{ for all } e \in E.$$

We claim that $m = 0$. If $m > 0$, since $\mathbf{a}$ is decreasing and $\mathbf{b}$ is decreasing, we would have

$$0 = a_m < a_0 \leq b_0 < b_m = 0.$$

This implies that

$$a_n < \cdots < a_1 < a_0 = 0 = b_0 < b_1 < \cdots < b_n.$$

Since $\mathbf{x} \in \sigma_{\mathcal{F}}$, the parts S of type (I), where $x_s = a_i$ for some $i \in I - 0$, must come after the part of type (IJ) in $\mathcal{S}$. Analogously, the parts S of type (II) must come before the part of type (IJ) in $\mathcal{S}$.

We conclude that the parts $S_1, \ldots, S_k$ are of type (J), S_{k+1} is of type (IJ), and $S_{k+2}, \ldots, S_{r+1}$ are of type (I). Their respective minimum elements $j_1, \ldots, j_k \in J$, $0 \in I \cap J$, $i_{k+2}, \ldots, i_{r+1}$ have decreasing x coordinates, which means that $j_1 \ldots j_k 0$ is decreasing and $0 i_{k+2} \ldots i_{r+1}$ is increasing. Thus $\mathcal{F}$ is a maximal chain whose Jordan-Hölder sequence has descent set $[k]$. As we saw in Section 3.2, there are $\mu^k(\mathsf{M})$ such chains.

Rereading the proof, the reader will see that the point $\mathbf{x}$ computed above does provide an intersection point of our tropical fans; it is not difficult to verify that it has index 1. This completes the proof that the intersection degree is $\mu^k(\mathsf{M})$. □

3.4 The Chow Ring in Terms of Tropical Intersection

Let $1 \leq r_1 \leq r_2 \leq r$. The (r_1, r_2)-*truncation* of M is the Minkowski weight $\Sigma_{\mathsf{M},[r_1,r_2]}$ on the braid fan Σ_E whose weight on the cone $\sigma_\mathcal{F}$ corresponding to a flag $\mathcal{F} = \{\emptyset \subsetneq F_{r_1} \subsetneq F_{r_1+1} \subsetneq \cdots \subsetneq F_{r_2} \subsetneq E\}$ is

$$w(\sigma_\mathcal{F}) = \begin{cases} |\mu(F_{r_1})| & \text{if each } F_i \text{ is a flat of rank } i \text{ in M, or} \\ 0 & \text{otherwise.} \end{cases}$$

Proposition 3.5 *[32, 34] Let* M *be a matroid of rank* $r+1$. *For* $1 \leq r_1 < r_2 \leq r+1$,

$$\alpha \cdot \Sigma_{\mathsf{M},[r_1,r_2]} = \Sigma_{\mathsf{M},[r_1,r_2-1]}, \qquad \beta \cdot \Sigma_{\mathsf{M},[r_1,r_2]} = \Sigma_{\mathsf{M},[r_1+1,r_2]}.$$

For $1 \leq k \leq r$,

$$\deg(\alpha \cdot \Sigma_{\mathsf{M},[k,k]}) = \mu^{k-1}, \qquad \deg(\beta \cdot \Sigma_{\mathsf{M},[k,k]}) = \mu^k.$$

In particular, each (r_1,r_2)-*truncation* $\Sigma_{\mathsf{M},[r_1,r_2]}$ *is a balanced fan.*

Proof Let us prove the statements for α; the proofs for β are similar and can be found in [32]. To show that $\alpha \cdot \Sigma_{\mathsf{M},[r_1,r_2]} = \Sigma_{\mathsf{M},[r_1,r_2-1]}$, we show that the Minkowski weight $\alpha \cdot \Sigma_{\mathsf{M},[r_1,r_2]}$ has the correct value on each cone of $\Sigma_{\mathsf{M},[r_1,r_2-1]}$, and has weight 0 on every other cone.

First, consider a cone $\sigma_\mathcal{G}$ in $\Sigma_{\mathsf{M},[r_1,r_2-1]}$ with $\mathcal{G} = \{G_{r_1} \subsetneq \cdots \subsetneq G_{r_2-1}\}$, and let $F_1, \ldots, F_m$ be the rank r_2 flats of M compatible with $\mathcal{G}$; that is, the flats covering $G_{r_2-1} = G$ in the lattice L_M. They correspond to the rank 1 flats $F_1 - G, \ldots, F_m - G$ of $\mathsf{M}[G,E] = \mathsf{M}/G$, which partition its ground set $E - G$, so

$$\mathbf{e}_{F_1} + \cdots + \mathbf{e}_{F_m} = (m-1)\,\mathbf{e}_G + \mathbf{e}_E = (m-1)\,\mathbf{e}_G \qquad \text{in } N_E.$$

The full-dimensional cones of $\Sigma_{\mathsf{M},[r_1,r_2]}$ containing $\sigma_\mathcal{G}$ are $\sigma_{\mathcal{G} \cup F_i}$ for $1 \leq i \leq m$, all with weight $|\mu(G_{r_1})|$. Let us choose an element $e \in G$ [7] and compute the weight of the divisor $\alpha \cdot \Sigma_{\mathsf{M},[r_1,r_2]}$ on this cone $\sigma_\mathcal{G}$, using $\alpha = \alpha_e$:

$$\begin{aligned}
\alpha \cdot \Sigma_{\mathsf{M},[r_1,r_2]}(\sigma_\mathcal{G}) &:= \sum_{\mathcal{F} \supset \mathcal{G}} w(\sigma_\mathcal{F}) \alpha(\mathbf{e}_{\mathcal{F}/\mathcal{G}}) - \alpha \left(\sum_{\mathcal{F} \supset \mathcal{G}} w(\sigma_\mathcal{F}) \mathbf{e}_{\mathcal{F}/\mathcal{G}} \right) \\
&= |\mu(G_{r_1})| \left(\sum_{j=1}^m \alpha(\mathbf{e}_{F_j}) - \alpha \left(\sum_{j=1}^m \mathbf{e}_{F_j} \right) \right) \\
&= |\mu(G_{r_1})| \left(m - \alpha((m-1)\mathbf{e}_G) \right) \\
&= |\mu(G_{r_1})|,
\end{aligned}$$

noting that $e \in G \subset F_j$ imply $\alpha(\mathbf{e}_G) = \alpha(\mathbf{e}_{F_j}) = 1$ for all j.

Now consider a cone $\sigma_\mathcal{G}$ of $\Sigma_{\mathsf{M},[r_1,r_2]}$ that is **not** in $\Sigma_{\mathsf{M},[r_1,r_2-1]}$; let the flats in $\mathcal{G} = \{G_{r_1} \subsetneq \cdots \subsetneq G_{s-1} \subsetneq G_{s+1} \subsetneq \cdots \subsetneq G_{r_2}\}$ have ranks $r_1, \ldots, \widehat{s}, \ldots, r_2$ where $s \neq r_2$. Let $F_1, \ldots, F_m$ be the flats of rank s that are compatible with M. Now let us

---

[7] One will obtain the same answer using $\alpha = \alpha_f$ for any other $f \in E$, but choosing $e \in G$ simplifies the computation.

choose an element $e \notin G_{s+1}$ (and hence $e \notin F_j$ for all j), and perform the following computation with $\alpha = \alpha_e$

$$\begin{aligned}
\alpha \cdot \Sigma_{\mathsf{M},[r_1,r_2]}(\sigma_{\mathcal{G}}) &= \sum_{\mathcal{F} \supset \mathcal{G}} w(\sigma_{\mathcal{F}}) \alpha(\mathbf{e}_{\mathcal{F}/\mathcal{G}}) - \alpha \left(\sum_{\mathcal{F} \supset \mathcal{G}} w(\sigma_{\mathcal{F}}) \mathbf{e}_{\mathcal{F}/\mathcal{G}} \right) \\
&= \left(\sum_{j=1}^{m} w(\sigma_{\mathcal{G} \cup F_j}) \alpha(\mathbf{e}_{F_j}) - \alpha \left(\sum_{j=1}^{m} w(\sigma_{\mathcal{G} \cup F_j}) \mathbf{e}_{F_j} \right) \right) \\
&= 0
\end{aligned}$$

because none of the $\mathbf{e}_{F_j}$s involve the e-th coordinate, so α takes a value of 0 on every term in this sum.

Finally, we compute $\alpha \cdot \Sigma_{\mathsf{M},[k,k]}$, which is supported on the origin $\bullet$. Summing over the flats G of rank k in M, we compute with $\alpha = \alpha_e$ for any $e \in E$:

$$\begin{aligned}
\alpha \cdot \Sigma_{\mathsf{M},[k,k]}(\bullet) &= \sum_G w(\mathbf{e}_G) \alpha(\mathbf{e}_{G/\bullet}) - \alpha \left(\sum_G w(\mathbf{e}_G) \mathbf{e}_{G/\bullet} \right) \\
&= \sum_G |\mu(G)| \alpha(\mathbf{e}_G) - \alpha \left(\sum_G |\mu(G)| \mathbf{e}_G \right) \\
&= \sum_{G \ni e} |\mu(G)| \\
&= |\mu(\mathrm{Tr}^{r+1-k} \mathsf{M})| = \mu^k,
\end{aligned}$$

because the balancing condition for $\Sigma_{\mathsf{M},[k,k]}$ gives $\sum_G |\mu(G)| \mathbf{e}_G = 0$ in N_E, and the last step follows from Weisner's theorem:

For any lattice L and any element $x \neq \hat{1}$, $\displaystyle\sum_{x \in L : x \vee e = \hat{1}} \mu(x) = 0$.

applied to the lattice of flats of $\mathrm{Tr}^{r+1-k} \mathsf{M}$ and the atom e. □

Proof 4 of Theorem 1.1: We use Proposition 3.5 to compute the degree of $\alpha^{r-k} \beta^k \cdot \Sigma_{\mathsf{M}} = \alpha^{r-k} \beta^k \cdot \Sigma_{\mathsf{M}[1,r]}$. For $k \neq r$ we have

$$\deg(\alpha \cdot \alpha^{r-k-1} \beta^k \cdot \Sigma_{\mathsf{M}[1,r]}) = \deg(\alpha \cdot \Sigma_{\mathsf{M}[k+1,k+1]}) = \mu^k$$

and for $k \neq 1$ we have

$$\deg(\beta \cdot \alpha^{r-k} \beta^{k-1} \cdot \Sigma_{\mathsf{M}[1,r]}) = \deg(\beta \cdot \Sigma_{\mathsf{M}[k,k]}) = \mu^k$$

as desired. □

This final proof of Theorem 1.1 is based on [32, 34].

4 Further Developments

We close with a small selection of recent results that highlight a few additional directions and techniques in the intersection theory of matroids. They all involve one of the most important functions of a matroid: the *Tutte polynomial*

$$T_{\mathsf{M}}(x,y) = \sum_{A \subseteq E} (x-1)^{r-r(A)}(y-1)^{|A|-r(A)}.$$

This is a very powerful invariant, because every matroid invariant that satisfies a deletion-contraction recurrence – for instance, the characteristic polynomial – can be expressed in terms of the $T_{\mathsf{M}}(x,y)$. Some of the results also involve the beta invariant $\beta(\mathsf{M})$, which is the coefficient of $x^1 y^0$ and of $x^0 y^1$ in $T_{\mathsf{M}}(x,y)$.

Our goal here is to give a brief description of the key combinatorial aspects of these constructions, but each one of them has an elegant geometric origin. To fully understand the motivation, as well as the relevant definitions, we invite the reader to consult the relevant references.

Matroid Fans and Symmetrized Minkowski Weights Berget, Spink, and Tseng [15] defined the *one-window symmetrized Minkowski weights* $\Phi_{r,k}$ on the permutahedral variety, and proved that

$$\deg[\Sigma_{\mathsf{M}} \cdot \Phi_{r,k}] = \text{coeff. of } q^k \text{ in } T_{\mathsf{M}}(1,q)$$

for $0 \leq k \leq n-r$. They computed these degrees by finding the stable intersection of the corresponding Minkowski weights, as described in Section 2.3. They introduced the combinatorial framework of "sliding sets" to describe the relevant intersection points.

Chern-Schwartz-MacPherson Cycles of a Matroid López de Medrano, Rincón, and Shaw [39] defined the k-th *Chern-Schwartz-MacPherson (CSM) cycle of a matroid* M to be the k-skeleton of the matroid fan Σ_{M} with weights

$$w(\sigma_{\mathcal{F}}) := (-1)^{r-k} \prod_{i=0}^{k} \beta(\mathsf{M}[F_i, F_{i+1}]), \quad \mathcal{F} = \{\emptyset = F_0 \subset F_1 \subset \cdots \subset F_k \subset F_{k+1} = E\}$$

where $\beta(\mathsf{M}[F_i, F_{i+1}])$ is the beta invariant of the minor $\mathsf{M}[F_i, F_{i+1}]$ [39]. They proved that this is a Minkowski weight on Σ_E, with degree

$$\deg_E(\mathrm{csm}_k(\mathsf{M}) \cdot \alpha^k) = \text{coefficient of } q^k \text{ in } \overline{\chi}_{\mathsf{M}}(q+1). \tag{4.1}$$

for $0 \leq k \leq r$.

They gave a deletion-contraction proof, in the context of Minkowski weights. This required describing the CSM cycles of M in terms of the CSM cycles of the deletion $\mathsf{M} \setminus e$ and the contraction M/e for an element e that is not a loop or coloop, using the relevant pushforward and pullback maps.

Ashraf and Backman [12] gave an alternative proof of (4.1) using stable intersections of Minkowski weights, relying on the Gioan-Las Vergnas refined activities expansion of the Tutte polynomial [29].

The Conormal Fan of a Matroid Ardila, Denham, and Huh [7, 6] introduced the *conormal fan* $\Sigma_{\mathsf{M},\mathsf{M}^\perp}$ of a matroid M. Its analysis required them to go beyond the permutahedral fan, introducing the *bipermutahedral fan* $\Sigma_{E,E}$.

The conormal fan $\Sigma_{\mathsf{M},\mathsf{M}^\perp}$ has support $|\Sigma_{\mathsf{M},\mathsf{M}^\perp}| = |\Sigma_\mathsf{M}| \times |\Sigma_{\mathsf{M}^\perp}|$ in $\mathrm{N}_E \times \mathrm{N}_E$, where $\mathsf{M}^\perp$ is the dual matroid. It is a subfan of the bipermutahedral fan $\Sigma_{E,E}$, whose Chow ring contains elements $\gamma, \delta \in A^1(\Sigma_{E,E})$, such that

$$\deg_{E,E}[\Sigma_{\mathsf{M},\mathsf{M}^\perp} \cdot \gamma^k \delta^{n-1-k}] = (-1)^{r-k} \text{coeff. of } q^k \text{ in } \overline{\chi}_\mathsf{M}(q+1) \qquad (4.2)$$

for $0 \leq k \leq r$.

They gave two proofs of (4.2), based on Brion's presentation of the Chow ring as in Section 2.1. One describes the lexicographic expansion of $\gamma^k \delta^{n-1-k}$ in the Chow ring of $\Sigma_{\mathsf{M},\mathsf{M}^\perp}$ in terms of the combinatorics of biflats and biflags [6], relying on work of LasVergnas on basis activities [38]. The other one shows that the combinatorially intricate CSM cycles of Σ_M are "shadows" of simpler cycles of $\Sigma_{\mathsf{M},\mathsf{M}^\perp}$ under the pushforward map of Minkowski weights:

$$\mathrm{csm}_k(\mathsf{M}) = (-1)^{r-k} \pi_*(\delta^{n-k-1} \cdot 1_{\mathsf{M},\mathsf{M}^\perp})$$

for $0 \leq k \leq r$, where $1_{\mathsf{M},\mathsf{M}^\perp}$ is the top-dimensional constant Minkowski weight on $\Sigma_{\mathsf{M},\mathsf{M}^\perp}$ [7]. The projection formula then shows that (4.1) implies (4.2).

Matroid Valuations in Intersection Theory The *matroid polytope* of a matroid M is

$$P_\mathsf{M} = \mathrm{conv}\{e_B : B \text{ is a basis of } \mathsf{M}\} \subset \mathbb{R}^E.$$

A *matroid valuation* is a function Φ from the set of matroids on E to an additive abelian group such that for any subdivision of a matroid polytope P_M into matroid polytopes $P_{\mathsf{M}_1}, \ldots, P_{\mathsf{M}_k}$ we have the inclusion-exclusion relation

$$\Phi(\mathsf{M}) = \sum_{i=1}^{k} (-1)^{\dim P_\mathsf{M} - \dim P_{\mathsf{M}_i}} \Phi(\mathsf{M}_i). \qquad (4.3)$$

This property seems restrictive but is surprisingly common [8, 10, 23]. For example, $H(\mathsf{M}) = \sum_{\sigma \in S_E} (\sigma, r_\mathsf{M}(\{\sigma(1)\}), r_\mathsf{M}(\{o(1), \sigma(2)\}), \ldots, r_\mathsf{M}(\{\sigma(1), \ldots, \sigma(n)\}))$ is valuative [8], and a very broad range of matroid valuations can be built from it. Notice that $H(\mathsf{M})$ determines M entirely.

This framework is relevant and useful in the intersection theory of matroids. For example, the Bergman fan Σ_M is a matroid valuation, when regarded as a Minkowski weight on Σ_E [39, Theorem 4.5] or as the piecewise polynomial (3.6) on Σ_E [8, Theorem 5.4], [14, Proposition 5.6]. More generally, the CSM cycles of a matroid are also valuative [39]. The multivariate *volume polynomial* of M, which is equivalent to the Chow ring $A(\mathsf{M})$, is also valuative [24].

The case of matroid invariants, which satisfy $f(M_1) = f(M_2)$ when $M_1 \cong M_2$, is best understood. Examples include the characteristic and Tutte polynomials and the beta invariant. Derksen and Fink described the universal valuative matroid invariant $G(\mathsf{M}) = \sum_{\sigma \in S_E} (r_\mathsf{M}(\{\sigma(1)\}), r_\mathsf{M}(\{\sigma(1), \sigma(2)\}), \ldots, r_\mathsf{M}(\{\sigma(1), \ldots, \sigma(n)\}))$; this is the symmetrization of $H(\mathsf{M})$ above. They also showed that any valuative matroid invariant f is determined by its value on *Schubert matroids*.

This gives a powerful way to prove an equation $f(M) = g(M)$ for all matroids M:
1. Prove that f and g are both valuative (e.g. using the techniques of [8, 10, 23]).
2. Prove that $f(M) = g(M)$ for all **Schubert** matroids M (e.g. by a combinatorial argument that uses the structure of Schubert matroids, or by a geometric argument that works for realizable matroids, which include Schubert matroids).

Tautological Matroid Classes Berget, Eur, Spink, and Tseng defined the *tautological Chern classes* $c_i(\mathcal{S}_M^\vee), c_i(\mathcal{Q}_M) \in A^i(\Sigma_E)$ in the permutahedral variety. They gave a valuative proof of the identity

$$\deg_E[c_r(\mathcal{S}_M^\vee)c_{n-r}(\mathcal{Q}_M)] = \beta(M) \tag{4.4}$$

in the spirit of the previous section. Once one understands the relevant definitions, the valuativity of both sides of (4.4) follows directly from the discussion there. They then proved (4.4) algebro-geometrically for all matroids realizable over $\mathbb{C}$ – which includes Schubert matroids – relying on earlier work of Speyer [47].

More generally, they showed that the intersections of the tautological Chern classes with powers of the classes α and β give the following reparameterization of the Tutte polynomial:

$$\sum \deg_E[\alpha^i\beta^j c_k(\mathcal{S}_M^\vee)c_l(\mathcal{Q}_M)]x^iy^jz^kw^l = \frac{(y+z)^{r+1}(x+w)^{n-r}}{x+y}T_M\left(\frac{x+y}{y+z}, \frac{x+y}{x+w}\right),$$

summing over all indices with $i + j + k + l = n$. They used the framework of Section 2.2, analyzing how the piecewise polynomials representing $\alpha^i\beta^j c_k(\mathcal{S}_M^\vee)c_l(\mathcal{Q}_M)$ behave under deletion-contraction. Our proof in Section 3.2 was inspired by theirs.

Tropical Critical Points of Affine Matroids The affine Bergman fan $\widehat{\Sigma}_M$ is the Bergman fan Σ_M with an added lineality space $\mathbb{R}\,\mathbf{e}_E$ in $\mathbb{R}^E$. An *affine matroid* (M, e) consts of a matroid M and a chosen element $e \in E$. The Bergman fan of (M, e) is $\widehat{\Sigma}_{(M,e)} = \{\mathbf{x} \in \mathbb{R}^{E-e} : (0, \mathbf{x}) \in \widehat{\Sigma}_M\}$. Ardila, Eur, and Penaguião gave two proofs of the following formula conjectured by Sturmfels [2]:

$$\deg[\widehat{\Sigma}_{(M,e)} \cdot (-\widehat{\Sigma}_{(M/e)^\perp})] = \beta(M). \tag{4.5}$$

Their first proof described the stable intersection of the fans $\widehat{\Sigma}_{(M,e)}$ and $-\widehat{\Sigma}_{(M/e)^\perp}$ explicitly, by developing the framework of *arboreal pairs of set partitions* and connecting to Ziegler's βnbc bases of ordered matroids [51]. Their second proof wrote down piecewise polynomials representing the left-hand sides of (4.4) and (4.5), and showed that their difference is a multiple of a linear function, and hence equal to 0 in the Chow ring of Σ_E. Thus (4.4) implies (4.5).

Acknowledgements

I would like to thank the organizers of the Clay Mathematics Institute and the 2024 British Combinatorics Conference for the invitation to deliver the Clay Lecture and write this accompanying survey. I am very thankful to many coauthors and friends with whom I have learned the material in this survey, including Carly Klivans, Chris Eur, Dusty Ross, Felipe Rincón, Graham Denham, Johannes Rau, June Huh,

Kris Shaw, Lauren Williams, Mont Cordero–Aguilar, and Raúl Penaguião. This work was partially supported by United States National Science Foundation grant DMS-2154279.

References

[1] Karim Adiprasito, June Huh, and Eric Katz, *Hodge theory for combinatorial geometries*, Ann. of Math. (2) **188** (2018), no. 2, 381–452. MR 3862944

[2] Daniele Agostini, Taylor Brysiewicz, Claudia Fevola, Lukas Kühne, Bernd Sturmfels, and Simon Telen, *Likelihood degenerations*, Adv. Math. **414** (2023), Paper No. 108863, 39, With an appendix by Thomas Lam. MR 4539061

[3] Lars Allermann and Johannes Rau, *First steps in tropical intersection theory*, Mathematische Zeitschrift **264** (2010), no. 3, 633–670.

[4] Federico Ardila, *Algebraic and geometric methods in enumerative combinatorics*, Handbook of enumerative combinatorics, Discrete Math. Appl. (Boca Raton), CRC Press, Boca Raton, FL, 2015, pp. 3–172. MR 3409342

[5] _____, *The geometry of geometries: matroid theory, old and new*, Proceedings of the International Congress of Mathematicians, 2022.

[6] Federico Ardila, Graham Denham, and June Huh, *Lagrangian combinatorics of matroids*, Algebr. Comb. **6** (2023), no. 2, 387–411. MR 4591592

[7] _____, *Lagrangian geometry of matroids*, Journal of the American Mathematical Society **36** (2023), no. 3, 727–794.

[8] Federico Ardila, Alex Fink, and Felipe Rincón, *Valuations for matroid polytope subdivisions*, Canad. J. Math. **62** (2010), no. 6, 1228–1245. MR 2760656

[9] Federico Ardila and Caroline J. Klivans, *The Bergman complex of a matroid and phylogenetic trees*, J. Combin. Theory Ser. B **96** (2006), no. 1, 38–49. MR 2185977 (2006i:05034)

[10] Federico Ardila and Mario Sanchez, *Valuations and the Hopf monoid of generalized permutahedra*, Int. Math. Res. Not. IMRN (2023), no. 5, 4149–4224. MR 4565665

[11] Federico Ardila-Mantilla, Christopher Eur, and Raul Penaguiao, *The tropical critical points of an affine matroid*, arXiv preprint arXiv:2212.08173 (2022).

[12] Ahmed Umer Ashraf and Spencer Backman, *Matroid Chern-Schwartz-MacPherson cycles and Tutte activities*, Proc. Amer. Math. Soc. **151** (2023), no. 6, 2303–2309. MR 4576299

[13] Matthew Baker, *Hodge theory in combinatorics*, Bulletin of the American Mathematical Society **55** (2018), no. 1, 57–80.

[14] Andrew Berget, Christopher Eur, Hunter Spink, and Dennis Tseng, *Tautological classes of matroids*, Invent. Math. **233** (2023), no. 2, 951–1039. MR 4607725

[15] Andrew Berget, Hunter Spink, and Dennis Tseng, *Log-concavity of matroid h-vectors and mixed eulerian numbers*, arXiv preprint arXiv:2005.01937 (2020).

[16] Louis J. Billera, *The algebra of continuous piecewise polynomials*, Adv. Math. **76** (1989), no. 2, 170–183. MR 1013666

[17] Anders Björner, *The homology and shellability of matroids and geometric lattices*, Matroid applications, Encyclopedia Math. Appl., vol. 40, Cambridge Univ. Press, Cambridge, 1992, pp. 226–283. MR 1165544

[18] Michel Brion, *Piecewise polynomial functions, convex polytopes and enumerative geometry*, Parameter spaces (Warsaw, 1994), Banach Center Publ., vol. 36, Polish Acad. Sci. Inst. Math., Warsaw, 1996, pp. 25–44. MR 1481477

[19] _____, *Equivariant Chow groups for torus actions*, Transform. Groups **2** (1997), no. 3, 225–267. MR 1466694

[20] David A. Cox, John B. Little, and Henry K. Schenck, *Toric varieties*, Graduate Studies in Mathematics, vol. 124, American Mathematical Society, Providence, RI, 2011. MR 2810322

[21] Vladimir Danilov, *The geometry of toric varieties*, Uspekhi Mat. Nauk **33** (1978), no. 2(200), 85–134, 247. MR 495499

[22] Jeshu Dastidar and Dustin Ross, *Matroid psi classes*, Selecta Math. (N.S.) **28** (2022), no. 3, Paper No. 55, 38. MR 4405747

[23] Harm Derksen and Alex Fink, *Valuative invariants for polymatroids*, Adv. Math. **225** (2010), no. 4, 1840–1892. MR 2680193

[24] Christopher Eur, *Divisors on matroids and their volumes*, J. Combin. Theory Ser. A **169** (2020), 105135, 31. MR 4011081

[25] _____, *Essence of independence: Hodge theory of matroids since June Huh*, Bulletin of the American Mathematical Society (2023).

[26] William Fulton, *Introduction to toric varieties*, Annals of Mathematics Studies, vol. 131, Princeton University Press, Princeton, NJ, 1993, The William H. Roever Lectures in Geometry. MR 1234037

[27] _____, *Intersection theory*, second ed., Ergebnisse der Mathematik und ihrer Grenzgebiete. 3. Folge. A Series of Modern Surveys in Mathematics [Results in Mathematics and Related Areas. 3rd Series. A Series of Modern Surveys in Mathematics], vol. 2, Springer-Verlag, Berlin, 1998. MR 1644323

[28] William Fulton and Bernd Sturmfels, *Intersection theory on toric varieties*, Topology **36** (1997), no. 2, 335–353. MR 1415592 (97h:14070)

[29] Emeric Gioan and Michel Las Vergnas, *The active bijection 2.a – decomposition of activities for matroid bases, and Tutte polynomial of a matroid in terms of beta invariants of minors*, arXiv preprint arXiv:1807.06516 (2018).

[30] Andrew Heron, *Matroid polynomials*, Combinatorics (Proc. Conf. Combinatorial Math., Math. Inst., Oxford, 1972), Inst. Math. Appl., Southend-on-Sea, 1972, pp. 164–202. MR 340058

[31] June Huh, *Milnor numbers of projective hypersurfaces and the chromatic polynomial of graphs*, J. Amer. Math. Soc. **25** (2012), no. 3, 907–927. MR 2904577

[32] _____, *Rota's conjecture and positivity of algebraic cycles in permutohedral varieties*, ProQuest LLC, Ann Arbor, MI, 2014, Thesis (Ph.D.)–University of Michigan. MR 3321982

[33] _____, *Combinatorial applications of the Hodge-Riemann relations*, Proceedings of the International Congress of Mathematicians—Rio de Janeiro 2018. Vol. IV. Invited lectures, World Sci. Publ., Hackensack, NJ, 2018, pp. 3093–3111. MR 3966524

[34] June Huh and Eric Katz, *Log-concavity of characteristic polynomials and the Bergman fan of matroids*, Math. Ann. **354** (2012), no. 3, 1103–1116. MR 2983081

[35] Anders Jensen and Josephine Yu, *Stable intersections of tropical varieties*, J. Algebraic Combin. **43** (2016), no. 1, 101–128. MR 3439302

[36] Eric Katz, *Tropical intersection theory from toric varieties*, Collect. Math. **63** (2012), no. 1, 29–44. MR 2887109

[37] Eric Katz and Sam Payne, *Piecewise polynomials, Minkowski weights, and localization on toric varieties*, Algebra & Number Theory **2** (2008), no. 2, 135–155.

[38] Michel Las Vergnas, *The Tutte polynomial of a morphism of matroids—5. Derivatives as generating functions of Tutte activities*, European J. Combin. **34** (2013), no. 8, 1390–1405. MR 3082209

[39] Lucía López de Medrano, Felipe Rincón, and Kristin Shaw, *Chern-Schwartz-MacPherson cycles of matroids*, Proc. Lond. Math. Soc. (3) **120** (2020), no. 1, 1–27. MR 3999674

[40] Diane Maclagan and Bernd Sturmfels, *Introduction to Tropical Geometry*, Graduate Studies in Mathematics, vol. 161, American Mathematical Society, Providence, RI, 2015.

[41] Peter McMullen, *The polytope algebra*, Advances in Mathematics **78** (1989), no. 1, 76–130.

[42] Grigory Mikhalkin, *Tropical geometry and its applications*, International Congress of Mathematicians. Vol. II, Eur. Math. Soc., Zürich, 2006, pp. 827–852. MR 2275625

[43] Grigory Mikhalkin and Johannes Rau, *Tropical geometry*, https://math.uniandes.edu.co/~j.rau/downloads/main.pdf, 2018.

[44] James G. Oxley, *Matroid theory*, Oxford Science Publications, The Clarendon Press, Oxford University Press, New York, 1992. MR 1207587

[45] Johannes Rau, *Intersections on tropical moduli spaces*, Rocky Mountain J. Math. **46** (2016), no. 2, 581–662. MR 3529085

[46] Gian-Carlo Rota, *Combinatorial theory, old and new*, Actes du Congrès International des Mathématiciens (Nice, 1970), Tome 3, Gauthier-Villars / Paris, 1971, pp. 229–233. MR 0505646

[47] David E. Speyer, *A matroid invariant via the K-theory of the Grassmannian*, Adv. Math. **221** (2009), no. 3, 882–913. MR 2511042

[48] Bernd Sturmfels, *Solving systems of polynomial equations*, CBMS Regional Conference Series in Mathematics, vol. 97, Conference Board of the Mathematical Sciences, Washington, DC; by the American Mathematical Society, Providence, RI, 2002. MR 1925796

[49] Angelo Vistoli, *Alexander duality in intersection theory*, Compositio Math. **70** (1989), no. 3, 199–225. MR 1002043

[50] Dominic Welsh, *Matroid theory*, L. M. S. Monographs, vol. No. 8, Academic Press [Harcourt Brace Jovanovich, Publishers], London-New York, 1976. MR 427112

[51] Günter M. Ziegler, *Matroid shellability, β-systems, and affine hyperplane arrangements*, J. Algebraic Combin. **1** (1992), no. 3, 283–300. MR 1194080

Department of Mathematics
San Francisco State University
1600 Holloway Avenue
San Francisco, CA 94132, USA
federico@sfsu.edu

Erdős Covering Systems

Paul Balister

Abstract

Introduced by Erdős in 1950, a covering system of the integers is a finite collection of infinite arithmetic progressions whose union is the set of all integers. Many beautiful questions and conjectures about covering systems have been posed over the past several decades, but until the last decade little was known about their properties. Most famously, the so-called minimum modulus problem of Erdős was resolved in 2015 by Hough, who proved that in every covering system with distinct moduli, the minimum modulus is at most a fixed constant. The ideas of Hough were simplified and extended in 2018 by Balister, Bollobás, Morris, Sahasrabudhe and Tiba, to give solutions (or progress towards solutions) to a number of related questions. We give a summary of this and other progress that has been made since.

1 Introduction

A *covering system* of the integers is a finite set of (infinite) arithmetic progressions that covers all the integers, that is a finite set $\mathcal{A} = \{A_1, \ldots, A_n\}$ where each A_i is an infinite arithmetic progression of the form $A_i = a_i + m_i \mathbf{Z}$ and $\bigcup_{i=1}^n A_i = \mathbf{Z}$. We shall usually write the arithmetic progression $a + m\mathbf{Z}$ as 'a mod m' and call m the *modulus* of the progression.

Of course any set of arithmetic progressions containing a covering system is also a covering system, so we will often require our covering systems to be *minimal*, that is, not strictly containing some other covering system. We have the trivial cover $\{\mathbf{Z}\} = \{0 \bmod 1\}$, and the only slightly less trivial covers of the form

$$\{0 \bmod m, 1 \bmod m, 2 \bmod m, \ldots, m-1 \bmod m\}$$

for any $m \geq 1$. To avoid these we often (but not always) demand that the m_i be distinct and that all $m_i > 1$.

At first sight it might not be obvious that a covering system exists with distinct $m_i > 1$, but a little thought yields some examples, e.g.,

$$\{1 \bmod 2, 2 \bmod 4, 1 \bmod 3, 2 \bmod 6, 0 \bmod 12\},$$

which is in fact (up to isomorphism) the smallest such example.

Covering systems were introduced in 1950 by Erdős in the process of answering a question of Romanov concerning a problem of de Polignac. In 1849, de Polignac [41] stated that every odd integer can be written as a sum of a prime and a power of two. This is easily seen to be false (consider 127). Later, however, Romanov (see [38]) proved that the set of numbers of the form $2^k + p$ where p is prime does at least have positive density (actually at least 0.107648 by [12]), and asked if all sufficiently large odd numbers are of this form. Erdős showed that this is not the case, and indeed, an entire arithmetic progression of odd numbers cannot be written in this way.

Theorem 1.1 (Erdős [13]) *There exists an infinite arithmetic progression consisting only of odd numbers, no term of which is of the form $2^k + p$ with p prime.*

Proof We note that $\{0 \bmod 2, 0 \bmod 3, 1 \bmod 4, 3 \bmod 8, 7 \bmod 12, 23 \bmod 24\}$ is a covering system and that the multiplicative order of $2 \bmod p$ is equal to 2, 3, 4, 8, 12 and 24 for the primes $p = 3, 7, 5, 17, 13$ and 241, respectively. Therefore if n is simultaneously congruent to $2^0 \bmod 3$, $2^0 \bmod 7$, $2^1 \bmod 5$, $2^3 \bmod 17$, $2^7 \bmod 13$ and $2^{23} \bmod 241$, then for any k, $n - 2^k$ is divisible by one of the primes 3, 7, 5, 17, 13 or 241. Indeed, for each of these primes we obtain divisibility for all k lying in the corresponding arithmetic progression. If in addition $n \equiv 27 \bmod 32$, say, then $n - 2^k$ cannot be equal to any of these primes, and so is composite. Applying the Chinese Remainder Theorem then gives an arithmetic progression of odd numbers n satisfying these congruences, none of which can be of the form $2^k + p$. □

Numerous other results can be proved using similar techniques. For example, Graham [31] showed that one can construct Fibonacci-like sequences $(a_n)_{n \geq 0}$, where a_0 and a_1 are relatively prime integers and $a_n = a_{n-1} + a_{n-2}$ for all $n \geq 2$, in which all the terms a_n are composite. There has also been much subsequent refinement of this result. Sierpiński [44] showed that there is an arithmetic progression of numbers k such that $k2^n + 1$ is composite for all $n \geq 0$, see also [24], and Erdős [19] showed that there exist infinitely many primes such that if any digit is changed the result is composite, see also [25].

2 Minimum Modulus and the Density of the Uncovered Set

Erdős noted that, in Theorem 1.1, if we could find a similar covering system covering **Z** multiple times, then one could find arithmetic sequences of n which were not of the form $2^k + d$ with d having fewer than some specified number of prime factors. However, in order to achieve this one would have to construct covering systems with larger and larger moduli as each modulus m can only be used a finite number of times (at most once for each prime factor of $2^m - 1$). This prompted him to ask the following question.

Question 2.1 (Minimum modulus problem) *Does there exist covering systems with distinct moduli, $m_1 < m_2 < \cdots < m_n$, with arbitrarily large values of the minimum modulus m_1?*

Erdős noted that it is possible to avoid the use of the modulus 2, and wrote "*It seems likely that for every c there exists such a system all the moduli of which are $> c$*". Later he offered $1000 for a resolution of this problem.

At first sight Erdős's belief seems justified as each progression $a \bmod m$ covers a proportion $\frac{1}{m}$ of the integers and $\sum_{m \geq m_1} \frac{1}{m} = \infty$ for any m_1. Indeed, over the years, various constructions have been given for covering systems with larger and larger minimum modulus [7, 37, 6, 29, 39, 40]. The best example to date is by Owens [40] and has minimum modulus $m_1 = 42$.

In 1980, Erdős and Graham [20] approached the problem from a different angle by considering the density of the uncovered set, that is, the set of integers that are not covered by any of the arithmetic progressions. In particular, they made the following conjecture.

Conjecture 2.2 (Erdős and Graham [20]) *For any $C > 1$ there exists an $n_0(C)$ and $\varepsilon(C) > 0$ such that the following holds. For any set of arithmetic progressions with distinct moduli $m_1, \ldots, m_k \in [n, Cn]$, where $n > n_0(C)$, the uncovered set $\mathbf{Z} \setminus \bigcup_{i=1}^{k} A_i$ has density at least $\varepsilon(C) > 0$, independently of n.*

This is in contrast to the above heuristic as $\sum_{m \in [n, Cn]} \frac{1}{m} \sim \log C$ can be much larger than 1 and hence integers are typically covered many times. On the other hand, we might consider the arithmetic progressions for large moduli as 'usually almost independent', which would suggest a nearly $1/C$ density of integers would remain uncovered.

The first significant progress on these problems was made by Filaseta, Ford, Konyagin, Pomerance and Yu [23] in 2007, who proved Conjecture 2.2 (in a very strong form) and took an important step towards solving Erdős' minimum modulus problem by showing that the sum of reciprocals of the moduli of a covering system with distinct differences grows rapidly with the minimum modulus. Building on their work, and in a remarkable breakthrough, Hough [34] resolved the minimum modulus problem in 2015, showing that in every covering system with distinct moduli, the minimum modulus is indeed bounded.

Theorem 2.3 (Hough [34]) *In any covering system with distinct moduli, $m_1 < m_2 < \cdots < m_n$, we must have $m_1 < 10^{16}$.*

The method used was later simplified and extended by Balister, Bollobás, Morris, Sahasrabudhe and Tiba [4] to reprove this (with an improved bound of $m_1 < 616000$) as well as proving several other results on covering systems. We give a description of this method, which is called the *distortion method*, in sections 5–6.

Theorem 2.3 has since been generalised by Cummings, Filaseta and Trifonov [10] and, independently, by Klein, Koukoulopoulos and Lemieux [36] to to show that, for each fixed k, m_k is also bounded. Indeed, [36] gives the following explicit bound.

Theorem 2.4 (Klein, Koukoulopoulos and Lemieux [36]) *There is an absolute constant c such that in any minimal covering system with moduli $m_1 < m_2 < \cdots < m_n$ we have $m_k < \exp(ck^2/\log(k+1))$.*

Also, in [10], a better bound of $m_1 \leq 118$ is obtained in the special case when all the moduli are square-free.

On the other hand, if only an approximate cover is needed, the situation is very different. The following beautiful question was posed by Filaseta, Ford, Konyagin, Pomerance and Yu [23].

Question 2.5 *Is it true that for each $C > 0$, there exist constants $M > 0$ and $\varepsilon > 0$ such that the following holds: for every covering system whose distinct moduli satisfy*

$$m_1, \ldots, m_n \geq M \quad \text{and} \quad \sum_{k=1}^{n} \frac{1}{m_k} < C,$$

the uncovered set has density at least ε?

Rather surprisingly, this turns out to be false in the strongest possible manner.

Theorem 2.6 (Balister, Bollobás, Morris, Sahasrabudhe and Tiba [4])
For every $M > 0$ and $\varepsilon > 0$, there exists a finite collection of arithmetic progressions $A_1, \ldots, A_n$, with distinct (and square-free) moduli $m_1, \ldots, m_n \geq M$, such that

$$\sum_{i=1}^{n} \frac{1}{m_i} < 1 \tag{2.1}$$

and with the density of the uncovered set $\mathbf{Z} \setminus \bigcup_{i=1}^{n} A_i$ less than ε.

Of course, if $\sum_{i=1}^{n} \frac{1}{m_i} = C < 1$, then the density of the uncovered set is forced to be at least $1 - C > 0$. We reproduce the proof of Theorem 2.6 in Section 8.

Indeed, stronger results are given in [4] showing that the above result holds if we replace the condition (2.1) with

$$\sum_{i=1}^{n} \frac{\mu(m_i)}{m_i} < C$$

where $\mu(m)$ is the multiplicative function defined by $\mu(p^i) = 1 + \frac{\lambda}{p}$ for every prime p and $i \geq 1$, and $C = C(\lambda)$. However, using the distortion method it can be shown to fail for any $C > 0$ if $\mu(p^i)$ is slightly larger.

Theorem 2.7 (Balister, Bollobás, Morris, Sahasrabudhe and Tiba [4])
Let $\varepsilon > 0$ and let μ be the multiplicative function defined by

$$\mu(p^i) = 1 + \frac{(\log p)^{3+\varepsilon}}{p}$$

for all primes p and integers $i \geq 1$. Then there exists $M > 0$ so that if $A_1, \ldots, A_n$ are arithmetic progressions with distinct moduli $m_1, \ldots, m_n \geq M$, and

$$\sum_{i=1}^{n} \frac{\mu(m_i)}{m_i} = C,$$

then the density of the uncovered set $\mathbf{Z} \setminus \bigcup_{i=1}^{n} A_i$ is at least $e^{-4C}/2$.

It should be noted that Theorem 2.7 immediately implies that the minimum modulus of a covering system is bounded, and in fact also proves Conjecture 2.2, as one can show that, for all C, $\sum_{m \in [n, Cn]} \frac{\mu(m)}{m}$ is bounded independently of n.

Another direction concerning how much of $\mathbf{Z}$ remains uncovered when we fail to have a covering system is to ask about the longest consecutive sequence of integers that is covered by a collection of arithmetic progressions.

Strengthening a conjecture of Stein [46], made in 1958, Erdős [15] conjectured that if $A_1, \ldots, A_n$ covers the set $[2^n] = \{1, \ldots, 2^n\}$ then it covers all of $\mathbf{Z}$. The family of progressions $A_i = 2^{i-1}$ mod 2^i for $i = 1, \ldots, n$ shows that 2^n cannot be decreased to $2^n - 1$. Erdős also mentioned this conjecture in several of his later papers, and in support of this conjecture, proved (see [17]) that there exists $N(n)$ such that if $A_1, \ldots, A_n$ covers $[N(n)]$ then it also covers $\mathbf{Z}$. However, the full conjecture was only proved in 1969 by Crittenden and Vanden Eynden [8, 9]. A short proof is given in [1].

Theorem 2.8 (Crittenden and Vanden Eynden) *If the arithmetic progressions $A_1, \ldots, A_n$ cover any set of 2^n consecutive integers, then they cover $\mathbf{Z}$.*

3 Divisibility Constraints

Another line of inquiry concerns covering systems with divisibility constraints on the moduli m_i. In [43], Schinzel related a problem on the irreducibility of certain integer polynomials to the existence of covering systems with moduli having certain divisibility properties. More specifically, consider the following assertion.

(A) For every integer polynomial $f(x) \in \mathbf{Z}[x]$ with $f \not\equiv 1$, $f(1) \neq -1$, $f(0) \neq 0$ there exists an infinite arithmetic progression of values of n such that the polynomial $f(x)x^n + 1$ is irreducible in $\mathbf{Z}[x]$.

Schinzel showed that (A) was equivalent to

(B) For every covering system with all $m_i > 1$ there exists two moduli m_i and m_j, $i \neq j$, with $m_j/m_i = p^a$ for some prime p and $a \geq 0$ with either $p > 2$, $p = 2$ and m_i odd, or $p = 2$ and $a_j \not\equiv a_i \bmod m_i/2$.

He also noted two implications. If (B) holds then it clearly implies the following.

Conjecture 3.1 (Schinzel) *In any covering system, there exists $i \neq j$ with $m_i \mid m_j$.*

Also, he showed that (B) is implied by the following conjecture of Erdős and Selfridge (see [32]).

Conjecture 3.2 (Erdős–Selfridge odd covering conjecture) *There is no covering system of congruences with distinct odd moduli $m_i > 1$.*

As recounted in [22], Erdős (who thought that such coverings are likely to exist) offered \$25 for a proof that there is no covering with these properties, and Selfridge (who expected the opposite) offered \$300 (later increased to \$2000) for a construction of such a covering.

The implication Conjecture 3.2 $\Rightarrow$ Conjecture 3.1 had previously been noted by Selfridge and a simple proof can be found in [21]. Schinzel's conjecture was later proved by the distortion method.

Theorem 3.3 (Balister, Bollobás, Morris, Sahasrabudhe and Tiba [4]) *Schinzel's conjecture holds: in every covering system, there exists $i \neq j$ with $m_i \mid m_j$.*

We sketch the proof of Theorem 3.3 in Section 7. The Erdős–Selfridge conjecture on the other hand is still wide open, but a slightly weaker version of this conjecture was proved in [35].

Theorem 3.4 (Hough and Nielsen [35]) *There is no covering system with distinct moduli $m_i > 1$ and no m_i divisible by either 2 or 3.*

Also, a square-free version of the conjecture was proved in [3].

Theorem 3.5 (Balister, Bollobás, Morris, Sahasrabudhe and Tiba [3]) *There is no covering system with distinct odd square-free moduli $m_i > 1$.*

We discuss these results further in Section 6. For further results connecting the irreducibility of polynomials with covering systems, see [26].

4 Counting Covering Systems

Another question posed by Erdős [14] is: "*How many minimal covering systems of size n are there?*" In order to answer this question, Erdős gave a simple proof that there are only finitely many minimal covering systems of size n, but the bound he obtained on their number was doubly exponential. A more reasonable upper bound follows from a result of Simpson [45], who proved in 1985 that the largest modulus in a minimal covering system of size n is at most 2^{n-1} (even with repeated moduli allowed). Note that this bound is best possible since $\{2^i - 1 \bmod 2^i : 0 \leq i \leq n-1\} \cup \{0 \bmod 2^{n-1}\}$ is a minimal covering system. This bound easily implies that there are at most $2^{O(n^2)}$ minimal covering systems of size n. In fact there are rather fewer such systems.

Theorem 4.1 (Balister, Bollobás, Morris, Sahasrabudhe and Tiba [5])
The number of minimal covering systems of size n (either with or without repeated moduli allowed) is

$$\exp\left(\left(\frac{4\sqrt{\tau}}{3} + o(1)\right)\frac{n^{3/2}}{(\log n)^{1/2}}\right)$$

as $n \to \infty$, where

$$\tau = \sum_{t=1}^{\infty} \left(\log \frac{t+1}{t}\right)^2 \approx 0.977189.$$

To give a lower bound that is of the right order of magnitude, we can describe a simple construction that gives a slightly weaker lower bound. Let $p_1 < \cdots < p_k$ be the first k primes, and for each $i \in [k]$, choose $p_i - 1$ arithmetic progressions $A_1^{(i)}, \ldots, A_{p_i-1}^{(i)}$ with the following properties: for each $j \in [p_i - 1]$, the modulus of $A_j^{(i)}$ is divisible by p_i and divides $Q_i := p_1 \ldots p_i$, and $A_j^{(i)}$ contains $j \cdot Q_{i-1}$. It is not difficult to show that, for each such choice, by adding the progression $0 \bmod Q_k$, we obtain distinct minimal covering systems of size $n = \sum_{i=1}^{k}(p_i - 1) + 1 \sim \frac{k^2}{2}\log k$. Since we have 2^{i-1} choices for the progression $A_j^{(i)}$ for each $i \in [k]$ and $j \in [p_i - 1]$, this implies that there are at least

$$\prod_{i=1}^{k} 2^{(i-1)(p_i-1)} = \exp\left(\Theta(k^3 \log k)\right) = \exp\left(\frac{\Theta(n^{3/2})}{(\log n)^{1/2}}\right)$$

minimal covering systems of size n.

This lower bound is not significantly changed if we require the moduli to be distinct. We just need the $p_i - 1$ choices of the modulus of $A_j^{(i)}$ among the 2^{i-1} possibilities to be distinct (and, for $i = k$, distinct from Q_k). As $2^{i-1} \geq p_i - 1$ for all i, and as 2^{i-1} grows much faster than $p_i - 1$, we get asymptotically about the same bound. It is also worth remarking here that this shows that there exists covering systems with distinct *square-free* moduli. Indeed the above construction shows one exists with all moduli $m_i \mid 2 \cdot 3 \cdot 5 \cdot 7$.

In [5] a more general construction is given that gives a lower bound with the right constant in the exponent. The hard part of course is to give a corresponding upper bound. For this a structural result is needed that shows that most such covering systems are 'close' to a (rather technical) generalisation of the systems given above.

The question of counting *exact covering systems*, that is, covering systems where each integer is covered exactly once, has also been considered. See for example [28, 30].

5 The Distortion Method

In this section we will give an outline of the distortion method, which is based on the proof of the minimum modulus problem by Hough, but significantly simplified by Balister, Bollobás, Morris, Sahasrabudhe and Tiba [4]. A simple exposition of this method in the case of square-free moduli is given in [2].

We wish to show that a collection of progressions $\{a_i \bmod m_i : 1 \le i \le n\}$ does *not* cover the integers. Clearly the collection covers $a \in \mathbf{Z}$ if and only if it covers the entire arithmetic progression $a \bmod Q$ where $Q = \mathrm{lcm}(m_1, \ldots, m_n)$. Thus we are reduced to the question of whether these progressions cover the finite set $\mathbf{Z}/Q\mathbf{Z}$. Write

$$Q = p_1^{\gamma_1} \cdots p_t^{\gamma_t},$$

where the p_k are distinct primes and $\gamma_k \ge 1$. Then by the Chinese Remainder Theorem $\mathbf{Z}/Q\mathbf{Z} \cong (\mathbf{Z}/p_1^{\gamma_1}\mathbf{Z}) \times \cdots \times (\mathbf{Z}/p_t^{\gamma_t}\mathbf{Z})$ and each progression $A_i = \{a_i \bmod m_i\}$ corresponds to some product of subsets of the form $A_{i,k} = \{a_i \bmod p_k^{\alpha_{i,k}}\} \subseteq \mathbf{Z}/p_k^{\gamma_k}\mathbf{Z}$, where $m_i = p_1^{\alpha_{i,1}} \cdots p_t^{\alpha_{i,t}}$.

At this point we can, to a large extent, dispense with the number theory and consider the following general setting. We have a finite product set

$$P = P_1 \times P_2 \times \cdots \times P_t,$$

with each $|P_i| \ge 2$ and we wish to know if we can cover it with product sets $A_1, \ldots, A_n$, where

$$A_i = A_{i,1} \times A_{i,2} \times \cdots \times A_{i,t} \subseteq P.$$

We note that in the simplest case when all m_i are square-free, each P_k is just $\mathbf{Z}/p_k\mathbf{Z}$ and each $A_{i,k}$ is either a singleton or the whole of P_k. For each i let $F(A_i) = \{k : A_{i,k} \ne P_k\}$ be the set of fixed (or restricted) coordinates of A_i. Then in our case $F(A_i) = \{k : p_k \mid m_i\}$ corresponds to the set of primes dividing the modulus m_i of A_i. To keep the notation reasonably consistent with the case of arithmetic progressions, we will define

$$m_{i,k} = \frac{|P_k|}{|A_{i,k}|} \qquad \text{and} \qquad m_i = \prod_{k=1}^{t} m_{i,k} = \frac{|P|}{|A_i|},$$

where, in the more general geometric case, we do not insist that the m_i and $m_{i,k}$ are integers. Then $F(A_i) = \{k : m_{i,k} > 1\}$.

We aim to show the A_i do not cover P by finding a random point $X \in P$ that is not covered with positive probability. The random variable X will not be uniform on P, but will be constructed so as to avoid the sets A_i as much as possible while still being approximately uniform elsewhere.

More specifically we construct the distribution of X in stages. Writing $X = (X_1, X_2, \ldots, X_t) \in P = P_1 \times \cdots \times P_t$, we construct the distribution of each X_k inductively so that X_k is dependent on the X_ℓ, $\ell < k$, but even conditioned on the previous X_ℓ, X_k will still be 'close' to uniform on P_k. However, the distribution of

X_k will be distorted so as to increase the probability that $(X_1, \ldots, X_t)$ is not covered by the A_i with $F(A_i) \subseteq \{1, \ldots, k\}$, but not so much that later progressions A_i with $F(A_i) \not\subseteq \{1, \ldots, k\}$ might be hit with too high a probability.

To this end, define
$$\mathcal{A}_k := \{A_i : \max F(A_i) = k\}$$
to be the collection of sets A_i where the largest restricted coordinate is k. In the original covering system this corresponds to all progressions where the largest prime dividing m_i is p_k. We let
$$B_k := \bigcup_{A \in \mathcal{A}_k} A \subseteq P$$
be the subset of P that is covered by these. For each k in turn we shall 'sieve out' the elements of B_k and define X_k so that the measure sieved out is small, without compromising too much the uniformity of the measure. Note that B_k is of the form $B'_k \times P_{k+1} \times \cdots \times P_t$ for some $B'_k \subseteq P_1 \times \cdots \times P_k$ as, by definition, none of the sets in $\mathcal{A}_k$ are restricted in coordinates $k+1, \ldots, t$. Thus $\mathbb{P}(X \in B_k)$ only depends on the distribution of $(X_1, \ldots, X_k)$ and will be unaffected by the distributions of X_ℓ, $\ell > k$.

Now suppose we have defined $X_1, \ldots, X_{k-1}$ and fix some $\delta_k \in [0, \frac{1}{2}]$. Define
$$\alpha_k = \alpha_k(X_1, \ldots, X_{k-1}) = \frac{|\{x_k : (X_1, \ldots, X_{k-1}, x_k) \in B'_k\}|}{|P_k|},$$
i.e., α_k is the proportion of extensions of $(X_1, \ldots, X_{k-1})$ to $(X_1, \ldots, X_{k-1}, x_k)$ that lie in B'_k. Equivalently we can think of α_k as the random variable
$$\alpha_k = \mathbb{P}\left((X_1, \ldots, X_{k-1}, U_k, \ldots, U_t) \in B_k \mid X_1, \ldots, X_{k-1}\right), \quad (5.1)$$
where U_j are independent random variables that are uniform in P_j.

The random variable X_k is then defined so that, for $x_k \in P_k$,
$$\mathbb{P}(X_k = x_k \mid X_1, \ldots, X_{k-1}) =$$
$$= \begin{cases} \max\left\{0, \dfrac{\alpha_k - \delta_k}{\alpha_k(1 - \delta_k)}\right\} \cdot \dfrac{1}{|P_k|}, & \text{if } (X_1, \ldots, X_{k-1}, x_k) \in B'_k; \\ \min\left\{\dfrac{1}{1-\alpha_k}, \dfrac{1}{1-\delta_k}\right\} \cdot \dfrac{1}{|P_k|}, & \text{if } (X_1, \ldots, X_{k-1}, x_k) \notin B'_k. \end{cases} \quad (5.2)$$

In other words, we minimise $\mathbb{P}(X \in B_k)$ while at the same time ensuring that
$$\mathbb{P}(X_k = x_k \mid X_1, \ldots, X_{k-1}) \leq \frac{1}{(1 - \delta_k)|P_k|} \quad (5.3)$$
is not much larger than for a uniform random variable for any $x_k \in P_k$. Indeed, if for a particular choice of $X_1, \ldots, X_{k-1}$ we have $\alpha_k \leq \delta_k$, then we completely avoid B_k with our choice of X_k. Otherwise we place as much mass as we can outside of B_k and distribute the rest uniformly in B_k. These two cases correspond to the first and second terms respectively in the max and min expressions in (5.2).

The rationale behind the restriction (5.3) is that increasing the probability too much in some places could make it possible that X is much more likely to lie in some future B_ℓ, $\ell > k$. Bounding the amount the probability distribution is distorted mitigates against this.

It is easily checked that (5.2) does indeed define a probability distribution for $(X_1, \ldots, X_k)$ as $\alpha_k |P_k|$ values of x_k result in $(X_1, \ldots, X_{k-1}, x_k) \in B_k'$ and

$$\max\left\{0, \frac{\alpha_k - \delta_k}{\alpha_k(1-\delta_k)}\right\} \cdot \frac{\alpha_k |P_k|}{|P_k|} + \min\left\{\frac{1}{1-\alpha_k}, \frac{1}{1-\delta_k}\right\} \cdot \frac{(1-\alpha_k)|P_k|}{|P_k|} = 1.$$

Our next task is to obtain reasonable upper bounds on $\mathbb{P}(X \in B_k)$. We first do this in terms of the 1st and 2nd moments of α_k. Later we will discuss how to bound these moments in practice.

Lemma 5.1

$$\mathbb{P}(X \in B_k) \leq \mathbb{E}[\alpha_k] \quad \text{and} \quad \mathbb{P}(X \in B_k) \leq \frac{\mathbb{E}[\alpha_k^2]}{4\delta_k(1-\delta_k)}.$$

Proof Conditioning on $X_1, \ldots, X_{k-1}$ we have

$$\mathbb{P}(X \in B_k \mid X_1, \ldots, X_{k-1}) = \max\left\{0, \frac{\alpha_k - \delta_k}{\alpha_k(1-\delta_k)}\right\} \cdot \alpha_k = \max\left\{0, \frac{\alpha_k - \delta_k}{1-\delta_k}\right\}.$$

This is clearly at most α_k as $\alpha_k - \delta_k \leq \alpha_k(1-\delta_k)$. It is also at most $\alpha_k^2/4\delta_k(1-\delta_k)$. Indeed, in general, $\max\{\alpha - \delta, 0\} \leq \alpha^2/4\delta$ for $\alpha, \delta > 0$ follows by rearranging the inequality $(\alpha - 2\delta)^2 \geq 0$. Hence

$$\mathbb{P}(X \in B_k \mid X_1, \ldots, X_{k-1}) \leq \alpha_k \quad \text{and} \quad \mathbb{P}(X \in B_k \mid X_1, \ldots, X_{k-1}) \leq \frac{\alpha_k^2}{4\delta_k(1-\delta_k)}.$$

Taking expectations over $X_1, \ldots, X_{k-1}$ then gives the result. □

In general, when using the first bound we usually set $\delta_k = 0$ and so X_k is simply uniform on P_k. The interesting case is when the second bound is more effective, and in this case we need to choose $\delta_k \in (0, \frac{1}{2}]$ suitably.

Now let $B = \bigcup_{k=1}^t B_k = \bigcup_{i=1}^n A_i \subseteq P$ be the set covered by the A_i. To show $B \neq P$ it is enough to show $\mathbb{P}(X \in B) < 1$. Thus, by the union bound, it is enough to show

$$\sum_{k=1}^t \mathbb{P}(X \in B_k) < 1$$

using the bounds given in Lemma 5.1. Of course we now have the problem of bounding $\mathbb{E}[\alpha_k]$ or $\mathbb{E}[\alpha_k^2]$. We start with a simple observation.

Lemma 5.2 *If $A = A_1 \times \cdots \times A_t \subseteq P$ then*

$$\mathbb{P}(X \in A) \leq \frac{|A|}{|P|} \prod_{k \in F(A)} \frac{1}{1-\delta_k}.$$

Proof Recall that $F(A) = \{k : A_k \neq P_k\}$ and expand $\mathbb{P}(X \in A)$ as

$$\mathbb{P}(X \in A) = \prod_{k=1}^{t} \mathbb{P}(X_k \in A_k \mid X_1 \in A_1, \ldots, X_{k-1} \in A_{k-1}).$$

The factors where $k \notin F(A)$ have $A_k = P_k$, and so are clearly equal to 1. On the other hand, for $k \in F(A)$,

$$\mathbb{P}(X_k = x_k \mid X_1, \ldots, X_{k-1}) \leq \frac{1}{1 - \delta_k} \cdot \frac{1}{|P_k|}$$

by (5.3). Thus

$$\mathbb{P}(X_k \in A_k \mid X_1 \in A_1, \ldots, X_{k-1} \in A_{k-1}) \leq \frac{1}{1 - \delta_k} \cdot \frac{|A_k|}{|P_k|}.$$

The result then follows as $\prod_{k \in F(A)} \frac{|A_k|}{|P_k|} = \frac{|A|}{|P|}$. □

Bounding $\mathbb{E}[\alpha_k]$ and $\mathbb{E}[\alpha_k^2]$ depends somewhat on the restrictions placed on the sets A_i, but the most general bounds are as follows.

Lemma 5.3

$$\mathbb{E}[\alpha_k] \leq \sum_{A_i \in \mathcal{A}_k} \frac{1}{m_{i,k}} \prod_{\ell \in F(A_i) \setminus \{k\}} \frac{1}{(1 - \delta_\ell) m_{i,\ell}} = \sum_{A_i \in \mathcal{A}_k} \frac{1}{m_i} \prod_{\ell \in F(A_i) \setminus \{k\}} \frac{1}{1 - \delta_\ell},$$

$$\mathbb{E}[\alpha_k^2] \leq \sum_{A_i, A_j \in \mathcal{A}_k} \frac{1}{m_{i,k} m_{j,k}} \prod_{\ell \in F(A_i) \cup F(A_j) \setminus \{k\}} \frac{1}{(1 - \delta_\ell) \max\{m_{i,\ell}, m_{j,\ell}\}}.$$

Proof We prove the second statement only as the first is much easier. Let U_j be independent uniform random variables in P_j and, for convenience write $(X_{<k}, U_{\geq k}) = (X_1, \ldots, X_{k-1}, U_k, \ldots, U_t)$. Then by (5.1) and the union bound we have

$$\alpha_k = \mathbb{P}\left((X_{<k}, U_{\geq k}) \in B_k \mid X_{<k}\right) \leq \sum_{A_i \in \mathcal{A}_k} \mathbb{P}\left((X_{<k}, U_{\geq k}) \in A_i \mid X_{<k}\right).$$

Hence

$$\mathbb{E}[\alpha_k^2] \leq \mathbb{E} \sum_{A_i, A_j \in \mathcal{A}_k} \mathbb{P}\left((X_{<k}, U_{\geq k}) \in A_i \mid X_{<k}\right) \mathbb{P}\left((X_{<k}, U_{\geq k}) \in A_j \mid X_{<k}\right)$$

$$\leq \mathbb{E} \sum_{A_i, A_j \in \mathcal{A}_k} \mathbb{P}(U_k \in A_{i,k}) \mathbb{1}_{\{X \in A_i'\}} \cdot \mathbb{P}(U_k \in A_{j,k}) \mathbb{1}_{\{X \in A_j'\}}$$

$$= \sum_{A_i, A_j \in \mathcal{A}_k} \mathbb{P}(U_k \in A_{i,k}) \mathbb{P}(U_k \in A_{j,k}) \mathbb{P}(X \in A_i' \cap A_j')$$

where $A_i' = A_{i,1} \times \cdots \times A_{i,k-1} \times P_k \times \cdots \times P_n$ is obtained from A_i by replacing $A_{i,k}$ by P_k. (As $A_i \in \mathcal{A}_k$ we have $A_{i,j} = P_j$ for $j > k$.)

Now $A'_i \cap A'_j$ is the product of the sets $A'_{i,\ell} \cap A'_{j,\ell}$ and we can bound $|A'_{i,\ell} \cap A'_{j,\ell}| \le \min(|A'_{i,\ell}|, |A'_{j,\ell}|)$. Thus, by Lemma 5.2, the definition $m_{i,k} = |P_k|/|A_{i,k}|$ and the definition of A'_i, we have

$$\mathbb{P}(X \in A'_i \cap A'_j) \le \prod_{\ell \in F(A_i) \cup F(A_j) \setminus \{k\}} \frac{1}{(1-\delta_\ell) \max\{m_{i,\ell}, m_{j,\ell}\}}.$$

Hence

$$\mathbb{E}[\alpha_k^2] \le \sum_{A_i, A_j \in \mathcal{A}_k} \frac{1}{m_{i,k} m_{j,k}} \prod_{\ell \in F(A_i) \cup F(A_j) \setminus \{k\}} \frac{1}{(1-\delta_\ell) \max\{m_{i,\ell}, m_{j,\ell}\}}.$$

as required. □

Applying Lemma 5.3 involves choosing the δ_i and calculating the bounds given the relevant restrictions on the m_i applicable to the problem at hand. However, before we look at specific cases, we note that it is also possible to bound the (undistorted) size of the uncovered set, $\mathbb{P}(U \notin B) = \frac{|P \setminus B|}{|P|}$, in terms of $\mathbb{P}(X \notin B) \ge 1 - \sum_k \mathbb{P}(X \in B_k)$. This is used, for example, in the proof of Theorem 2.7 where we wish to have an explicit bound on the density of the uncovered set.

Theorem 5.4 *If* $\mathbb{P}(X \notin B) > 0$ *then*

$$\mathbb{P}(U \notin B) \ge \mathbb{P}(X \notin B) \exp\Big(- \frac{2}{\mathbb{P}(X \notin B)} \sum_i \frac{1}{m_i} \prod_{\ell \in F(A_i)} \frac{1}{1-\delta_\ell} \Big).$$

Proof We first note that for $x \notin B$,

$$\frac{\mathbb{P}((X_1, \ldots, X_{k-1}, X_k, U_{k+1}, \ldots, U_t) = x)}{\mathbb{P}((X_1, \ldots, X_{k-1}, U_k, U_{k+1}, \ldots, U_t) = x)} = \min\Big\{ \frac{1}{1-\alpha_k}, \frac{1}{1-\delta_k} \Big\}$$

by (5.2), where $\alpha_k = \alpha_k(x)$ is now considered as a function of x. Now if $\alpha_k \le \frac{1}{2}$ then $1 - \alpha_k \ge e^{-2\alpha_k}$. Otherwise, as $\delta_k \in [0, \frac{1}{2}]$, we have $1 - \delta_k \ge \frac{1}{2} \ge e^{-2\alpha_k}$. Hence

$$\frac{\mathbb{P}(X = x)}{\mathbb{P}(U = x)} \ge \prod_{k=1}^t \min\Big\{ \frac{1}{1-\alpha_k}, \frac{1}{1-\delta_k} \Big\} \le \exp\Big(2 \sum_{k=1}^t \alpha_k\Big).$$

Now, by the convexity of e^{-z} we have

$$\mathbb{P}(U \notin B) \ge \sum_{x \notin B} \exp\Big(- 2 \sum \alpha_k(x) \Big) \mathbb{P}(X = x)$$

$$= \mathbb{E}\Big[\exp\Big(- 2 \sum \alpha_k(X) \Big) \mathbb{1}_{\{X \notin B\}} \Big]$$

$$= \mathbb{P}(X \notin B) \mathbb{E}\Big[\exp\Big(- 2 \sum \alpha_k(X) \Big) \mid X \notin B \Big]$$

$$\ge \mathbb{P}(X \notin B) \exp\Big(- 2 \mathbb{E}\Big[\sum \alpha_k(X) \mid X \notin B \Big] \Big)$$

$$\ge \mathbb{P}(X \notin B) \exp\Big(- \frac{2}{\mathbb{P}(X \notin B)} \sum \mathbb{E}[\alpha_k] \Big).$$

The result follows from Lemma 5.3 as

$$\mathbb{E}[\alpha_k] \le \sum_{A_i \in \mathcal{A}_k} \frac{1}{m_i} \prod_{\ell \in F(A_i) \setminus \{k\}} \frac{1}{1-\delta_\ell} \le \sum_{A_i \in \mathcal{A}_k} \frac{1}{m_i} \prod_{\ell \in F(A_i)} \frac{1}{1-\delta_\ell}.$$

□

6 Applying the Distortion Method

In this section we describe practical methods of applying the distortion method to some of the questions concerning covering systems described in the earlier sections. Write p_k for the kth prime, so that $p_1 = 2$, $p_2 = 3$, etc. Recall that $P_k = \mathbf{Z}/p_k^{\gamma_k}\mathbf{Z}$, and each modulus m_i has the factorisation $m_i = \prod m_{i,k}$ where $m_{i,k} = p_k^{\alpha_{i,k}}$ for some $0 \leq \alpha_{i,k} \leq \gamma_k$.

Recall that from Lemma 5.3 we have the general bound

$$\mathbb{E}[\alpha_k^2] \leq \sum_{A_i, A_j \in \mathcal{A}_k} \frac{1}{m_{i,k} m_{j,k}} \prod_{\ell \in F(A_i) \cup F(A_j) \setminus \{k\}} \frac{1}{(1-\delta_\ell) \max\{m_{i,\ell}, m_{j,\ell}\}}$$

We can write each m_i with $A_i \in \mathcal{A}_k$ as $m_i = m_i' p_k^{\alpha_i}$, where $p_k \nmid m_i'$, in which case this bound becomes

$$\mathbb{E}[\alpha_k^2] \leq \sum_{A_i, A_j \in \mathcal{A}_k} \frac{1}{p_k^{\alpha_i+\alpha_j}} \frac{1}{\mathrm{lcm}(m_i', m_j')} \prod_{p_\ell | \mathrm{lcm}(m_i', m_j')} \frac{1}{1-\delta_\ell}. \qquad (6.1)$$

We wish to find effective ways of estimating the bound given in (6.1). For this, fix k and write each $m_i = m_i' m_i''$, where m_i' is divisible only by primes p_ℓ, $\ell \leq k$, the p_k-smooth part of m_i, and m_i'' is only divisible by primes p_ℓ, $\ell > k$. Set

$$S_k = \{m' : \exists i \colon m_i' = m' \text{ and } m_i'' > 1\}$$

to be the p_k-smooth parts of the m_i that are not themselves p_k-smooth. Note that for any k

$$S_k \subseteq \{mp_k^\alpha : m \in S_{k-1}, 0 \leq \alpha \leq \gamma_k\}. \qquad (6.2)$$

Now define

$$\kappa_k = \sum_{m, m' \in S_k} \frac{1}{\mathrm{lcm}(m, m')} \prod_{p_\ell | \mathrm{lcm}(m, m')} \frac{1}{1-\delta_\ell}. \qquad (6.3)$$

Then by (6.1) we have the inequalities

$$\mathbb{E}[\alpha_k^2] \leq \left(\sum_{i,j=1}^{\gamma_k} \frac{1}{p_k^{i+j}} \right) \cdot \kappa_{k-1}$$

and, by (6.2),

$$\kappa_k \leq \sum_{i,j=0}^{\gamma_k} \sum_{m,m' \in S_{k-1}} \frac{1}{\mathrm{lcm}(m,m') p_k^{\max\{i,j\}}} \prod_{p_\ell | \mathrm{lcm}(m,m') p_k^{\max\{i,j\}}} \frac{1}{1-\delta_\ell}$$

$$= \left(1 + \sum_{\substack{i,j=0 \\ (i,j) \neq (0,0)}}^{\gamma_k} \frac{1}{p_k^{\max\{i,j\}}} \cdot \frac{1}{1-\delta_k} \right) \cdot \kappa_{k-1}.$$

Now one can check that

$$\sum_{i,j=1}^\infty \frac{1}{p^{i+j}} = \left(1 + \frac{1}{p} + \frac{1}{p^2} + \cdots \right)^2 = \frac{1}{(p-1)^2}$$

Erdős Covering Systems

and
$$\sum_{\substack{i,j=0 \\ (i,j)\neq(0,0)}}^{\infty} \frac{1}{p^{\max\{i,j\}}} = \frac{3}{p} + \frac{5}{p^2} + \frac{7}{p^3} + \cdots = \frac{3p-1}{(p-1)^2}.$$

For convenience, define for each $k \in \mathbf{N}$,

$$a_k = \frac{3p_k - 1}{(p_k - 1)^2} \quad \text{and} \quad b_k = \frac{1}{(p_k - 1)^2}. \tag{6.4}$$

Then, applying Lemma 5.2, we obtain the general bounds

$$\mathbb{P}(X \in B_k) \leq \frac{\kappa_{k-1} b_k}{4\delta_k(1-\delta_k)}, \qquad \kappa_k \leq \left(1 + \frac{a_k}{1-\delta_k}\right)\kappa_{k-1}. \tag{6.5}$$

We note for future reference that if we were interested in square-free moduli only, then $\gamma_k = 1$ and we can set

$$a_k = \frac{3}{p_k} \quad \text{and} \quad b_k = \frac{1}{p_k^2} \tag{6.6}$$

in place of (6.4).

The bounds given in (6.5) allow us to inductively bound $\sum \mathbb{P}(X \in B_k)$, and if we can show the sum is less than 1 we know that the system of arithmetic progressions does not cover the integers. Two problems remain. The first is that usually we have no upper bound on the size of the primes dividing the moduli. The second is that we still need to choose the δ_k appropriately.

To simplify the calculations, we define a sequence of numbers $f_k = f_k(\mathcal{A})$, which will (roughly speaking) encode how "well" we are doing after k steps of our sieve. Define

$$\mu_k = 1 - \sum_{j \leq k} \mathbb{P}(X \in B_j) \tag{6.7}$$

and

$$f_k = \frac{\kappa_k}{\mu_k}. \tag{6.8}$$

Lemma 6.1 *Let $k > k_0$, and assume that $\mu_{k-1} > 0$. If $b_k f_{k-1} < 4\delta_k(1-\delta_k)$, then $\mu_k > 0$, and*

$$f_k \leq \left(1 + \frac{a_k}{1-\delta_k}\right)\left(1 - \frac{b_k f_{k-1}}{4\delta_k(1-\delta_k)}\right)^{-1} f_{k-1}. \tag{6.9}$$

Proof We have

$$\frac{f_k}{f_{k-1}} = \left(\frac{\kappa_k}{\kappa_{k-1}}\right)\left(\frac{\mu_k}{\mu_{k-1}}\right)^{-1} = \left(\frac{\kappa_k}{\kappa_{k-1}}\right)\left(1 - \frac{\mathbb{P}(X \in B_k)}{\mu_{k-1}}\right)^{-1}$$

$$\leq \left(1 + \frac{a_k}{1-\delta_k}\right)\left(1 - \frac{b_k \kappa_{k-1}}{4\delta_k(1-\delta_k)\mu_{k-1}}\right)^{-1},$$

by (6.5). The result then follows noting that $\kappa_{k-1}/\mu_{k-1} = f_{k-1}$. □

Theorem 6.2 *Let $k \geq 10$. If $\mu_k > 0$ and $f_k(\mathcal{A}) \leq (\log k + \log\log k - 3)^2 k$, then the system of arithmetic progressions $\mathcal{A}$ does not cover $\mathbf{Z}$.*

Proof Set $\delta_j = \frac{1}{2}$ for all $j \geq k$. By Lemma 6.1, it is enough to show that $b_j f_{j-1} < 1$ and $f_j \leq (\log j + \log \log j - 3)^2 j$ for all $j \geq k$ by induction on j. Indeed, this would imply that $\mu_j > 0$ for all j, and so, finally, after all t rounds of the sieve, we have $\mathbb{P}(X \in \bigcup_i A_i) \geq 1 - \sum_{k=1}^{t} \mathbb{P}(X \in B_k) = \mu_t > 0$.

To show $f_j \leq (\log j + \log \log j - 3)^2 j$ one bounds $p_j \geq j(\log j + \log \log j - 1)$ for $j \geq 2$ using the main result of [11] and then deduces the bound on f_j from that of f_{j-1} via Lemma 6.1 and some (rather unpleasant) algebraic manipulation. See [4] for details. □

The criterion given in Theorem 6.2 is reasonably good for large k but, by combining Theorem 6.2 and Lemma 6.1, we can deduce better bounds on f_k that imply that the uncovered set is non-empty. To do so, observe first that, given f_{k-1}, the bound on f_k given by Lemma 6.1 is just a function of f_{k-1} and a *single* δ_k. Elementary calculus shows that the optimal choice of δ_k occurs when

$$\delta_k = \frac{1 + a_k}{1 + \sqrt{1 + 4a_k(1 + a_k)/(b_k f_{k-1})}}. \tag{6.10}$$

This expression for δ_k allows for fast numerical computation of the bounds on the f_k. Indeed, for each k one can calculate an almost optimal value f_k^* such that if $f_k \leq f_k^*$ then the uncovered set is non-empty. We simply find an f_k^* such that, starting with this value for k and repeatedly applying Lemma 6.1 with the choice (6.10) of δ_k, we get a bound on $f_{k'}$ for some much larger k' which satisfies the conditions of Theorem 6.2.

In Table 1 we give suitable bounds f_k^* by performing this calculation, which was implemented as follows: starting with a potential value of f_1, we ran the iteration given in (6.9) using the value of δ_k given in (6.10) until either the conditions of Theorem 6.2 were satisfied, or the condition $b_i f_i - 1 < 4\delta_i(1 - \delta_i)$ failed. (Also, at each stage, f_k was increased very slightly to compensate for any errors in the floating point arithmetic.) The optimal value f_1^* was determined by binary search. The other bounds f_k^* were read off by taking the largest successful f_1^* and listing the corresponding bounds on f_k (rounded down to ensure bound is rigorous). The bounds f_k^{*sf} were calculated in a similar manner for the case of square-free moduli (so using (6.6) instead of (6.4)). To summarise we have the following.

Corollary 6.3 *If $f_k(\mathcal{A}) \leq f_k^*$ for some $k \in \mathbb{N}$, where f_k^* are given in Table 1, then the system of arithmetic progressions $\mathcal{A}$ does not cover $\mathbb{Z}$. If all moduli are square-free then $f_k(\mathcal{A}) \leq f_k^{*sf}$ is sufficient.*

As an example, we prove the slight strengthening of Theorem 3.4 given in [4].

Theorem 6.4 *Let $\mathcal{A}$ be a finite collection of arithmetic progressions with distinct moduli $m_1, \ldots, m_n$ that covers the integers, and let $Q = \text{lcm}(m_1, \ldots, m_n)$ be the least common multiple of the moduli. Then either $2 \mid Q$, or $9 \mid Q$, or $15 \mid Q$.*

In other words, either there is an even m_i, an m_i divisible by $3^2 = 9$, or there are m_i, m_j (possibly equal) with $3 \mid m_i$ and $5 \mid m_j$. We remark that we are unable to prove that a *single* m_i has $15 \mid m_i$ in this last case.

k	p_k	f_k^*	f_k^{*sf}	f_k^T
1	2	0.31785912	0.77410271	-
2	3	1.26099737	2.01063772	-
3	5	3.00788829	4.02433030	-
4	7	5.86093846	7.08291937	-
5	11	9.03208215	10.4496911	-
6	13	13.3034434	14.8801623	-
7	17	17.9968785	19.7240000	-
8	19	23.9097336	25.7500269	-
9	23	30.3872205	32.3252543	-
10	29	36.7237215	38.7675845	0.18664351
100	541	1691.36578	1695.78046	981.161534
1000	7919	42420.7852	42427.6802	34110.2723
10000	104729	802133.868	802143.238	710761.490
100000	1299709	12989056.4	12989068.2	12004260.9
1000000	15485863	191138582	191138597	180668612

Table 1: Upper bounds f_k^* on f_k that ensure that the system does not cover the integers. Slightly weaker bounds f_k^{*sf} are sufficient if all moduli are square-free. The bounds $f_k^T := (\log k + \log \log k - 3)^2 k$ given by Theorem 6.2 are listed for comparison. The bounds f_k^* are slightly improved from [4] as more extensive calculations were performed for this survey. All the bounds given are rounded down in the last place.

Proof If no modulus is divisible by 3 we are reduced to proving Theorem 3.4. In this case $S_2 = \{1\}$ as no modulus can have a factor of 2 or 3. Then $\kappa_2 = 1$ and $\mu_2 = 1$ (by (6.3) and (6.7)), so $f_2 = \kappa_2/\mu_2 = 1 \leq f_2^* = 1.26\ldots$, and the progressions do not cover the integers by Corollary 6.3.

If some modulus is divisible by 3 then no modulus is divisible by 5. In this case $S_3 = \{1,3\}$ as no modulus can have a factor of 2, 5, or 9. In this case we assume $\delta_1 = \delta_2 = \delta_3 = 0$ so X_1, X_2, X_3 are uniform. Then $\mu_3 \geq 1 - \frac{1}{3} = \frac{2}{3}$ (the only 5-smooth modulus is 3) and $\kappa_3 = 1 + \frac{3}{3} = 2$, so $f_3 = \kappa_3/\mu_3 \leq 3 \leq f_3^* = 3.007\ldots$, and the progressions do not cover the integers by Corollary 6.3. $\square$

We note that for the Erdős–Selfridge conjecture (Conjecture 3.2), we would have $S_1 = \{1\}$ and $f_1 = \kappa_1 = \mu_1 = 1$. Unfortunately, $f_1^* < 1$, so the Erdős–Selfridge conjecture does not follow from the results given above. Even for square-free moduli, we do not quite get the conjecture as $f_1^{*sf} < 1$ as well. However, the square-free version of the Erdős–Selfridge is proved, see [3]. The extra idea used there is that we can optimise the probability distribution of X_k for the primes $3, 5, 7, 11$. Roughly speaking, one considers every possible set of $\mathbf{Z}/(3 \cdot 5 \cdot 7 \cdot 11)\mathbf{Z}$ that is the union of arithmetic progressions, one for each factor of $3 \cdot 5 \cdot 7 \cdot 11$. Then (using linear programming) one finds a probability distribution on the uncovered set of $\mathbf{Z}/(3 \cdot 5 \cdot 7 \cdot 11)\mathbf{Z}$ that minimises f_5. This is still not quite enough to prove the result: the proof also uses the idea of removing congruences mod p for primes $p \leq 73$ first, as this can be done wlog, and effectively reduces the sizes of the sets $P = \mathbf{Z}/p\mathbf{Z}$ by one (and complicates the criterion one wishes to achieve as it is now not exactly

f_5 that one wants to minimise). Additionally, the number of possible such subsets of $\mathbf{Z}/(3 \cdot 5 \cdot 7 \cdot 11)\mathbf{Z}$ is astronomical, and several other ideas are needed to reduce the problem to something more computationally tractable. Nevertheless, it is worth noting that this shows the methods described above are not truly optimal, and further improvements can sometimes be made by suitable choice of the distribution of the first few X_k.

As another illustration of the power of this technique, we give a simple proof (without explicit bound) of the minimum modulus problem, Theorem 2.3.

Proof Suppose we have a collection of progressions, all with moduli $m_i \geq M$ for some large M to be determined. Pick a large k and set $\delta_j = 0$ for $j \leq k$. Then freely allowing any modulus with a prime factor $> p_k$ gives

$$\kappa_k \leq \prod_{\ell \leq k} \left(1 + \frac{3p_\ell - 1}{(p_\ell - 1)^2}\right) \leq \exp\left(\sum_{\ell \leq k} \frac{3p_\ell - 1}{(p_\ell - 1)^2}\right)$$

$$\leq \exp\left(\sum_{\ell \leq k} \frac{3}{p_\ell} + \sum_{\ell \leq k} \frac{O(1)}{p_\ell^2}\right)$$

$$\leq \exp(3\log\log p_k + O(1)) = \Theta((\log k)^3),$$

where in the last line we have used the estimates $\sum_{p < N} \frac{1}{p} = \log\log N + O(1)$ and $\log p_k \sim \log k$.

On the other hand

$$\sum_{m \ p_k\text{-smooth}} \frac{1}{m} = \prod_{\ell \leq k} \left(1 - \frac{1}{p_\ell}\right)^{-1} < \infty,$$

so

$$\mu_k \geq 1 - \sum_{\substack{m \geq M \\ m \ p_k\text{-smooth}}} \frac{1}{m} > \frac{1}{2}$$

for sufficiently large M (depending on k). But then for sufficiently large k, $f_k = \kappa_k/\mu_k = \Theta((\log k)^3)$ satisfies the conditions of Theorem 6.2 and so the system of progressions does not cover the integers for sufficiently large M. □

As a final example in this section we consider the question of multiple coverings of the integers (see [33]). We say a set of covering systems have *mutually distinct* moduli, if the moduli in each system are distinct, and no modulus used in one system is used in any of the others.

Theorem 6.5 *There does not exist a set of* 10 *covering systems with mutually distinct moduli* > 1.

Proof We introduce an extra 'prime' p_0 corresponding to a factor P_0 with $|P_0| = 10$ in the geometric setting. Each modulus can now be assumed to have this extra 'prime' as a factor, with the congruence mod p_0 (i.e., projection into P_0) determining which of the ten covering systems it belongs to. We set $\delta_k = 0$ for primes up to $p_8 = 19$. A simple union bound over all progressions with 19-smooth moduli gives

$$\mu_8 \geq 1 - \frac{1}{10} \sum_{19\text{-smooth } m} \frac{1}{m} = 1 - \frac{1}{10}\left(\prod_{p \leq 19}(1 - \tfrac{1}{p})^{-1} - 1\right) \geq 0.515.$$

(The -1 is because we do not include the trivial congruence mod 1.) For κ_8 we have the bound

$$\kappa_8 \le \frac{1}{10} \prod_{p \le 19} \left(1 + \frac{3p-1}{(p-1)^2}\right) \le 12.3.$$

Hence $f_8 \le 12.3/0.515 < 23.9 < f_8^*$, so the systems do not cover $P_0 \times \mathbf{Z}$. □

In fact the proof shows the slightly stronger statement that one cannot find a system of progressions with distinct moduli > 1 that covers each integer at least 10 times. This is because the sets A_i in the geometric setting do not have to be product sets. One only needs that they have the correct measures in each factor when fixing the later coordinates.

In contrast to Theorem 6.5, one can construct three covering systems with mutually distinct moduli [33], although we know of no set of four such covering systems. We note that a similar calculation for square-free moduli shows that we cannot have 4 covering systems with mutually distinct square-free moduli. For this it is enough to consider $f_1 = \kappa_1/\mu_1 \le (2.5/4)/(1-1/8) < 0.72 < f_1^{*sf}$.

7 Schinzel's Conjecture

In this section we will describe the proof, given in [4], of Schinzel's Conjecture, which we restate here (in an equivalent form) for convenience.

Theorem 7.1 *If $1 < m_1 < m_2 < \cdots < m_k$ are the moduli of a finite collection of arithmetic progressions that covers the integers, then $m_i \mid m_j$ for some $i < j$.*

We will argue by contradiction, assuming that we have a set $M = \{m_1, \ldots, m_k\}$ of moduli of a covering system of the integers that forms an antichain under divisibility, i.e., $m_i \nmid m_j$ for any $i \ne j$.

The reason for giving this example is that the result does not quite follow from the method given in Section 6 as the sets S_k used to define κ_k are unrestricted and so κ_k is large. However, if we write $m_i = m_i' m_i''$ with m_i' the p_k-smooth part of m_i, then for any *fixed* value of m_i'', the set of m_i' allowed must form an antichain under divisibility, i.e., $m_i' \nmid m_j'$ for any i, j with $m_i'' = m_j''$. This then restricts the bound on $\mathbb{E}[\alpha_k^2]$ given in (6.1) in a similar manner to the previous examples. Indeed, it is enough to redefine κ_k so that (6.5) still holds, as that is all we needed in all the subsequent proofs.

We first observe that we may assume that none of the moduli m_i are prime powers. Indeed, we may assume that the covering is minimal, so the removal of any A_i results in a set of progressions that do not cover $\mathbf{Z}$. If $m_i = p^j$ for some prime p and $j > 0$, then the prime can appear at most to the $(j-1)$st power in any other moduli. Thus the other progressions fail to cover some congruence class mod Q/p, where $Q = \mathrm{lcm}\{m_1, \ldots, m_n\}$. But this congruence class cannot be covered by A_i as $m_i \nmid Q/p$, a contradiction.

Call an antichain of natural numbers *p-smooth* if all its elements are p-smooth, i.e., have no prime factor greater than p. We will need the following simple lemmas about 5-smooth antichains.

Lemma 7.2 *If A is a 5-smooth antichain containing no prime power, then*
$$\sum_{m \in A} \frac{1}{m} \leq \frac{1}{3}.$$

Proof Let us write the 5-smooth antichain A as a union of sets of the form $\{5^i m : m \in A_i\}$, where each A_i is a (possibly empty) 3-smooth antichain. We note that any 3-smooth antichain A_i must be of the form
$$A_i = \{2^{a_1} 3^{b_1}, 2^{a_2} 3^{b_2}, \ldots, 2^{a_k} 3^{b_k}\}$$
for some $a_1 > a_2 > \cdots > a_k$ and $b_1 < b_2 < \cdots < b_k$.

Since A contains no prime power, neither can A_0. It is then clear by successively reducing the a_i and b_i as much as possible that $\sum_{m \in A_0} \frac{1}{m}$ is minimised by taking $(a_1, \ldots, a_k) = (k, \ldots, 1)$ and $(b_1, \ldots, b_k) = (1, \ldots, k)$. Furthermore, $\sum_{m \in A_0} \frac{1}{m} = 2^{-k} - 3^{-k}$ is then maximised by taking $k = 1$ and $A_0 = \{6\}$. Hence $\sum_{m \in A_0} \frac{1}{m} \leq \frac{1}{6}$.

For A_i, $i > 0$, the fact that A contains no prime powers implies $1 \notin A_i$. Again, it is easy to check that $\sum_{m \in A_i} \frac{1}{m} \leq \frac{5}{6}$ with equality when $A_i = \{2, 3\}$. However, if $A_i = \{2, 3\}$ then $A_j = \emptyset$ for all $j > i$. Thus we look for the next best choice of A_i which is $\{2, 9\}$, so $\sum_{m \in A_i} \frac{1}{m} \leq \frac{11}{18}$ when $A_j \neq \emptyset$ for some $j > i$.

Thus if $A_1 = \{2, 3\}$ then
$$\sum_{m \in A} \frac{1}{m} = \sum_{m \in A_0} \frac{1}{m} + \frac{1}{5} \sum_{m \in A_1} \frac{1}{m} \leq \frac{1}{6} + \frac{1}{5} \cdot \frac{5}{6} = \frac{1}{3}.$$

On the other hand, if $A_1 \neq \{2, 3\}$ then
$$\sum_{m \in A} \frac{1}{m} = \sum_{i=0}^{\infty} \frac{1}{5^i} \sum_{m \in A_i} \frac{1}{m} \leq \frac{1}{6} + \frac{1}{5} \cdot \frac{11}{18} + \sum_{i=2}^{\infty} \frac{1}{5^i} \cdot \frac{5}{6} = \frac{119}{360} < \frac{1}{3}.$$

as required. $\square$

Lemma 7.3 *If A and B are two 3-smooth antichains, then*
$$\sum_{a \in A} \sum_{b \in B} \frac{1}{\mathrm{lcm}(a, b)} \leq \frac{31}{36},$$
except in the cases $A = B = \{1\}$ and $A = B = \{2, 3\}$. If A and B are two 5-smooth antichains, then
$$\sum_{a \in A} \sum_{b \in B} \frac{1}{\mathrm{lcm}(a, b)} \leq \frac{17}{10}.$$

Proof Suppose first that A and B are 3-smooth antichains with $|A| \leq |B|$, and that $|B| = k \geq 5$. Then we can write $B = \{2^{a_1} 3^{b_1}, 2^{a_2} 3^{b_2}, \ldots, 2^{a_k} 3^{b_k}\}$ with $a_1 > \cdots > a_k$ and $b_1 < \cdots < b_k$. Reducing the exponents can only increase the sum $\sum_{a \in A} \sum_{b \in B} \frac{1}{\mathrm{lcm}(a,b)}$, so we may assume $B = \{2^{k-1}, 2^{k-2} 3^1, \ldots, 3^{k-1}\}$. Then
$$\sum_{a \in A} \sum_{b \in B} \frac{1}{\mathrm{lcm}(a, b)} \leq \sum_{a \in A} \sum_{b \in B} \frac{1}{b} = |A| \sum_{b \in B} \frac{1}{b} \leq 6k(2^{-k} - 3^{-k}) < \frac{31}{36}.$$

The lemma therefore reduces to a finite check of families with $\max\{|A|, |B|\} \leq 4$, and in fact it is sufficient to consider the antichains $\{1\}$, $\{2,3\}$, $\{2,9\}$, $\{3,4\}$, $\{4,6,9\}$, and $\{8, 12, 18, 27\}$. The lemma now follows from a trivial case analysis, which can be done by hand.

Now consider 5-smooth antichains. We decompose A and B as a union of sets $5^i \cdot A_i$ and $5^j \cdot B_j$, where A_i and B_j are 3-smooth antichains, as in the proof of Lemma 7.2. Suppose first that there is no pair (i,j) with $A_i = B_j = \{1\}$ or $A_i = B_j = \{2,3\}$. Then

$$\sum_{a \in A} \sum_{b \in B} \frac{1}{\text{lcm}(a,b)} = \sum_{i=0}^{\infty} \sum_{j=0}^{\infty} \frac{1}{5^{\max\{i,j\}}} \sum_{a \in A_i} \sum_{b \in B_j} \frac{1}{\text{lcm}(a,b)}$$

$$\leq \frac{31}{36}\left(1 + \frac{3}{5} + \frac{5}{5^2} + \frac{7}{5^3} + \cdots\right) = \frac{31}{36} \cdot \frac{15}{8} < \frac{17}{10} - \frac{1}{12}.$$

Next, suppose that $A_i = B_j = \{1\}$ for some pair (i,j), and observe that $A_{i'} = B_{j'} = \emptyset$ for every $i' > i$ and $j' > j$, so the pair (i,j) is unique. If there is no pair (s,t) with $A_s = B_t = \{2,3\}$, then the bound above increases by at most $(1 - \frac{31}{36})5^{-\max\{i,j\}}$, and this is less that $\frac{1}{12}$ if $\max\{i,j\} \geq 1$. On the other hand, if $A_0 = B_0 = \{1\}$, then (since $A_i = B_i = \emptyset$ for all $i > 0$) we have $A = B = \{1\}$, and so $\sum_{a \in A} \sum_{b \in B} \frac{1}{\text{lcm}(a,b)} = 1 < \frac{17}{10}$.

We may therefore assume that $A_i = B_j = \{2,3\}$ for some pair (i,j), which implies that $A_{i'}, B_{j'} \subseteq \{1\}$ for every $i' > i$ and $j' > j$, and (as above) at most one of the sets in each sequence is non-empty. The bound above increases by at most

$$\left(\frac{7}{6} - \frac{31}{36}\right) \frac{1}{5^{\max\{i,j\}}} + \left(1 - \frac{31}{36}\right) \frac{1}{5^{\max\{i,j\}+1}} = \frac{1}{3} \cdot \frac{1}{5^{\max\{i,j\}}} < \frac{1}{12}$$

if $\max\{i,j\} \geq 1$. However, if $A_0 = B_0 = \{2,3\}$, then it is easy to see that $\sum_{a \in A} \sum_{b \in B} \frac{1}{\text{lcm}(a,b)}$ is maximised by taking $A = B = \{2,3,5\}$, and in that case it is equal to $\frac{17}{10}$. □

Having completed the easy preliminaries, we are ready to prove Schinzel's Conjecture.

Proof We set $\delta_1 = \delta_2 = \delta_3 = 0$. Observe that, by Lemma 7.2, the total measure of $B_1 \cup B_2 \cup B_3$ is at most $\frac{1}{3}$, so $\mu_3 \geq \frac{2}{3}$. Now we deduce bounds on $\mathbb{E}[\alpha_k^2]$ for $k \geq 3$. Let S_{k-1} be the set of integers whose prime factors all lie between 7 and p_{k-1}, and for each $m \in S_{k-1}$ and $j \geq 1$, define

$$M(m, j) = \{a : amp_k^j \in M \text{ and } a \text{ is 5-smooth}\}$$

where M is the set of moduli of our covering system. By (6.1), we have

$$\mathbb{E}[\alpha_k^2] \leq \sum_{\substack{j_1, j_2 \geq 1 \\ m_1, m_2 \in S_{k-1}}} \frac{1}{p_i^{j_1+j_2} \text{lcm}(m_1, m_2)} \prod_{p_\ell | \text{lcm}(m_1, m_2)} \frac{1}{1 - \delta_\ell} \sum_{\substack{a \in M(m_1, j_1) \\ b \in M(m_2, j_2)}} \frac{1}{\text{lcm}(a,b)}.$$

Since $M(m, j)$ is a 5-smooth antichain, it follows from Lemma 7.3 that

$$\mathbb{E}[\alpha_k^2] \leq \frac{17}{10} \sum_{j_1, j_2 \geq 1} \frac{1}{p_i^{j_1+j_2}} \sum_{m_1, m_2 \in S_{k-1}} \frac{1}{\text{lcm}(m_1, m_2)} \prod_{p_\ell | \text{lcm}(m_1, m_2)} \frac{1}{1 - \delta_\ell}.$$

But then defining

$$\kappa_k = \frac{17}{10} \sum_{m_1,m_2 \in S_{k-1}} \frac{1}{\operatorname{lcm}(m_1,m_2)} \prod_{p_\ell \mid \operatorname{lcm}(m_1,m_2)} \frac{1}{1-\delta_\ell}$$

we deduce that

$$\mathbb{E}[\alpha_k^2] \leq \frac{1}{(p_k-1)^2} \cdot \kappa_{k-1}$$

and, by the same calculation as in Section 6, that for $k \geq 4$

$$\kappa_k \leq \left(1 + \frac{3p_k - 1}{(p_k-1)^2(1-\delta_k)}\right) \cdot \kappa_{k-1}.$$

Hence (6.5) holds. Now $\kappa_3 = \frac{17}{10}$ and, recalling from above that $\mu_3 \geq \frac{2}{3}$, we obtain $f_3 \leq \frac{17}{10} \cdot \frac{3}{2} = 2.55 < f_3^*$, so the system of arithmetic progressions does not cover the integers. □

8 An Efficient Almost-Covering

In this section we will reproduce the proof from [4] of Theorem 2.6 showing that we can construct collections of arithmetic progressions with distinct square-free moduli $m_i \geq M$, $\sum \frac{1}{m_i} < 1$, covering all but ε density of the integers.

Proof We will choose a collection $P_1, \ldots, P_N$ of disjoint sets of primes, and define

$$Q_i := \prod_{j \leq i} \prod_{p \in P_j} p \quad \text{and} \quad D_i := \{p \cdot Q_{i-1} : p \in P_i\}$$

for each $i \in [N]$, where $Q_0 := 1$. We will show that, for a suitable choice of $P_1, \ldots, P_N$, the set $D = D_1 \cup \cdots \cup D_N$ has the following properties:

$$\sum_{m \in D} \frac{1}{m} \leq 1 + \frac{\varepsilon}{3}, \qquad (8.1)$$

and there exists a collection of arithmetic progressions, with distinct moduli in D, such that the uncovered set has density at most $\varepsilon/3$. By removing a few of the progressions from this family, we will obtain the claimed collection.

We construct the sets $P_1, \ldots, P_N$ of primes as follows. First, let us fix some positive constants c_0, c and δ such that

$$c_0 := 1 + \frac{\varepsilon}{3}, \quad \text{and} \quad c := \frac{\delta}{1-e^{-\delta}} \in (1, c_0).$$

Indeed, $\delta/(1-e^{-\delta})$ is a continuous increasing function of δ which tends to 1 as $\delta \to 0$, so for sufficiently small δ we have $1 < \delta/(1-e^{-\delta}) < c_0$. Assume (without loss of generality) that $M > 3/\varepsilon$, and choose N sufficiently large so that $e^{-\delta N} < \varepsilon/3$. Now let P_1 be any set of primes such that $p \geq M$ for every $p \in P_1$, and

$$\delta \leq \sum_{p \in P_1} \frac{1}{p} \leq \delta + \frac{c_0 - c}{N}.$$

In general, if we have already constructed $P_1, \ldots, P_j$, then let P_{j+1} be any set of primes disjoint from $P_1 \cup \cdots \cup P_j$ such that $p \geq M$ for every $p \in P_{j+1}$, and

$$\delta e^{-\delta j} \leq \sum_{p \in P_{j+1}} \frac{1}{p \cdot Q_j} \leq \delta e^{-\delta j} + \frac{c_0 - c}{N}. \tag{8.2}$$

As the sum $\sum 1/p$ over prime p diverges, it is clear that sets $P_1, \ldots, P_N$ exist with these properties. It follows that

$$\sum_{m \in D} \frac{1}{m} = \sum_{j=0}^{N-1} \sum_{p \in P_{j+1}} \frac{1}{p \cdot Q_j} \leq \sum_{j=0}^{N-1} \left(\delta e^{-\delta j} + \frac{c_0 - c}{N} \right) \leq c_0 = 1 + \frac{\varepsilon}{3},$$

where in the final inequality we used the identity $\sum_{j=0}^{\infty} \delta e^{-\delta j} = \delta(1 - e^{-\delta})^{-1} = c$.

Now, to construct the arithmetic progressions, simply choose (for each $m \in D$ in turn) any arithmetic progression with modulus m that has the largest intersection with the (as yet) uncovered set. To be more precise, for each $j \in [N]$ let $\mathcal{A}_j$ denote the collection of arithmetic progressions whose modulus lies in D_j, and write ε_j for the density of the uncovered set $R_j := \mathbf{Z} \setminus \bigcup_{l \leq j} \bigcup_{A \in \mathcal{A}_l} A$. Now observe that R_j is a union of $\varepsilon_j Q_j$ congruence classes mod Q_j and so if $m = p \cdot Q_j \in D_{j+1}$, we can pick a congruence class mod m lying in one of the $\varepsilon_j Q_j$ congruence classes mod Q_j that lie in R_j. In this case the congruence mod m will remove a fraction $1/p$ of the elements in this congruence class. Indeed, it will remove a fraction $1/p$ of any remaining elements in this congruence class, even after we have removed any previous congruences mod $m' \in D_{j+1}$, as the prime p does not divide these other moduli. Thus picking a congruence class mod Q_j that is least covered by previous progressions shows that we can find a congruence class mod m that covers at least a fraction $1/(p\varepsilon_j Q_j)$ of the as yet uncovered set. It follows that

$$\varepsilon_{j+1} \leq \varepsilon_j \prod_{p \in P_{j+1}} \left(1 - \frac{1}{p \varepsilon_j Q_j} \right) \leq \varepsilon_j \cdot \exp\left(-\frac{1}{\varepsilon_j} \sum_{p \in P_{j+1}} \frac{1}{p \cdot Q_j} \right)$$

for each $0 \leq j \leq N - 1$ (where $\varepsilon_0 := 1$), and hence, by (8.2),

$$\varepsilon_{j+1} \leq \varepsilon_j \cdot \exp\left(-\frac{\delta e^{-\delta j}}{\varepsilon_j} \right).$$

It now follows immediately by induction that $\varepsilon_j \leq e^{-\delta j}$ for every $j \in [N]$, and in particular $\varepsilon_N \leq e^{-\delta N} \leq \varepsilon/3$, by our choice of N.

We have therefore constructed a collection of arithmetic progressions whose set D of (distinct and square-free) moduli satisfies (8.1), and whose uncovered set has density at most $\varepsilon/3$. To complete the construction, simply choose a maximal subset $D' \subseteq D$ such that $\sum_{m \in D'} \frac{1}{m} < 1$, and observe that the density of the set uncovered by by the progressions with $m \in D'$ is at most

$$\frac{\varepsilon}{3} + \sum_{m \in D \setminus D'} \frac{1}{m} \leq \frac{\varepsilon}{3} + \frac{\varepsilon}{3} + \frac{1}{M} \leq \varepsilon,$$

as required. $\square$

References

[1] P. Balister, B. Bollobás, R. Morris, J. Sahasrabudhe and M. Tiba, *Covering intervals with arithmetic progressions*, Acta Math. Hungar. **161** (2020), 197–200.

[2] _____, *Erdős covering systems*, Acta Math. Hungar. **161** (2020), 540–549.

[3] _____, *The Erdős–Selfridge problem with square-free moduli*, Algebra Number Theory **15** (2021), 609–626.

[4] _____, *On the Erdős covering problem: the density of the uncovered set*, Invent. Math. **228** (2022), 377–414.

[5] _____, *The structure and number of Erdős covering systems*, J. Eur. Math. Soc., in press.

[6] S. L. G. Choi, *Covering the set of integers by congruence classes of distinct moduli*, Math. Comp. **25** (1971), 885–895.

[7] R. F. Churchhouse, *Covering sets and systems of congruences*, In: *Computers in Mathematical Research* North-Holland, Amsterdam, 1968, pp. 20–36.

[8] R. B. Crittenden and C. L. Vanden Eynden, *A proof of a conjecture of Erdős*, Bull. Amer. Math. Soc. **75** (1969), 1326–1329.

[9] _____, *Any n arithmetic progressions covering the first 2^n integers cover all integers*, Proc. Amer. Math. Soc. **24** (1970), 475–481.

[10] M. Cummings, M. Filaseta and O. Trifonov, *An upper bound for the minimum modulus in a covering system with squarefree moduli*, ArXiv preprint arXiv:2211.08548 (2022).

[11] P. Dusart, *The kth prime is greater than $k(\log k + \log \log k - 1)$ for $k \geq 2$*, Math. Comp. **68** (1999), 411–415.

[12] C. Elsholtz and J-C. Schlage-Puchta, *On Romanov's constant*, Math. Z. **288** (2018), 713–724.

[13] P. Erdős, *On integers of the form $2^k + p$ and some related problems*, Summa Brasil. Math. **2** (1950), 113–123.

[14] _____, *On a problem concerning congruence-systems (in Hungarian)*, Mat. Lapok **3** (1952), 122–128.

[15] _____, *Remarks on number theory, IV. Extremal problems in number theory, I. (in Hungarian)*, Mat. Lapok **13** (1962), 228–255.

[16] _____, *Some recent advances and current problems in number theory*, Lectures on Modern Mathematics, Vol. III, Wiley, New York, 1965, pp. 196–244.

[17] _____, *Extremal problems in number theory*, Proc. Sympos. Pure Math., Vol. VIII, Amer. Math. Soc., Providence, R.I., 1965, pp. 181–189.

[18] _____, *Problems and results on combinatorial number theory III*, In: Number Theory Day, Lecture Notes in Math., vol. 626, Springer, Berlin, 1977, pp. 43–72.

[19] _____, *Solution to problem 1029: Erdős and the computer*, Math. Mag. **52** (1979), 180–181.

[20] P. Erdős and R. L. Graham, *Old and new problems and results in combinatorial number theory*, Monographies de L'Enseignement Mathématique, vol. 28, Université de Genève, L'Enseignement Mathématique, Geneva, 1980.

[21] J. Fabrykowski, T. Smotzer, *Covering Systems of Congruences*, Math. Mag. **78** (2005), pp. 228–231.

[22] M. Filaseta, K. Ford and S. Konyagin, *On an irreducibility theorem of A. Schinzel associated with coverings of the integers*, Illinois J. Math. **44** (2000), 633–643.

[23] M. Filaseta, K. Ford, S. Konyagin, C. Pomerance and G. Yu, *Sieving by large integers and covering systems of congruences*, J. Amer. Math. Soc. **20** (2007), 495–517.

[24] M. Filaseta, C. Finch and M. Kozek, *On powers associated with Sierpiński numbers, Riesel numbers and Polignac's conjecture*, J. Number Theory **128** (2008), 1916–1940.

[25] M. Filaseta and J. Juillerat, *Consecutive primes which are widely digitally delicate*, Integers **21A** (2021), A12.

[26] M. Filaseta and J. Harrington, *A polynomial investigation inspired by work of Schinzel and Sierpiński*, Acta Arith. **155** (2012), 149–161.

[27] M. Filaseta and W. Harvey, *Covering subsets of the integers by congruences*, Acta Arith. **182** (2018), 43–72.

[28] J. Friedlander, *On exact coverings of the integers*, Israel J. Math. **12** (1972), 299–305.

[29] D. J. Gibson, *A covering system with least modulus 25*, Math. Comp. **78** (2009), 1127–1146.

[30] I. P. Goulden, A. Granville, L. B. Richmond and J. Shallit, *Natural exact covering systems and the reversion of the Möbius series*, Ramanujan J. **50** (2019), 211–235.

[31] R. L. Graham, *A Fibonacci-like sequence of composite numbers*, Math. Mag. **37** (1964), 322–324.

[32] R. K. Guy, *Unsolved problems in number theory*, Problem Books in Mathematics, Springer-Verlag, New York, 2004.

[33] J. Harrington, *Two questions concerning covering systems*, Int. J. Number Theory **11** (2015), 1739–1750.

[34] R. D. Hough, *Solution of the minimum modulus problem for covering systems*, Ann. Math. **181** (2015), 361–382.

[35] R. D. Hough and P. P. Nielsen, *Covering systems with restricted divisibility*, Duke Math. J. **168** (2019), 3261–3295.

[36] J. Klein, D. Koukoulopoulos and S. Lemieux, *On the j-th smallest modulus of a covering system with distinct moduli*, Int. J. Number Theory, in press.

[37] C. E. Krukenberg, *Covering sets of the integers*, Ph.D. dissertation, University of Illinois, Urbana-Champaign, 1971.

[38] E. Landau, *Über einige neuere Fortschritte der additiven Zahlentheorie*, Cambridge tracts in mathematics and mathematical physics, vol. 35, Cambridge University Press, Cambridge, 1937.

[39] P. P. Nielsen, *A covering system whose smallest modulus is 40*, J. Number Theory **129** (2009), 640–666.

[40] T. Owens, *A Covering System with Minimum Modulus 42*, Master's thesis, Brigham Young University, 2014, available at https://scholarsarchive.byu.edu/etd/4329.

[41] A. de Polignac, *Six propositions arithmologiques déduites du crible d'Ératosthéne*, Nouvelles Annales de Mathématiques **1** (1849), no. 8, 423–429.

[42] Š. Porubský and J. Schönheim, *Covering systems of Paul Erdős. Past, present and future*, In: Paul Erdős and his mathematics, I, János Bolyai Math. Soc., Budapest, 2002, 581–627.

[43] A. Schinzel, *Reducibility of polynomials and covering systems of congruences*, Acta Arith. **13** (1967), 91–101.

[44] W. Sierpiński, *Sur un problème concernant les nombres $k \cdot 2^n + 1$*, Elem. Math. **15** (1960), 73–74.

[45] R. J. Simpson, *Regular coverings of the integers by arithmetic progressions*, Acta Arith. **45** (1985), 145–152.

[46] S. K. Stein, *Unions of arithmetic sequences*, Math. Ann. **134** (1958), 289–294.

Mathematical Institute
University of Oxford
Oxford OX2 6GG, UK
paul.balister@maths.ox.ac.uk

The Cluster Expansion in Combinatorics

Matthew Jenssen

Abstract

The cluster expansion is a classical tool from statistical physics used to study the phase diagram of interacting particle systems. Recently, the cluster expansion has seen a number of applications in combinatorics and the field of approximate counting/sampling. In this article, we give an introduction to the cluster expansion and survey some of these recent developments.

1 Introduction

A central theme in both statistical physics and combinatorics is to understand the relationship between structure and randomness. It is therefore natural that these two fields are interlinked and indeed both fields have independently developed powerful tools and methodologies to study the very same phenomena. Most of the topics covered in this article can be viewed from the perspective of the following question:

How does global structure emerge in systems governed only by local interactions?

From the physics perspective, perhaps the first image that springs to mind is that of a gas of interacting particles freezing into a solid, a move from disorder to order. One of the major achievements of statistical physics has been to formulate a rigorous way to understand this type of phenomena through the notion of *phase transition*. Phase transitions are also familiar in combinatorics, the most immediate example perhaps being the emergence of the giant component in the Erdős-Renyi random graph. There are a number of other classical problems in combinatorics that exemplify the notion of phase transition, even though this language is often not used in their study. Examples include understanding the typical structure of

1. Triangle-free graphs on n vertices and m edges;
2. Antichains of size m in the Boolean lattice $\mathcal{P}([n])$;
3. Sum-free sets of size m in an Abelian group G.

In each of the above examples, as m increases, the typical structure of the above combinatorial objects exhibit a 'disorder to order' phase transition. For example, in the case of triangle free graphs, if $m = cn^{3/2}$, the global structure of a typical triangle-free graph on n vertices and m edges depends drastically on c: for c small a typical graph has a max-cut of size close to $m/2$ whereas for c large a typical graph has a cut of size close to m and so 'looks bipartite'.

One of the most powerful and oldest tools in the rigorous study of phase transitions is the *cluster expansion*. It was introduced by Mayer [49] in the 1930s in his study of the phenomenon of condensation and remains widely used today. The purpose of this article is to give an introduction to the cluster expansion and survey some of its connections and applications to combinatorics and algorithms. The topics covered in this survey are by no means exhaustive and are naturally a reflection of the author's own particular perspective on the field.

The protagonist of this article will be the *hard-core* model, one of the oldest and simplest models of a gas. The hard-core model is a probability distribution on *independent sets* in a graph and so its relevance to combinatorics is also immediate.

1.1 Introducing the Hard-Core Model

Given a graph G, let $\mathcal{I}(G)$ denote the collection of independent sets in G. The hard-core model on G at 'activity' $\lambda > 0$ is the probability distribution on $\mathcal{I}(G)$ given by

$$\mathbb{P}_{G,\lambda}(I) = \frac{\lambda^{|I|}}{Z_G(\lambda)} \qquad (1.1)$$

for $I \in \mathcal{I}(G)$, where the normalising constant

$$Z_G(\lambda) = \sum_{I \in \mathcal{I}(G)} \lambda^{|I|}$$

is known as the *hard-core model partition function*. In combinatorics $Z_G(\lambda)$ is commonly known as the *independence polynomial*. The hard-core model originated in statistical physics as as a simple model of a gas. The vertices of the graph G are to be thought of as 'sites' that can be occupied by particles and neighbouring sites cannot both be occupied. This constraint is meant to model a system of particles with a 'hard-core' that cannot overlap.

As λ increases, the hard-core model is biased towards larger independent sets. To the combinatorialist, perhaps the most natural incarnation of the hard-core model is given by the case $\lambda = 1$. Indeed $\mathbb{P}_{G,1}$ is the uniform distribution on independent sets in G and $Z_G(1)$ is the number of independent sets in G. Given this, one might wonder why one should bother studying the hard-core model at general λ. This survey will hopefully convince the reader that there are several reasons for doing so. One immediate reason is that there is a wealth of probabilistic information hidden within the *derivatives* of the partition function. For example, letting $\mathbb{E}_{G,\lambda}(|I|)$ denote the expected size of an independent set drawn from the hard-core model, then

$$\mathbb{E}_{G,\lambda}(|I|) = \sum_{I \in \mathcal{I}(G)} |I| \frac{\lambda^{|I|}}{Z_G(\lambda)} = \lambda \frac{Z'_G(\lambda)}{Z_G(\lambda)} = \lambda (\log Z_G(\lambda))' \qquad (1.2)$$

and higher moments of $|I|$ can be accessed via higher derivatives. In fact, we go further and contend that it's fruitful to study a *multivariate* generalisation of the hard-core model. Here each vertex v of the graph G is given an activity parameter $\lambda_v > 0$ and we let $\boldsymbol{\lambda} = (\lambda_v : v \in V(G))$ be the vector of vertex activities. We then have the following generalisation of (1.1) given by

$$\mathbb{P}_{G,\boldsymbol{\lambda}}(I) = \frac{\prod_{v \in I} \lambda_v}{Z_G(\boldsymbol{\lambda})} \qquad (1.3)$$

where

$$Z_G(\boldsymbol{\lambda}) = \sum_{I \in \mathcal{I}(G)} \prod_{v \in I} \lambda_v.$$

Once again we will see several reasons for why this is a useful perspective. An immediate reason is that it allows us to access correlation statistics through partial derivatives. For example, given an independent set I drawn from the distribution (1.3), and vertices u, v of G one easily checks that

$$\mathbb{P}(u \in I, v \in I) - \mathbb{P}(u \in I)\mathbb{P}(v \in I) = \lambda_u \lambda_v \frac{\partial^2}{\partial \lambda_u \partial \lambda_v} \log Z_G(\boldsymbol{\lambda}). \quad (1.4)$$

To conclude this short introduction, we record a simple yet powerful identity. For $U \subseteq V(G)$, let

$$Z_U(\boldsymbol{\lambda}) := \sum_{U \subseteq I \in \mathcal{I}(G)} \prod_{v \in I} \lambda_v = Z_G(\boldsymbol{\lambda} \mathbf{1}_U).$$

By considering independent sets that contain a fixed vertex v in G and those that do not we have

$$Z_G(\boldsymbol{\lambda}) = Z_{G-v}(\boldsymbol{\lambda}) + \lambda_v Z_{G-v-N(v)}(\boldsymbol{\lambda}). \quad (1.5)$$

This is often referred to as *the fundamental identity* in the literature.

To illustrate the connection between statistical physics and combinatorics and to further motivate the study of the multivariate hard-core model, we can think of no better place to start than with the remarkable work of Scott and Sokal [62] which shows that a pair of the most influential tools from these respective fields are in fact equivalent.

1.2 Notation

Given $n \in \mathbb{N}$, we let $[n]$ denote the set $\{1, \ldots, n\}$. For a graph $G = (V, E)$ we will often write uv to denote an edge $\{u, v\} \in E$ and write $u \sim v$ to indicate that u is adjacent to v. For a vertex $v \in V$, we let $N(v) = \{u \in V : uv \in E\}$ denote the vertex neighbourhood of v in G. Given two vectors $\boldsymbol{x}, \boldsymbol{y} \in \mathbb{R}^n$ we write $\boldsymbol{x} \geq \boldsymbol{y}$ to denote that $x_i \geq y_i$ for all $i \in [n]$. For a vector $\boldsymbol{\lambda} \in \mathbb{C}^n$, we let $|\boldsymbol{\lambda}|$ denote the vector whose ith coordinate is $|\lambda_i|$. For $\boldsymbol{x} \in \mathbb{C}^n$ and $\boldsymbol{R} \in \mathbb{R}_{>0}^n$, we let $D(\boldsymbol{x}; \boldsymbol{R}) = \{\boldsymbol{\lambda} \in \mathbb{C}^n : |\boldsymbol{x} - \boldsymbol{\lambda}| < \boldsymbol{R}\}$ the open polydisc centred at $\boldsymbol{x}$ with radius $\boldsymbol{R}$, and we let $\bar{D}(\boldsymbol{x}; \boldsymbol{R})$ denote the closed polydisc.

2 A Tale of Two Theorems: Zero-Free Regions and the Lovász Local Lemma

In the previous section, we introduced the hard-core model as a probability distribution on the independent sets of a graph G. Of course, the probability distribution defined at (1.3) only makes sense for $\boldsymbol{\lambda} \geq 0$. On the other hand, the partition function $Z_G(\boldsymbol{\lambda})$ makes perfect sense as a multivariate polynomial over the complex numbers. If we're interested in the hard-core model as a probability distribution, why bother with complex numbers? This question by now has several compelling answers, the earliest of which comes from the work of Lee and Yang [71] in 1952 in their study of phase transitions. Here we think of a phase transition as a point at which some physical quantity (e.g. the expected fraction of occupied vertices in the hard-core model as in (1.2)) depends non-analytically (or even discontinuously) on some control parameter (e.g. the activity λ). This is of course impossible for any sensible model on a finite graph, instead we think of a sequence graphs G_n (e.g. boxes in $\mathbb{Z}^d$ of increasing volume) and consider 'infinite volume limits' such as $f(\lambda) = \lim_{n \to \infty} \frac{1}{V(G_n)} \log Z_{G_n}(\lambda)$. Absence of zeros of Z_{G_n} in the complex plane can be used to show that $f(\lambda)$ is real-analytic in some interval which implies absence of phase transition in that interval.

We begin with a seminal result of Dobrushin on zero-freeness of the multivariate hard-core model.

Theorem 2.1 (Dobrushin [15, 16]) *Let $G = (V, E)$ be a graph. Let $\boldsymbol{R} = (R_v)_{v \in V}$ with $\boldsymbol{R} \geq 0$. Suppose that there exist constants $0 \leq r_v \leq 1$ such that*

$$R_v \leq r_v \prod_{u \sim v} (1 - r_u)$$

for all $v \in V$. Then for all $\boldsymbol{\lambda}$ satisfying $|\boldsymbol{\lambda}| \leq \boldsymbol{R}$,

$$|Z_G(\boldsymbol{\lambda})| \geq Z_G(-\boldsymbol{R}) \geq \prod_{v \in V} (1 - r_v) > 0.$$

The reader may have noticed the striking resemblance of this theorem with the Lovász Local Lemma [19].

Theorem 2.2 (Lovász Local Lemma) *Let V be a finite set and let $(A_v)_{v \in V}$ be a collection of events in some probability space with dependency graph $G = (V, E)$. Suppose that there exist constants $0 \leq p_v \leq 1$ such that*

$$\mathbb{P}(A_v) \leq p_v \prod_{u \sim v} (1 - p_u)$$

for all $v \in V$. Then

$$\mathbb{P}\left(\bigcap_{v \in V} \bar{A}_v\right) \geq \prod_{v \in V} (1 - p_v) > 0.$$

In a remarkable paper, Scott and Sokal [62] showed that this resemblance is no coincidence. In fact, they show that Theorems 2.1 and 2.2, proved 20 years apart for different purposes in separate fields, are indeed *equivalent*. Their proof builds on the beautiful work of Shearer [64] who was the first to exploit the connection between the Lovàsz Local Lemma and the multivariate hard-core model (more on this later). To discuss these results in more detail, we require some terminology.

Definition 2.3 Let $G = (V, E)$ be a graph. Call $\boldsymbol{R} = (R_v)_{v \in V} \in [0, \infty)^V$ *zero-free for G* if $Z_G(\boldsymbol{\lambda}) \neq 0$ for all $\boldsymbol{\lambda} \in \mathbb{C}^V$ such that $|\boldsymbol{\lambda}| \leq \boldsymbol{R}$.

We remark that the above definition is not standard. Following [63] we also make the following definition.

Definition 2.4 Let $G = (V, E)$ be a graph. Call $\boldsymbol{q} = (q_v)_{v \in V} \in [0, \infty)^V$ *good for G* if the following holds: if $(A_v)_{v \in V}$ is a collection of events in some probability space with dependency graph $G = (V, E)$, and $\mathbb{P}(A_v) \leq q_v$ for all $v \in V$, then $\mathbb{P}\left(\bigcap_{v \in V} \bar{A}_v\right) > 0$.

Theorems 2.1 and 2.2 raise the following two natural questions: given a graph G which vectors are good for G and which vectors are zero-free for G? Scott and Sokal [62] showed that these questions are one and the same.

Theorem 2.5 (Scott and Sokal [62]) *Let $G = (V, E)$ be a graph and let $q \in [0, \infty)^V$. Then q is good for G if and only if q is zero-free for G.*

We note that if q contains a coordinate $q_v \geq 1$, then q cannot be good for G since we can take an event A_v with $\mathbb{P}(A_v) = 1$. We can therefore restrict our attention to $q \in [0, 1)^V$. Theorem 2.5 shows that we can do the same when considering zero-free vectors which is a less obvious statement.

We give a brief account of the proof of Theorem 2.5 and refer the reader to [62] (see also [63]) for more detail, insight and generalisations.

We begin with a preliminary lemma. Given a graph G and set $J \subseteq V(G)$, let

$$Z_G(\boldsymbol{\lambda}; J) = \sum_{J \subseteq I \in \mathcal{I}(G)} \prod_{v \in I} \lambda_v,$$

so in particular $Z_G(\boldsymbol{\lambda}; \emptyset) = Z_G(\boldsymbol{\lambda})$ and $Z_G(\boldsymbol{\lambda}; J) = 0$ if J is not an independent set.

Lemma 2.6 *Let G be a graph. If $\boldsymbol{R}$ is zero-free for G then*

$$(-1)^{|J|} Z_G(-\boldsymbol{R}; J) \geq 0.$$

for all $J \subseteq V(G)$ with equality if and only if $J \notin \mathcal{I}(G)$.

Proof We may assume that $J \in \mathcal{I}(G)$. We say a vertex v *extends* J if $v \notin J$ and $J \cup \{v\}$ is an independent set in G. Let $\boldsymbol{R}_J$ denote the vector with coordinates

$$(\boldsymbol{R}_J)_v = R_v \cdot \mathbf{1}_{\{v \text{ extends } J\}},$$

for $v \in V(G)$. Then

$$Z_G(-\boldsymbol{R}; J) = (-1)^{|J|} \left(\prod_{v \in J} R_v \right) Z_G(-\boldsymbol{R}_J).$$

Since $Z_G(0) = 1$ and $-\boldsymbol{R}_J$ lies in the polydisc $D(0; \boldsymbol{R})$ which contains no zeros of Z_G, we conclude by continuity of Z_G that $Z_G(-\boldsymbol{R}_J) > 0$. The result follows. □

It turns out that the converse of Lemma 2.6 also holds, and this is just one in a long list of equivalent statements for zero-freeness proved in [62] (Theorem 2.10, which the authors term 'the fundamental theorem').

Proof of Theorem 2.5. Let q be zero-free for $G = (V, E)$. We will define a probability space containing a collection of events $(B_v)_{v \in V}$ which minimises the quantity $\mathbb{P}\left(\bigcap_{v \in V} B_v\right)$ subject to the constraints (i) $\mathbb{P}(B_v) \leq q_v$ for all $v \in V$, and (ii) G is a dependency graph for $(B_v)_{v \in V}$. To this end, define the probability measure $\mathbb{P}$ on the σ-algebra generated by $(B_v)_{v \in V}$ where for $J \subseteq V$,

$$\mathbb{P}\left(\bigcap_{v \in J} B_v\right) = \begin{cases} \prod_{v \in J} q_v & \text{if } J \in \mathcal{I}(G), \\ 0 & \text{otherwise.} \end{cases} \quad (2.1)$$

This indeed defines a measure since by inclusion-exclusion and Lemma 2.6,

$$\mathbb{P}\left(\bigcap_{v\in J} B_v \cap \bigcap_{v\notin J} \bar{B}_v\right) = \sum_{I\supseteq J}(-1)^{|I|-|J|}\mathbb{P}\left(\bigcap_{v\in I} B_v\right) \tag{2.2}$$

$$= \sum_{I\supseteq J, I\in\mathcal{I}(G)}(-1)^{|I|-|J|}\prod_{v\in I} q_v \tag{2.3}$$

$$= (-1)^{|J|}Z_G(-\boldsymbol{q}; J) \tag{2.4}$$

$$\geq 0, \tag{2.5}$$

(the fact that its a probability measure follows by taking $J = \emptyset$ in (2.1)). We note that $\mathbb{P}(B_v) = q_v$ by definition and it's easily checked G is a dependency graph for $(B_v)_{v\in V}$. Note also that by (2.4) with $J = \emptyset$ we have $\mathbb{P}\left(\bigcap_{v\in V}\bar{B}_v\right) = Z_G(-\boldsymbol{q})$ and $Z_G(-\boldsymbol{q}) > 0$ by Lemma 2.6.

Suppose now that $(A_v)_{v\in V}$ is a collection of events (in some other probability space) with dependency graph G, and $\mathbb{P}(A_v) \leq q_v$ for all $v \in V$. We will show that $\mathbb{P}\left(\bigcap_{v\in V}\bar{A}_v\right) \geq \mathbb{P}\left(\bigcap_{v\in V}\bar{B}_v\right)$. For $J \subseteq V$, let

$$P_J = \mathbb{P}\left(\bigcap_{v\in J}\bar{A}_v\right) \quad \text{and} \quad Q_J = \mathbb{P}\left(\bigcap_{v\in J}\bar{B}_v\right).$$

We show by induction on $|J|$ that P_J/Q_J monotone increasing in J (note that for all J, $Q_J \geq \mathbb{P}\left(\bigcap_{v\in V}\bar{B}_v\right) = Z_G(-\boldsymbol{q}) > 0$). By inclusion-exclusion,

$$Q_J = \sum_{I\subseteq J}(-1)^{|I|}\mathbb{P}\left(\bigcap_{v\in I} B_v\right) = \sum_{I\subseteq J, I\in\mathcal{I}(G)}(-1)^{|I|}\prod_{v\in I} q_v = Z_J(-\boldsymbol{q}).$$

Suppose now that $v \notin J$. By the identity (1.5) applied to $Z_J(-\boldsymbol{q})$ we have

$$Q_J - Q_{J\cup\{v\}} = q_v Q_{J\setminus N(v)}.$$

On the other hand,

$$P_J - P_{J\cup\{v\}} = \mathbb{P}\left(A_v \cap \bigcap_{v\in J}\bar{A}_v\right) \leq \mathbb{P}\left(A_v \cap \bigcap_{v\in J\setminus N(v)}\bar{A}_v\right) = q_v P_{J\setminus N(v)},$$

using the fact that G is a dependency graph for $(A_v)_{v\in V}$. It follows that

$$P_{J\cup\{v\}}Q_J - Q_{J\cup\{v\}}P_J \geq (P_J - q_v P_{J\setminus N(v)})Q_J - (Q_J - q_v Q_{J\setminus N(v)})P_J$$
$$= q_v(Q_{J\setminus N(v)}P_J - P_{J\setminus N(v)}Q_J)$$
$$\geq 0,$$

where for the final inequality we used that $P_J/Q_J \geq P_{J\setminus N(v)}/Q_{J\setminus N(v)}$ by the inductive hypothesis. P_J/Q_J is therefore monotone increasing in J as claimed and so $P_V/Q_V \geq P_\emptyset/Q_\emptyset = 1$. We conclude that $\boldsymbol{q}$ is good for G.

Suppose now that $\boldsymbol{q}$ is not zero-free. We can then choose $\boldsymbol{q}' \leq \boldsymbol{q}$ minimal so that $Z_G(-\boldsymbol{q}') = 0$. Define a set of events $(B'_v)_{v\in V}$ and a probability measure as in (2.1) with q'_v replacing q_v (by minimality $\boldsymbol{q}'$ must be a limit point of zero-free vectors and so, by continuity of Z_G, this does indeed define a measure as before). It follows that $\mathbb{P}(B'_v) \leq q'_v \leq q_v$ for all $v \in V$, G is a dependency graph for $(B'_v)_v \in V$ and $\mathbb{P}\left(\bigcap_{v \in V} B'_v\right) = Z_G(-\boldsymbol{q}') = 0$. Thus $\boldsymbol{q}$ is not good for G. □

Remark The proof of Theorem 2.5 showed that if $(A_v)_v \in V$ is a collection of events with dependency graph $G = (V, E)$ and $\mathbb{P}(A_v) \leq q_v$ for all $v \in V$ where $\boldsymbol{q} = (q_v)_{v \in V}$ is zero-free for G, then

$$\mathbb{P}\left(\bigcap_{v \in V} \bar{A}_v\right) \geq Z_G(-\boldsymbol{q}) \tag{2.6}$$

a fact first discovered by Shearer [64] with a strategy similar to the proof presented above. It's instructive to consider two extreme examples here.

1. If G has no edges, then the events A_v are mutually independent and $Z_G(-\boldsymbol{q}) = \prod_{v \in V}(1 - q_v)$ so that the lower bound in (2.6) is in fact an equality.

2. If G is the complete graph then $Z_G(-\boldsymbol{q}) = 1 - \sum_{v \in V} q_v$, which matches the union bound (and of course we cannot do any better in the absence of further information about the events A_v).

At this point, we're left with the following natural question: given a graph G, what is the set of good (equivalently zero-free) vectors for G? In most applications, the only salient information we have about G is some bound on its *maximum degree*. Moreover, it is often more convenient to work with good vectors that are uniform on their support. We might therefore rephrase the question and ask the following instead: what is the supremum of those r such that $(r, r, \ldots, r)$ is good for all graphs of maximum degree Δ? Following Shearer [64], we denote this supremum by $f(\Delta)$. Theorems 2.1 and 2.2 give a lower bound on $f(\Delta)$. Indeed, setting $p_v = 1/(1 + \Delta)$ for all v in Theorem 2.2 we see that

$$f(\Delta) \geq \frac{\Delta^\Delta}{(\Delta + 1)^{\Delta+1}}. \tag{2.7}$$

In a seminal paper, Shearer determined $f(\Delta)$ exactly. We give the proof of his lower bound on $f(\Delta)$ and follow up with a discussion of the matching upper bound.

Theorem 2.7 (Shearer [64]) *For $\Delta \geq 2$,*

$$f(\Delta) \geq \frac{(\Delta - 1)^{\Delta-1}}{\Delta^\Delta}. \tag{2.8}$$

Proof Let $\lambda_s(\Delta) := \frac{(\Delta-1)^{\Delta-1}}{\Delta^\Delta}$. We rephrase Shearer's proof in the language of zero-freeness. It suffices to show that if $G = (V, E)$ is a graph of maximum degree Δ and $\boldsymbol{\lambda} \in \mathbb{C}^V$ is such that $|\lambda_v| \leq \lambda_s(\Delta)$ for all $v \in V$, then $Z_G(\boldsymbol{\lambda}) \neq 0$. The proof relies crucially on the fundamental identity (1.5) which, after rephrasing, states that for $S \subseteq V$ and $v \in S$,

$$R_{S,v} := \frac{Z_S(\boldsymbol{\lambda})}{Z_{S-v}(\boldsymbol{\lambda})} - 1 = \lambda_v \frac{Z_{S-v-N(v)}(\boldsymbol{\lambda})}{Z_{S-v}(\boldsymbol{\lambda})}. \tag{2.9}$$

We prove the following statement by induction on $|S|$: if $v \in S$ has at most $\Delta - 1$ neighbours in S, then $|R_{S,v}| < 1/\Delta$. If $|S| = 1$, then

$$|R_{S,v}| = |\lambda_v| \leq \lambda_s(\Delta) < \frac{1}{\Delta}.$$

Suppose then that $|S| \geq 2$, let $S' = S - v$ and let the $\{u_1, \ldots, u_k\}$ denote the neighbours of v in S ($k \leq \Delta - 1$). By (2.9)

$$R_{S,v} = \lambda_v \prod_{i=1}^k \frac{Z_{S'-u_1-\ldots-u_i}(\boldsymbol{\lambda})}{Z_{S'-u_1-\ldots-u_{i-1}}(\boldsymbol{\lambda})} = \lambda_v \prod_{i=1}^k \frac{1}{1 + R_{S'-u_1-\ldots-u_{i-1},u_i}}. \tag{2.10}$$

We note that since u_i has v as a neighbour, it has at most $\Delta - 1$ neighbours in the set $S' - u_1 - \ldots - u_{i-1}$. By the inductive hypothesis we conclude that

$$|R_{S,v}| < \lambda_s(\Delta) \left(\frac{1}{1 - 1/\Delta} \right)^{\Delta-1} = \frac{1}{\Delta},$$

which concludes the induction. If v has Δ neighbours $\{u_1, \ldots, u_\Delta\}$ in S then we may still express $R_{S,v}$ as in (2.9) with $k = \Delta$ so that

$$|R_{S,v}| < \lambda_s(\Delta) \left(\frac{1}{1 - 1/\Delta} \right)^{\Delta} = \frac{1}{\Delta - 1}.$$

In particular, $|R_{G,v}| < 1/(\Delta - 1) \leq 1$ and so $Z_G(\boldsymbol{\lambda}) \neq 0$ as desired. □

2.1 The Δ-Regular Tree and a Look Ahead

We note that the bounds (2.7) and (2.8) are both asymptotically equal to $1/(e\Delta)$ as $\Delta \to \infty$. However, miraculously, Shearer's bound is tight. To show this it suffices to exhibit a sequence of graphs (G_n) of maximum degree Δ such that the (univariate) polynomial $Z_{G_n}(\lambda)$ has a root r_n converging to $\lambda_s(\Delta)$ as $n \to \infty$. The sequence we consider comes from truncations of the 'infinite regular tree', a central object in the study of the hard-core model. More precisely, let $\mathbb{T}_\Delta$ denote the infinite $(\Delta - 1)$-ary tree with root vertex r (so in particular v_0 has degree $\Delta - 1$ and all other vertices have degree Δ). For $n \geq 0$, we let $T_n = T_{\Delta,n}$ denote the subgraph of $\mathbb{T}_\Delta$ induced by all vertices at graph distance at most n from r.

This sequence is particularly nice from the perspective of the hard-core model due to its recursive structure. In particular, the fundamental identity (1.5) applied to T_n yields,

$$Z_{T_n}(\lambda) = \left(Z_{T_{n-1}}(\lambda) \right)^{\Delta-1} + \lambda \left(Z_{T_{n-2}}(\lambda) \right)^{(\Delta-1)^2} \tag{2.11}$$

valid for $n \geq 0$ if we set $Z_{T_{-1}} = Z_{T_{-2}} = 1$. By defining

$$R_n(\lambda) = \frac{Z_{T_n}(\lambda)}{(Z_{T_{n-1}}(\lambda))^{\Delta-1}} - 1 = \frac{\lambda \left(Z_{T_{n-2}}(\lambda)\right)^{(\Delta-1)^2}}{(Z_{T_{n-1}}(\lambda))^{\Delta-1}}$$

we can reduce the second-order recursion (2.11) to the first order recursion

$$R_n(\lambda) = \frac{\lambda}{(1 + R_{n-1}(\lambda))^{\Delta-1}}. \tag{2.12}$$

We note that $Z_{T_n}(\lambda) = 0$ if and only if $R_n(\lambda) = -1$. From here it's not too difficult to show that Z_{T_n} has a negative real root that approaches $-\lambda_s(\Delta)$ from the left as $n \to \infty$. We point the reader to [64] (see also [62, Example 3.6]) for the details.

The recursion (2.12) is also useful for probing the probabilistic properties of the hard-core model on the tree. The ratio $R_n(\lambda)$ is equal to $p_n(\lambda)/(1 - p_n(\lambda))$ where $p_n(\lambda)$ is the probability that the root of T_n is occupied in a sample from the hard-core model at activity λ. Suppose that we want to understand whether a typical sample from the hard-core model on T_n is *ordered* or *disordered*. One way to formalise this is to understand whether the model exhibits long-range correlations: suppose we place boundary conditions (specifications of occupied/unoccupied) on the leaves of T_n, does this significantly influence the probability that the root is occupied? Two extreme boundary conditions one can consider are those where we force all leaves to be occupied or unoccupied. We note that

$$\mathbb{P}_{T_n,\lambda}(r \text{ is occupied} \mid \text{all leaves of } T_n \text{ are unoccupied}) = p_{n-1}(\lambda),$$

and

$$\mathbb{P}_{T_n,\lambda}(r \text{ is occupied} \mid \text{all leaves of } T_n \text{ are occupied}) = p_{n-2}(\lambda).$$

We might then ask: for which $\lambda > 0$ is it the case that $p_{n-1}(\lambda)/p_{n-2}(\lambda) \to 1$ as $n \to \infty$? A folklore result (see e.g. [67]) states that there is a critical value $\lambda_c(\Delta)$ such that for $\lambda \leq \lambda_c(\Delta)$ the answer is 'yes' and for $\lambda > \lambda_c(\Delta)$ the answer is 'no'. Moreover, we have the expression

$$\lambda_c(\Delta) = \frac{(\Delta-1)^{\Delta-1}}{(\Delta-2)^{\Delta}}.$$

This 'phase transition' and it can be characterised in a number of ways:

1. If $\lambda \leq \lambda_c(\Delta)$ there exists a unique Gibbs measure[1] on the infinite tree $\mathbb{T}_\Delta$, whereas for $\lambda > \lambda_c(\Delta)$ there are multiple. $\lambda_c(\Delta)$ is therefore often referred to as the *tree uniqueness threshold*.

2. The limiting *free energy* $f(\lambda) := \lim_{n \to \infty} \frac{1}{|V(T_n)|} \log Z_{T_n}(\lambda)$ is real analytic on $[0, \infty)$ except at the point $\lambda = \lambda_c(\Delta)$.

3. When $\lambda < \lambda_c(\Delta)$ we have $|p_n(\lambda) - p_{n-1}(\lambda)| = e^{-\Theta(n)}$ and this is no longer true when $\lambda > \lambda_c(\Delta)$.

---

[1] a probability measure on independent sets of $\mathbb{T}_\Delta$ satisfying certain consistency conditions.

4. There exists an open set $D \subseteq \mathbb{C}$ containing the real interval $[0, \lambda_c(\Delta))$ such that $Z_{T_n}(\lambda) \neq 0$ for all $n \geq 0$ and $\lambda \in D$. Moreover $\lambda_c(\Delta)$ is a limit point of the set of zeros of the polynomials Z_{T_n}.

Exploring the connections between these analytic and probabilistic characterisations of phase transition has been a highly fruitful line of research. In Section 5, we will see that $\lambda_c(\Delta)$ can also be characterised *algorithmically*.

At this point let us highlight that $\lambda_c(\Delta) \sim \frac{e}{\Delta}$ as $\Delta \to \infty$ whereas $\lambda_s(\Delta) \sim \frac{1}{e\Delta}$. Item 4 hints at the fact that Shearer's theorem (Theorem 4) does not close the book on the topic of zero-free regions of the hard-core model. Shearer's theorem determines the maximal zero-free disc for the tree, but the obstruction to growing the disc further comes from roots on the negative real axis. From a probabilistic perspective this is rather unsatisfactory since when λ is negative, the hard-core model lacks a probabilistic interpretation. We return to the this topic in Section 5.1.

3 The Cluster Expansion

The cluster expansion is an important and classical tool from statistical physics which has been extensively used in the study of phase transitions (or absence thereof). Recently, the cluster expansion has proven to be a powerful tool in combinatorics and the study of algorithms. These applications will be our main focus.

We introduce the cluster expansion for the multivariate hard-core model $Z_G(\boldsymbol{\lambda})$ introduced in Section 1.1. Simply put, the cluster expansion is a Taylor expansion of the logarithm of the partition function $Z_G(\boldsymbol{\lambda})$. Observe that since $Z_G(\boldsymbol{\lambda})$ is an entire function (indeed it's a polynomial) and $Z_G(\mathbf{0}) = 1$, Z_G is zero-free in some neighbourhood of $\mathbf{0}$ and $\log Z_G$ can be expanded as a convergent Taylor series:

$$\log Z_G(\boldsymbol{\lambda}) = \sum_n a_{\mathbf{n}}(G) \boldsymbol{\lambda}^{\mathbf{n}}, \qquad (3.1)$$

where the sum is over vectors in $\mathbb{N}_0^V$ and $\boldsymbol{\lambda}^{\mathbf{n}} = \prod_{v \in V} \lambda_v^{n_v}$. This is the cluster expansion, also known as the 'Mayer expansion' following the work of Mayer [49], Mayer and Mayer [51] and Mayer and Montroll [50] in the 1930s and 40s. Groeneveld [30] was one of the first to give rigorous results on its convergence properties.

To make the connection to the previous section explicit, we record the following elementary fact from complex analysis.

Fact 3.1 *The series (3.1) is absolutely convergent for $|\boldsymbol{\lambda}| \leq \boldsymbol{R}$ if and only if $\boldsymbol{R}$ is zero-free for G.*

Each of the theorems of the previous section therefore give a lower bound on the radius of convergence of the cluster expansion for graphs G of maximum degree Δ. Early results bounding the radius of convergence of the cluster expansion (e.g. [30, 57, 60]) involved complicated combinatorial arguments for bounding the terms of the expansion directly. This then immediately implies zero-freeness via the forward implication of Fact 3.1. Dobrushin exploited the reverse implication: he established zero freeness directly and deduced cluster expansion convergence. Like Theorem 2.8, Theorem 2.1 has a simple inductive proof (essentially equivalent to the inductive proof of the Local Lemma!).

So, why care about the cluster expansion? One reason is that it gives us a way to analyse the statistics hidden inside the (partial) derivatives of $\log Z_G$ such as those in (1.2) and (1.4). Moreover, the coefficients $a_n(G)$ have a remarkable combinatorial interpretation which make them particularly easy to work with (see Theorem 3.2 below). Perhaps the most illuminating reason is that it gives a way to study *perturbations* of some ideal measure (e.g. a product measure). To illustrate this, consider the case where G has no edges and $\boldsymbol{\lambda} \geq 0$. In this case

$$Z_G(\boldsymbol{\lambda}) = \prod_{v \in V} (1 + \lambda_v)$$

and the hard-core measure $\mathbb{P}_{G,\boldsymbol{\lambda}}$ (defined at (1.3)) is seen to be a product measure (its independent site percolation where vertex v is occupied with probability $\lambda_v/(1+\lambda_v)$). Now if G has edges this is no longer the case as there are non-trivial interactions between vertices. However, if the coordinates of $\boldsymbol{\lambda}$ are small then we should think of $\mathbb{P}_{G,\boldsymbol{\lambda}}$ as a small perturbation of a product measure, and intuitively the measure should behave similarly. The cluster expansion gives a way to make this intuition rigorous.

3.1 The Coefficients of the Cluster Expansion

The coefficients of (3.1) have a useful combinatorial interpretation which we introduce now.

Let $\mathcal{G}_k$ denote the set of all graphs on vertex set $[k]$ and let $\mathcal{C}_k$ denote the set of all *connected* graphs in $\mathcal{G}_k$. For vertices $v_i v_j \in V$ let us write $v_i \sim v_j$ if v_i is adjacent to v_j in G or $v_i = v_j$. Given $\boldsymbol{v} = (v_1, \ldots, v_k) \in \boldsymbol{V}$ define the *Ursell function* $\phi = \phi_G$ by

$$\phi(\boldsymbol{v}) = \frac{1}{k!} \sum_{H \in \mathcal{C}_k} \prod_{\{i,j\} \in E(H)} (-\mathbf{1}_{v_i \sim v_j}).$$

A key property of the Ursell function is that $\phi(\boldsymbol{v})$ in non-zero only if the subgraph of G induced by the entries of $\boldsymbol{v}$ is *connected*. We call such a $\boldsymbol{v}$ a *cluster* and let $\mathcal{C}(G)$ denote the set of all such clusters. The term 'cluster expansion' usually refers to the combinatorial formulation of the expansion (3.1) given by the following theorem.

Theorem 3.2 *The expansion* (3.1) *can be rewritten as*

$$\log Z_G(\boldsymbol{\lambda}) = \sum_{\boldsymbol{v} \in \mathcal{C}(G)} \phi(\boldsymbol{v}) \prod_{v \in \boldsymbol{v}} \lambda_v. \qquad (3.2)$$

Proof We begin by rewriting the partition function as

$$Z_G(\boldsymbol{\lambda}) = 1 + \frac{1}{k!} \sum_{k \geq 1} \sum_{\boldsymbol{v} \in V^k} \prod_{v \in \boldsymbol{v}} \lambda_v \prod_{1 \leq i < j \leq k} (1 - \mathbf{1}_{v_i \sim v_j}),$$

where we note that the product is simply the indicator that the elements of $\boldsymbol{v}$ induce an independent set of size k in G. Expanding the product we have

$$\prod_{1 \leq i < j \leq k} (1 - \mathbf{1}_{v_i \sim v_j}) = \sum_{H \in \mathcal{G}_k} \prod_{\{i,j\} \in E(H)} (-\mathbf{1}_{v_i \sim v_j}),$$

and so
$$Z_G(\boldsymbol{\lambda}) = 1 + \sum_{k\geq 1} \frac{1}{k!} \sum_{H\in \mathcal{G}_k} \sum_{\boldsymbol{v}\in V^k} \prod_{v\in \boldsymbol{v}} \lambda_v \prod_{\{i,j\}\in E(H)} (-\mathbf{1}_{v_i\sim v_j}). \tag{3.3}$$

For $H \in \mathcal{G}_k$, let
$$Q(H) = \sum_{\boldsymbol{v}\in V^k} \prod_{v\in \boldsymbol{v}} \lambda_v \prod_{\{i,j\}\in E(H)} (-\mathbf{1}_{v_i\sim v_j}).$$

We note that if H has connected components $H_1,\ldots,H_\ell$ then $Q(H) = \prod_{i=1}^\ell Q(H_i)$. We may then write

$$\sum_{H\in\mathcal{G}_k} Q(H) = \sum_{\ell=1}^k \frac{1}{\ell!} \sum_{\substack{(m_1,\ldots,m_\ell)\in\mathbb{N}^\ell \\ m_1+\ldots+m_\ell=k}} \frac{n!}{m_1!\cdots m_\ell!} \sum_{H_1\in\mathcal{C}_{m_1}}\cdots \sum_{H_\ell\in\mathcal{C}_{m_\ell}} \prod_{i=1}^\ell Q(H_i).$$

Here the second sum is over all *ordered* integer partitions of k into ℓ parts. The multinomial coefficient $\frac{k!}{m_1!\cdots m_\ell!}$ is the number of ways of partitioning the ground set $[k]$ into ℓ labelled parts of size $m_1,\ldots,m_\ell$. The factor of $1/\ell!$ unlabels the resulting vertex partitions. Given a vertex partition of $[k] = V_1 \cup \ldots V_\ell$, we then consider all possible ways of placing a connected graph H_i on vertex set V_i. We now observe that, at least formally,

$$\sum_{k\geq 1}\sum_{\ell=1}^k \sum_{\substack{(m_1,\ldots,m_\ell)\in\mathbb{N}^\ell \\ m_1+\ldots+m_\ell=k}} (\cdot) = \sum_{\ell\geq 1}\sum_{k\geq \ell} \sum_{\substack{(m_1,\ldots,m_\ell)\in\mathbb{N}^\ell \\ m_1+\ldots+m_\ell=k}} (\cdot) = \sum_{\ell\geq 1} \sum_{(m_1,\ldots,m_\ell)\in\mathbb{N}^\ell} (\cdot). \tag{3.4}$$

Putting everything together we arrive at
$$Z_G(\boldsymbol{\lambda}) = 1 + \sum_{\ell\geq 1} \frac{1}{\ell!} \sum_{(m_1,\ldots,m_\ell)\in\mathbb{N}^\ell} \prod_{i=1}^\ell \left\{ \frac{1}{m_i!} \sum_{H_i\in\mathcal{C}_{m_i}} Q(H_i) \right\}$$
$$= 1 + \sum_{\ell\geq 1} \frac{1}{\ell!} \left(\sum_{m\geq 1}\sum_{\boldsymbol{v}\in V^m} \phi(\boldsymbol{v}) \prod_{v\in \boldsymbol{v}} \lambda_v \right)^\ell$$
$$= \exp\left\{ \sum_{\boldsymbol{v}\in\mathcal{C}(G)} \phi(\boldsymbol{v}) \prod_{v\in \boldsymbol{v}} \lambda_v \right\},$$

as desired. $\square$

Remark 1. Comparing the expansions (3.1) and (3.2) we see that for $\boldsymbol{n}\in \mathbb{N}_0^V$, $a_{\boldsymbol{n}}$ is equal to the sum of $\phi(\boldsymbol{v})$ over all $\boldsymbol{v}$ with exactly n_v coordinates equal to v. Fixing one such vector $\boldsymbol{u}$ we then have

$$a_{\boldsymbol{n}} = \frac{\sum_v n_v}{\prod_v (n_v)!} \phi(\boldsymbol{u}),$$

since $\phi(\boldsymbol{v})$ is invariant under permutations of the coordinates of $\boldsymbol{v}$. In particular, the expansion (3.1) therefore converges absolutely if and only if (3.2) converges absolutely.

The Cluster Expansion in Combinatorics

2. In the proof of Theorem (3.1), we assumed that the series (3.2) was absolutely convergent so that the manipulations of infinite series such as (3.4) was justified.

3. It is amusing to compare $Z_G(\boldsymbol{\lambda})$ and $\log Z_G(\boldsymbol{\lambda})$ in the following two forms. On the one hand, from (3.3) we have

$$Z_G(\boldsymbol{\lambda}) = 1 + \sum_{k \geq 1} \frac{1}{k!} \sum_{H \in \mathcal{G}_k} \sum_{\boldsymbol{v} \in V^k} \prod_{v \in \boldsymbol{v}} \lambda_v \prod_{\{i,j\} \in E(H)} (-1_{v_i \sim v_j}).$$

On the other hand, writing out the definition of the Ursell function in (3.2) we see that

$$\log Z_G(\boldsymbol{\lambda}) = \sum_{k \geq 1} \frac{1}{k!} \sum_{H \in \mathcal{C}_k} \sum_{\boldsymbol{v} \in V^k} \prod_{v \in \boldsymbol{v}} \lambda_v \prod_{\{i,j\} \in E(H)} (-1_{v_i \sim v_j}).$$

By replacing the sum over $\mathcal{G}_k$ by a sum over $\mathcal{C}_k$ in the expression for $Z_G(\boldsymbol{\lambda})$ (and subtracting 1) we get an expression for $\log Z_G(\boldsymbol{\lambda})$! One heuristic for why this is the case is that the operation of exponentiation 'combines' the connected graphs in all possible ways.

At this point it may be helpful to look at some examples.

Example 3.3 Let $G = (\{v\}, \emptyset)$ be the graph with a single vertex $\{v\}$. The clusters in G are all vectors of the form $\boldsymbol{v} = (v, \ldots, v)$ (of length ≥ 1). If $\boldsymbol{v} = (v, \ldots, v)$ has length k, then

$$\phi(\boldsymbol{v}) = \frac{1}{k!} \sum_{H \in \mathcal{C}_k} (-1)^{|H|} \qquad (3.5)$$

where $|H|$ denotes the number of edges in H. By Theorem 3.2, the cluster expansion of $\log Z_G$ is then

$$\log Z_G(\lambda) = \sum_{k=1}^{\infty} \left(\frac{1}{k!} \sum_{H \in \mathcal{C}_k} (-1)^{|H|} \right) \lambda^k.$$

On the other hand, in this case we know that $Z_G(\lambda) = (1 + \lambda)$ and so by Taylor expansion

$$\log Z_G(\lambda) = \sum_{k=1}^{\infty} \frac{(-1)^{k-1}}{k} \lambda^k.$$

Equating the coefficients of these two expansions we deduce the the curious combinatorial identity

$$\sum_{H \in \mathcal{C}_k} (-1)^{|H|} = (-1)^{k-1}(k-1)!.$$

Example 3.4 Let $G = (V, E)$ be a graph on n vertices and consider the univariate partition function $Z_G(\lambda)$. In this setting, it is often useful to express the coefficients of the cluster expansion of $\log Z_G(\lambda)$ in terms of *subgraph counts* of G. First observe that by Example 3.3, the contribution to the cluster expansion of $\log Z_G$ from clusters of the form $(v, v, \ldots, v)$ is $n \log(1+\lambda)$ and so we can set this aside and restrict our attention to non-constant clusters. We note that (u, v), $u \neq v$, is a cluster if and only if $uv \in E$ in which case $\phi(u, v) = -1/2$. Moreover there are $2e(G)$ such clusters. A non-constant vector $\boldsymbol{v} = (u, v, w)$ is a cluster if and only if the induced graph on $\{u, v, w\}$ is (i) a triangle, (ii) a path of length 2, or (iii) an edge. $\phi(\boldsymbol{v}) = 1/3, 1/6, 1/3$ in cases (i), (ii), (iii) respectively. Moreover there are $6K_3(G), 6P_2(G), 6e(G)$ clusters of type (i), (ii), (iii) where $P_2(G), K_3(G)$ denote the number of induced copies of P_2 and K_3 in G respectively. We conclude that

$$\log\left(\frac{Z_G(\lambda)}{(1+\lambda)^n}\right) = -e(G)\lambda^2 + (2K_3(G) + P_2(G) + 2e(G))\lambda^3 + \ldots$$

We note that when $\lambda > 0$, the ratio $\frac{Z_G(\lambda)}{(1+\lambda)^n}$ has a natural probabilistic interpretation. Suppose we select a random subset $S \subseteq V$, where each $v \in V$ is selected independently with probability $p = \lambda/(1+\lambda)$. Then $\frac{Z_G(\lambda)}{(1+\lambda)^n}$ is the probability that S is an independent set. To the combinatorialist, the first tool that may come to mind here is Janson's inequality [33] which in this case gives

$$\log\left(\frac{Z_G(\lambda)}{(1+\lambda)^n}\right) \leq -e(G)\left(\frac{\lambda}{1+\lambda}\right)^2 + P_2(G)\left(\frac{\lambda}{1+\lambda}\right)^3.$$

In this setting, the cluster expansion can be viewed as a refinement of Janson's inequality. The advantage of Janson's inequality of course is that it is valid for all $\lambda > 0$.

3.2 Polymer Models

In the physics literature, multivariate hard-core models often go under the name of *polymer models*. The language of polymer models turns out to be convenient for many applications and so we introduce it here.

Let $\mathcal{P}$ be a finite set whose elements we call 'polymers'. We equip $\mathcal{P}$ with a complex-valued weight $w(\gamma)$ for each polymer S as well as a symmetric and reflexive incompatibility relation between polymers. We write $\gamma \nsim \gamma'$ if polymers γ and γ' are incompatible. We refer to the triple $(\mathcal{P}, w, \nsim)$ as a polymer model. Let Ω be the collection of sets of pairwise compatible polymers from $\mathcal{P}$, including the empty set of polymers. The polymer model partition function is then

$$\Xi = \sum_{\Gamma \in \Omega} \prod_{\gamma \in \Gamma} w(\gamma), \tag{3.6}$$

where the contribution from the empty set is 1. We note that a polymer model is simply a multivariate hard-core model on the graph with vertex set $\mathcal{P}$ where γ is adjacent to γ' if and only if γ, γ' are incompatible. In applications, polymers are often taken to be connected subgraphs of some ambient graph G. It is slightly cumbersome to talk about a multivariate hard-core model on an auxiliary graph whose vertices are subgraphs of G. The language of polymer models is more natural here.

It will also be useful to rephrase/repeat some of the language around cluster expansion in this context. For a vector $\Gamma = (\gamma_1, \ldots, \gamma_k)$ of polymers, the *incompatibility graph*, H_Γ, is the graph with vertex set Γ (considered as a multiset) and an edge between any two incompatible polymers. We call Γ a cluster if H_Γ is a connected graph. We let $\mathcal{C} = \mathcal{C}(\mathcal{P})$ denote the set of all possible clusters. For $\Gamma = (\gamma_1, \ldots, \gamma_k) \in \mathcal{C}$ we have the Ursell function

$$\phi(\Gamma) = \frac{1}{k!} \sum_{\substack{A \subseteq E(H_\Gamma) \\ \text{spanning, connected}}} (-1)^{|A|}.$$

The cluster expansion then reads

$$\log \Xi = \sum_{\Gamma \in \mathcal{C}} \phi(\Gamma) \prod_{\gamma \in \Gamma} w(\gamma).$$

A particularly convenient sufficient condition for the convergence of the cluster expansion is given by a theorem of Kotecký and Preiss.

Theorem 3.5 (Kotecký and Preiss [44]) *Let $f : \mathcal{P} \to [0, \infty)$ and $g : \mathcal{P} \to [0, \infty)$ be two functions. Suppose that for all polymers $\gamma \in \mathcal{P}$,*

$$\sum_{\gamma' \not\sim \gamma} |w(\gamma')| e^{f(\gamma') + g(\gamma')} \leq f(\gamma). \tag{3.7}$$

Then the cluster expansion converges absolutely. Moreover, if we let $g(\Gamma) = \sum_{\gamma \in \Gamma} g(\gamma)$ and write $\Gamma \not\sim \gamma$ if there exists $\gamma' \in \Gamma$ so that $\gamma \not\sim \gamma'$, then for all polymers γ,

$$\sum_{\substack{\Gamma \in \mathcal{C} \\ \Gamma \not\sim \gamma}} \left| \phi(\Gamma) \prod_{\gamma \in \Gamma} w(\gamma) \right| e^{g(\Gamma)} \leq f(\gamma). \tag{3.8}$$

One could simply take $g \equiv 0$ in (3.7) in order to establish convergence of the cluster expansion. However, allowing g to take non-zero values (thus strengthening (3.7)) allows us to give strong tail bounds on the cluster expansion via (3.8). This allows one to show that certain truncations of the cluster expansion serve as good approximations to the logarithm of the partition function. We note also that one often defines $f(\gamma)$ to be $|\gamma|$ where $|\gamma|$ is some notion of 'size' of the polymer (e.g. number of vertices in γ if polymers are subgraphs of an ambient graph). This makes Theorem 3.5 particularly useful when polymers have some underlying structure.

4 Cluster Expansion and Enumeration

In this section we will discuss applications of the cluster expansion to combinatorial enumeration. Take, for example, the problem of counting the independent sets in a graph G. We can express this count as $Z_G(1)$ where Z_G is the (univariate) hard-core model partition function. At this point, it seems like cluster expansion has nothing to say here: for any graph G with more than one edge, the radius of convergence of the cluster expansion of $\log Z_G$ is at most $1/2$! However, fortunately, this is not the end of the story. To see why, we begin with a classical result of Korshunov and Sapozhenko on counting independent sets in the hypercube.

4.1 Independent Sets in the Hypercube

Let Q_d denote the discrete hypercube of dimension d: the graph with vertex set $\{0,1\}^d$ with edges between vectors that differ in exactly one coordinate. Let $i(G) = Z_G(1)$ denote the number of independent sets of G.

Theorem 4.1 (Korshunov and Sapozhenko [43])

$$i(Q_d) = (1 + o(1)) \cdot 2\sqrt{e} \cdot 2^{2^{d-1}}$$

as $d \to \infty$.

A beautiful and influential proof of Theorem 4.1 was later given by Sapozhenko in [61], an early example of the method of graph containers. See [28] for an excellent exposition of this proof.

Enumeration results such as this, often come with detailed probabilistic and structural information on the class of objects being enumerated. In proving Theorem 4.1, Korshunov and Sapozhenko showed that a typical independent set of Q_d is highly *unbalanced*. To make this precise, recall that Q_d is a bipartite graph with vertex classes $\mathcal{O}, \mathcal{E}$ (vertices of odd, even Hamming weight respectively). Korshunov and Sapozhenko observed that for $k \geq 0$ fixed (or slowly growing as a function of d), the number of independent sets I such that $\min\{|I \cap \mathcal{O}|, |I \cap \mathcal{E}|\} = k$ is asymptotically equal to

$$2 \cdot \frac{2^{-k}}{k!} \cdot 2^{2^{d-1}} . \tag{4.1}$$

Summing these contributions over all k we have the lower bound $i(Q_d) \geq (1 + o(1)) \cdot 2\sqrt{e} \cdot 2^{2^{d-1}}$. The difficult part of the proof of Theorem 4.1 is showing that this lower bound is asymptotically tight. Once this is proved, we can return to (4.1) to deduce that if $k \geq 0$ is fixed and I is a uniformly chosen independent set in Q_d then,

$$\mathbb{P}(\min\{|I \cap \mathcal{O}|, |I \cap \mathcal{E}|\} = k) = (1 + o(1)) \frac{2^{-k}}{k!} e^{-1/2} . \tag{4.2}$$

Let us call the vertices in the smaller of the sets $I \cap \mathcal{O}, I \cap \mathcal{E}$ the *defect vertices* of I. Then (4.2) is telling us that the number of defect vertices in I is asymptotically Poisson with mean $1/2$. The non-defect vertices can also be understood precisely: note that if we condition on $I \cap \mathcal{O} = S$ for some $S \subseteq \mathcal{O}$ (to be thought of as defect vertices), then $I \cap \mathcal{E}$ is simply a uniformly random subset of the unblocked vertices $\mathcal{E} \setminus N(S)$.

This perspective was further developed by Galvin [26] who refined the understanding of the distribution of the defect vertices and generalised the work of Sapozhenko [61] to the setting of the hard-core model. In particular, Galvin showed that for the hard-core model on Q_d at activity $\lambda = 1 + s/d$ with s constant, the number of defect vertices is asymptotically distributed as a Poisson random variable with mean $e^{-s/2}/2$. Prior to Galvin's work, Kahn [41], in his influential work on the entropy approach to the hard-core model, showed that for constant λ, a typical independent set drawn from hard-core model on Q_d has $o(2^d)$ defect vertices.

The above discussion suggests that we can understand the hard-core model on Q_d at activity λ as a *perturbation* of the measure $\frac{1}{2}\mu_{\mathcal{O}} + \frac{1}{2}\mu_{\mathcal{E}}$, where $\mu_{\mathcal{O}}, \mu_{\mathcal{E}}$ is the measure on $\mathcal{O}, \mathcal{E}$ where we select each vertex independently with probability $p = \lambda/(1+\lambda)$ (the hard-core model on the edgeless graphs $Q_d[\mathcal{O}], Q_d[\mathcal{E}]$). We refer to the measures $\mu_{\mathcal{O}}, \mu_{\mathcal{E}}$ as 'ground states'. To understand the perturbation from these ground states, we need to understand the distribution of the defect vertices. We are now back in the situation where we have a perturbation of some ideal measure and this is where cluster expansion re-enters the picture. The key here is that the global independent set is highly structured (mostly contained in $\mathcal{O}$ or $\mathcal{E}$), however the defects are highly *unstructured* and can therefore be understood in detail.

Recently, the author and Perkins [37] showed that the distribution of the defect vertices can be described precisely in terms of a polymer model. Moreover this polymer model is *unstructured* (its weights are small) and can therefore be analysed via cluster expansion. This allows for a particularly detailed description of the hard-core model on Q_d and gives detailed asymptotics for the partition function $Z_{Q_d}(\lambda)$. Specialising to $\lambda = 1$, for example, we have the following refinement of Theorem 4.1.

Theorem 4.2 (J. and Perkins [37])

$$i(Q_d) = 2\sqrt{e} \cdot 2^{2^{d-1}} \left(1 + \frac{3d^2 - 3d - 2}{8 \cdot 2^d} + \frac{243d^4 - 646d^3 - 33d^2 + 436d + 76}{384 \cdot 2^{2d}} + O\left(d^6 \cdot 2^{-3d}\right) \right)$$

as $d \to \infty$.

More generally, one has a formula and an algorithm for computing the asymptotics of $i(Q_d)$ to arbitrary order in 2^{-d} by taking further terms in a cluster expansion. Taking further terms in this expansion corresponds to accounting for independent sets with increasingly complex defects.

4.1.1 The Defect Polymer Model
In this section we go into a little more detail about the proof of Theorem 4.2 and describe the polymer model (see Section 3.2) used to describe the distribution of defect vertices in a sample from the hard-core model on Q_d at activity λ.

Suppose we fix $S \subseteq \mathcal{O}$ (to be thought of as a set of defect vertices of an independent set, so in particular $|S| \leq |\mathcal{O}|/2$) and consider the weight of all independent sets I such that $I \cap \mathcal{O} = S$:

$$\sum_{I \in \mathcal{I}(Q_d): I \cap \mathcal{O} = S} \lambda^{|I|} = \lambda^{|S|}(1+\lambda)^{|\mathcal{E}| - |N(S)|}.$$

The equality comes from the fact that we have forced the elements of S to be occupied and this leaves $|\mathcal{E}| - |N(S)|$ unblocked vertices in $\mathcal{E}$ that can be occupied or unoccupied. This motivates us to define

$$w(S) = \lambda^{|S|}(1+\lambda)^{-|N(S)|},$$

the *weight* of S. A key feature of this weight function is that it factorises over sets of vertices with disjoint neighbourhoods. We call a set $S \subseteq \mathcal{O}$ *2-linked* if it cannot be partitioned into disjoint subsets S_1, S_2 such that $N(S_1) \cap N(S_2) = \emptyset$ (in other words S induces a connected subgraph of the squared graph Q_d^2). A set $S \subseteq \mathcal{O}$ can be decomposed into maximal 2-linked components $\gamma_1, \ldots, \gamma_k$ in which case

$$w(S) = \prod_{i=1}^{k} w(\gamma_i).$$

We therefore arrive naturally at the following polymer model: polymers are 2-linked subsets of $\mathcal{O}$ of size at most $|\mathcal{O}|/2$; the weight $w(\gamma)$ of a polymer γ is $\lambda^{|\gamma|}(1+\lambda)^{-|N(\gamma)|}$; and the two polymers γ, γ' are compatible if and only if $N(\gamma), N(\gamma')$ are disjoint. This polymer model comes with the associated partition function $\Xi_\mathcal{O}$ as in (3.6). We may also define a polymer model on subsets of $\mathcal{E}$ in identical fashion with partition function $\Xi_\mathcal{E}$. The first step to proving Theorem 4.2 is to show that the partition function $Z_{Q_d}(\lambda)$ is very well approximated by

$$(1+\lambda)^{2^{d-1}} (\Xi_\mathcal{O} + \Xi_\mathcal{E}). \tag{4.3}$$

The intuition here is that $\Xi_\mathcal{O}, \Xi_\mathcal{E}$ capture the contribution of independent sets with defects in $\mathcal{O}, \mathcal{E}$ respectively. Now the aim is to understand $\Xi_\mathcal{O}, \Xi_\mathcal{E}$ in detail via cluster expansion. The difficult part here is proving that the cluster expansions of these polymer models converge. Here the container tools developed by Sapozhenko [61] and Galvin [26] combine remarkably well with the convergence criteria for the cluster expansion from the physics literature (e.g. Theorem 2.1 and Theorem 3.5).

To end this section, we use the cluster expansion to give an alternative viewpoint on the factor of $\sqrt{e}$ appearing in Theorem 4.1. The simplest clusters in the odd polymer model are those consisting of a single polymer γ where γ is a polymer consisting of a single vertex. There are 2^{d-1} such clusters and they each contribute $w(\gamma) = \lambda(1+\lambda)^{-d}$ to the cluster expansion of $\log \Xi_\mathcal{O}$. We therefore have

$$\log \Xi_\mathcal{O} = 2^{d-1} \lambda (1+\lambda)^{-d} + R_\lambda,$$

where R_λ is the remainder of the cluster expansion. When $\lambda = 1$, $R_\lambda = o(1)$ and so we have $\log \Xi_\mathcal{O} = 1/2 + o(1)$ which, in conjunction with the step described at (4.3), recovers Theorem 4.1. Taking further terms of the cluster expansion, we obtain Theorem 4.2.

Independent sets in the hypercube exemplify a phenomenon that is ubiquitous in combinatorics: often a natural measure on a class of combinatorial objects can be understood as a perturbation of some idealised measure. In any such situation one can attempt to model the perturbation as a polymer model. The next three sections contain a few more examples of this perspective in action.

4.2 Homomorphisms From the Hypercube

The problem of counting independent sets in a graph G can be placed within a rich class of counting problems: counting homomorphisms from G to a fixed graph H. We call a function $f : V(G) \to V(H)$ a *homomorphism* if the map preserves edges, that is, $f(e) \in E(H)$ for all $e \in E(G)$. Let $\text{Hom}(G, H)$ denote the set of homomorphisms from G to H. The set $\mathcal{I}(G)$ of independent sets in G is easily seen to be in bijection with $\mathrm{Hom}(G, H)$ where $H = $ ⌬, an edge with a loop at one vertex[2]. Another example of particular interest in combinatorics comes from taking $H = K_q$ the clique on q vertices. In this case, $\mathrm{Hom}(G, H)$ is the set of proper q-colourings of the vertices of G.

Building on the work of Sapozhenko [61], Galvin asymptotically determined the number of proper 3-colourings of Q_d.

Theorem 4.3 (Galvin [25])

$$|\mathrm{Hom}(Q_d, K_3)| \sim 6e2^{2^{d-1}}.$$

In a similar spirit to Theorem 4.1, Galvin shows that almost every 3-colouring of Q_d is 'close' to a 'ground state'. Here a ground state colouring is one that is monochromatic on one side of the bipartition of Q_d. More generally, given a graph H we can define a class of ground state homomorphisms in $\mathrm{Hom}(Q_d, H)$. For $A, B \subset V(H)$, we write $A \sim B$ if $\{a, b\} \in E(H)$ for all $a \in A$ and $b \in B$ and call such a pair (A, B) a *pattern*. We call a pattern (A, B) *dominant* if it maximises the quantity $|A||B|$ over all patterns. In particular, the dominant patterns in $H = K_q$ are those (A, B) that partition the vertex set of K_q (the set of colours) as evenly as possible. Let us call $f \in \mathrm{Hom}(Q_d, H)$ a ground state homomorphism if $f(\mathcal{O}) \subseteq A$, $f(\mathcal{E}) \subseteq B$ for some dominant pattern (A, B). In analogy with Section 4.1, given $f \in \mathrm{Hom}(Q_d, H)$ we choose the ground state homomorphism g that agrees with f the most (breaking ties arbitrarily) and call vertices v such that $f(v) \neq g(v)$ *defect vertices*. Using a delicate entropy argument Engbers and Galvin [18] proved the following result.

Theorem 4.4 (Engbers and Galvin [18]) *For every H there is an $\epsilon > 0$ such that all but a $2^{-\Omega(d)}$ proportion of homomorphisms in $\mathrm{Hom}(Q_d, H)$ have at most $(2 - \epsilon)^d$ defect vertices.*

In fact, Engbers and Galvin prove much more, and their results apply to weighted homomorphisms from the discrete torus $\mathbb{Z}_m^d$ of even side length m: the graph on vertex set $\{0, \ldots, m-1\}^n$ where two vertices x, y are adjacent if and only if there exists a coordinate $i \in [n]$ such that $x_i = y_i \pm 1 \pmod{m}$ and $x_j = y_j$ for all $j \neq i$ (the case $m = 2$ returns the cube Q_d). Theorem 4.4 provides an excellent starting point for attempts to asymptotically enumerate $\mathrm{Hom}(Q_d, H)$. Indeed it shows that it suffices to focus on the case where there are 'few' defect vertices. Using this starting point, Kahn and Park [42] delicately combined the container method with entropy tools to enumerate 4 colourings of the cube.

Theorem 4.5 (Kahn and Park [42])

$$|\mathrm{Hom}(Q_d, K_4)| \sim 6e2^{2^d}.$$

As in Theorem 4.3, Kahn and Park showed that a typical 4-colouring of Q_d has only a constant number of defects. Again we have a situation where a measure over combinatorial objects is a perturbation of some idealised measure (in this case the

---

[2] More generally, the hard-core model can be described in terms of 'weighted' graph homomorphisms, but for simplicity we restrict ourselves to the unweighted case in this section.

uniform distribution on ground state colourings) and so cluster expansion methods can be brought to bear. The author and Keevash showed that for any integer q the distribution of defects in a uniformly random q colouring of Q_d can be described by a mixture of polymer models with a convergent cluster expansion. In addition to detailed probabilistic and structural information on the space of q-colourings of Q_d, detailed asymptotics for the number of such colourings become available e.g.

Theorem 4.6 (J. and Keevash [35])

$$|\mathrm{Hom}(Q_d, K_5)| \sim 20\sqrt[3]{e} \cdot 6^{2^{d-1}} \exp\left\{\left(\frac{4}{3}\right)^d\right\}$$

$$|\mathrm{Hom}(Q_d, K_6)| \sim 20 \cdot 3^{2^d} \exp\left\{\left(\frac{4}{3}\right)^d\right\}$$

$$|\mathrm{Hom}(Q_d, K_7)| \sim 70 \cdot e^{d/2} 12^{2^{d-1}} \exp\left\{\left(\frac{3}{2}\right)^{d-1} + \frac{1}{2}\left(\frac{4}{3}\right)^{d-1} + \frac{d^2 - d - 54}{108}\left(\frac{9}{8}\right)^{d-1}\right\}$$

$$|\mathrm{Hom}(Q_d, K_8)| \sim 70 \cdot 4^{2^d} \exp\left\{\left(\frac{3}{2}\right)^d + \frac{d^2 + 41d - 54}{108}\left(\frac{9}{8}\right)^d\right\}$$

The cluster expansion provides an algorithm for computing the asymptotics of $|\mathrm{Hom}(Q_d, K_q)|$ for any q. As q grows one needs to take increasingly many terms of the cluster expansion to obtain an asymptotic formula. The results of [35] apply more generally to (weighted) homomorphisms from the discrete torus $\mathbb{Z}_m^d$, m even, to any target graph H. As usual, the most challenging part of the proofs is to show that the polymer models describing defects has a convergent cluster expansion. In [35] a combination of the container method, entropy tools, and algebraic properties of the torus are used to establish convergence.

A natural next step would to study homomorphisms from $\mathbb{Z}_m^d$ where m is odd. When m is odd $\mathbb{Z}_m^d$ is no longer bipartite and the set of ground states can have a much more complicated structure. We repeat a question left open in [35].

Question 4.7 *Is there a simple asymptotic formula for the number of independent sets in $\mathbb{Z}_3^d$?*

4.3 Triangle-Free Graphs

The problems of the previous two sections exemplify the common phenomenon in combinatorics where almost all elements of some family of objects is (close to) a subset of an *extremal* element in that family. For example, in Section 4.1 we saw that almost all independent sets in Q_d are close to a subset of one of the two maximal independent sets $\mathcal{O}, \mathcal{E}$.

Perhaps the most well-known result in extremal combinatorics is Mantel's theorem [48]: the number of edges in a triangle-free graph is at most that of a complete balanced bipartite graph on the same number of vertices. A classical result of Erdős, Kleitman, Rothschild [20] shows that almost all triangle-free graphs are bipartite and nearly balanced. Let $\mathcal{T}(n)$ be the set of triangle-free graphs on n vertices and $\mathcal{B}(n)$ be the set of bipartite graphs on n vertices.

Theorem 4.8 (Erdös, Kleitman, Rothschild [20]) *Almost all triangle-free graphs are bipartite. That is,*

$$|\mathcal{T}(n)| \sim |\mathcal{B}(n)|$$

It is not difficult to see that almost all bipartite graphs on n vertices are highly balanced (with part sizes differing by at most $O(n \log n)^{1/4}$). Theorem 4.8 therefore shows that a typical triangle-free graph is a random subgraph of a nearly balanced complete bipartite graph on n vertices; or in other words, typical triangle-free graphs exhibit the same rigid global structure as the extremal example provided by Mantel's theorem [48].

A recent trend in combinatorics has been to study sparse analogues of classical combinatorial results. In this context, it is natural to ask whether the structural behaviour exhibited in Theorem 4.8 persists for *sparse* triangle-free graphs. To make this question precise, let $\mathcal{T}(n,m)$ be the set of triangle-free graphs on n vertices and m edges and let $\mathcal{B}(n,m)$ be the set of bipartite graphs. In a celebrated paper, Osthus, Prömel, and Taraz [53], proved that there is a sharp threshold in m for the property that a typical element of $\mathcal{T}(n,m)$ is bipartite with high probability.

Theorem 4.9 (Osthus, Prömel, and Taraz [53]) *For every $\epsilon > 0$,*

1. *If $m \geq (1+\epsilon)\frac{\sqrt{3}}{4}n^{3/2}\sqrt{\log n}$, then almost every graph in $\mathcal{T}(n,m)$ is bipartite; that is,*

$$|\mathcal{T}(n,m)| \sim |\mathcal{B}(n,m)|.$$

2. *If $n/2 \leq m \leq (1-\epsilon)\frac{\sqrt{3}}{4}n^{3/2}\sqrt{\log n}$, then almost every graph in $\mathcal{T}(n,m)$ is not bipartite; that is,*

$$|\mathcal{B}(n,m)| = o(|\mathcal{T}(n,m)|).$$

In other words, the rigid global structural property of a typical triangle-free graph being bipartite persists, as the edge density is lowered, until $m \approx \frac{\sqrt{3}}{4}n^{3/2}\sqrt{\log n}$, and thus in this range of densities the asymptotic enumeration and typical structure problems reduce to the simpler problem of understanding bipartite graphs. What happens when m drops below $\frac{\sqrt{3}}{4}n^{3/2}\sqrt{\log n}$? A classical result of Łuczak [47] shows that a typical element of $\mathcal{T}(n,m)$ is almost bipartite all the way down to $m \geq Cn^{3/2}$ for a large constant C.

Theorem 4.10 (Łuczak [47]) *Given $\delta > 0$, there exists a constant $C > 0$ so that if $m \geq Cn^{3/2}$ then almost all elements of $\mathcal{T}(n,m)$ can be made bipartite by deleting at most δm edges.*

In the language of the previous two sections, Theorem 4.10 says that a typical triangle-free graph in $\mathcal{T}(n,m)$ is a perturbation of a 'ground state' where ground state graphs are bipartite (and nearly balanced). Let us call the edges not contained in the max-cut of a graph G, the *defect edges* of G. This raises the question of what the distribution of the defect edges is. The author, Perkins and Potukuchi [38] studied this problem in detail using the cluster expansion. We show that for $m \geq (1+\epsilon)\frac{1}{4}n^{3/2}\sqrt{\log n}$ the distribution of the defect edges is an Erdős-Renyi random graph up to $o(1)$ in total variation distance. When $Cn^{3/2} \leq m \leq (1-\epsilon)\frac{1}{4}n^{3/2}\sqrt{\log n}$, the distribution is that of an *exponential random graph* conditioned on being triangle-free and having not too large maximum degree (roughly speaking, if we allow the max degree to grow too large, then we violate the max-cut condition). This understanding of the typical structure of triangle-free graphs allows for detailed enumeration results. For example, in [38] the following is proved. Let $\lambda_0 = \lambda_0(n,m) = 4m/n^2$ and

$$\lambda = \lambda(n,m) = \lambda_0 + \lambda_0^2 + (n\lambda_0^2 - 1)\lambda_0 e^{-\lambda_0^2 n/2}. \tag{4.4}$$

Theorem 4.11 (J., Perkins, Potukuchi [38]) *Fix $\epsilon > 0$ and suppose $m \geq (1+\epsilon)\frac{1}{4}n^{3/2}\sqrt{\log n}$. Then*

$$|\mathcal{T}(n,m)| \sim \frac{1}{\sqrt{2}\lambda^{m+1}n}\binom{n}{n/2}(1+\lambda)^{n^2/4}\exp\left\{\lambda e^{-\lambda^2 n/2 + \lambda^3 n}\frac{n^2}{4} + \lambda^5 e^{-\lambda^2 n}\frac{n^4}{8}\right\},$$

where $\lambda = \lambda(n,m)$ is as in (4.4).

Asymptotics are available for smaller m, but the formulae become rather unwieldy.

For m far enough below $n^{3/2}$ edges, the asymptotic enumeration problem has also been solved, though by entirely different methods. Unlike in Theorems 4.8, 4.9 and 4.11, the asymptotics in this regime are not driven by a rigid global structure like bipartiteness, but rather by a lack of global structure i.e. disorder. When $m \leq n^{3/2-\epsilon}$, the asymptotics of $|\mathcal{T}(n,m)|$ have been determined in a series of papers [21, 34, 52, 59, 68, 70]. The first step was the result of Erdős and Renyi showing that with $m = \Theta(n)$ the distribution of the number of triangles in the random graph $G(n,m)$ is asymptotically Poisson, and thus the proportion of all graphs on n vertices with m edges that are triangle free is $\sim \exp(-\mu)$, where μ is the expected number of triangles in $G(n,m)$. Using Janson's Inequality, Janson, Luczak and Rucinski then showed that for $m = o(n^{6/5})$, the Poisson behaviour persists and the probability in $G(n,m)$ of seeing no triangles is still asymptotic to $\exp(-\mu)$. This approach was pushed further, to $m \leq n^{3/2-\epsilon}$ for any fixed $\epsilon > 0$ by Wormald [70] and Stark and Wormald [68] (see also [52]), and here the asymptotic formula for the probability of triangle freeness is the exponential of a sum whose number of terms grows as ϵ gets smaller.

The transition from order to disorder in triangle-free graphs is still poorly understood. In [38] we make the following conjecture.

Conjecture 4.12 *There exists $c^* > 0$ and a continuous function $\delta : (c^*, \infty) \to (0, 1/2]$ so that the following holds.*

1. *If $c < c^*$ and $m \sim cn^{3/2}$, then whp a graph G drawn uniformly from $\mathcal{T}(n,m)$ has a max cut of size $(1/2 + o(1))m$.*

2. *If $c > c^*$ and $m \sim cn^{3/2}$, then whp a graph G drawn uniformly from $\mathcal{T}(n,m)$ has a max cut of size $(1/2 + \delta(c) + o(1))m$.*

4.4 Further Applications of Cluster Expansion and Counting

Using the polymer models strategy of [37] described in Section 4.1, the author Perkins and Potukuchi [39] determined the asymptotics of the number of independent sets of size $\lfloor \beta 2^{d-1} \rfloor$ in the discrete hypercube $Q_d = \{0,1\}^d$ for any fixed $\beta \in (0,1)$ as $d \to \infty$, extending a result of Galvin [27] for $\beta \in (1 - 1/\sqrt{2}, 1)$. Similar polymer models were used by Balogh, Garcia and Li [1] to find detailed asymptotics for the number of independent sets in the subgraph H_d of Q_d induced by its largest two layers when d is odd. Building on the polymer model approach of [35], Li, McKinley and Park [45] determined asymptotics for the number q-colourings of H_d with d odd, q even. In [13] the author, Davies, and Perkins applied the cluster expansion to the extremal problem of determining which Δ-regular graphs maximise the number of matchings and independent sets of a fixed size.

5 The Algorithmic Perspective

Thus far we have used the (multivariate) hard-core model to illustrate connections between combinatorics and statistical physics. In recent years, the hard-core model has also played a starring role in the field of approximate counting/sampling.

We begin with a definition. Given $\epsilon > 0$, we say that $\hat{Z}$ is an ϵ-relative approximation to Z if
$$e^{-\epsilon}\hat{Z} \leq Z \leq e^{\epsilon}\hat{Z}.$$

There are two natural computational problems arise when considering the hard-core model on a graph G:

1. Given $\epsilon > 0$, compute an ϵ-relative approximation to the hard-core model partition function $Z_G(\boldsymbol{\lambda})$.

2. Given $\epsilon > 0$, sample a random independent set in G whose distribution is at most ϵ in total variation distance from the hard-core distribution $\mathbb{P}_{G,\boldsymbol{\lambda}}$.

We would like to perform both of these tasks *efficiently*, for example in time polynomial in $|V(G)|$ and $1/\epsilon$. A deterministic algorithm that performs Task 1 with this guarantee is called a *fully polynomial time approximation scheme* (FPTAS). A randomised algorithm that does the same with probability at least $2/3$ is called a *fully polynomial randomised approximation scheme* (FPRAS) respectively. An algorithm that performs Task 2 with the above running time guarantees is called a *polynomial time sampling scheme*. For sampling, it is also common to ask for a run-time that is polynomial in $|V(G)|$ and $\log(1/\epsilon)$. We highlight that Task 1 makes sense for arbitrary complex valued $\boldsymbol{\lambda}$ whereas Task 2 only makes sense in the physical setting $\boldsymbol{\lambda} \geq 0$.

We begin with a pair of seminal results on approximating the hard-core model partition function.

Theorem 5.1 (Weitz [69]) *If* $\lambda < \lambda_c(\Delta) = \lambda_c(\Delta) = \frac{(\Delta-1)^{\Delta-1}}{(\Delta-2)^{\Delta}}$, *then there is an FPTAS and a polynomial time sampling scheme for the hard-core model on graphs of maximum degree* Δ.

The following is a combination of results of Sly [65], Galanis, Ge, Štefankovič, Vigoda, and Yang [22], and Sly and Sun [66].

Theorem 5.2 *Let $\Delta \geq 3$. If $\lambda > \lambda_c(\Delta)$, then there is no FPRAS for the hard-core model on graphs of maximum degree Δ unless $NP = RP$.*

Taken together, Theorems 5.1 and 5.2 exhibit a remarkable computational phase transition for the hardness of approximating the hard-core model which coincides with the order-disorder phase transition on the infinite regular tree. We therefore have an algorithmic characterisation of $\lambda_c(\Delta)$ which we can add to our list in Section 2.1.

The algorithm of Weitz [69] pioneered what is now known as the 'correlation decay method'. Given a graph G of max degree Δ, Weitz relates the probability that a vertex is occupied by the hard-core model, to the probability that a vertex is occupied in a certain tree of maximum degree Δ. Correlation decay properties of the hard-core model on the tree (related to item 3 in the list of Section 2.1) are then used to show that this probability can be computed efficiently.

Perhaps the most widely used approach for approximate counting and sampling is the Markov Chain Monte Carlo (MCMC) approach. The idea is to construct a Markov chain with the desired stationary distribution μ. One can then obtain an approximate sample from μ by running the chain for long enough. If the chain mixes rapidly, and each step of the chain is easy to implement, then once can approximately sample from μ efficiently.

A more recent approach to approximate counting and sampling is based on zero-freeness of partition functions and is known as 'Barvinok's interpolation method'. We describe this approach in the next section. We then discuss the connection between Barvinok's method and the cluster expansion and discuss some algorithmic applications of the cluster expansion.

5.1 Barvinok's Interpolation Method

Barvinok's interpolation method [2, 3] is based on the simple yet powerful idea that to approximate a univariate of a polynomial $p(z)$, one can Taylor expand $\log p(z)$ and take the first few terms of the expansion. The function $\log p(z)$ has a Taylor series about $z = 0$ with radius of convergence R if and only if $p(z)$ is *zero-free* in the disc $D(0; R) \subseteq \mathbb{C}$. This then provides an algorithmic motivation for finding zero-free regions of partition functions.

The engine behind Barvinok's method is the following lemma.

Lemma 5.3 *Let $p(z)$ be a polynomial of degree n which is zero-free in the disc $D(0; R)$ for some $R > 0$. Let*

$$T_m(z) = \sum_{j=0}^{m} \frac{(\log p)^{(j)}(0)}{j!} z^j$$

be the Taylor series of $\log p$ truncated at order m. If $|z| \leq (1-\delta)R$ for some $\delta > 0$, then

$$|\log p(z) - T_m(z)| \leq n \frac{(1-\delta)^m}{(m+1)\delta}.$$

Proof Let $z_1, \ldots, z_n \in \mathbb{C}$ denote the roots of p (none of which are 0 by assumption). We may then write

$$p(z) = p(0) \prod_{j=1}^{n} \left(1 - \frac{z}{z_j}\right)$$

and so

$$\log p(z) = \log p(0) + \sum_{j=1}^{n} \log\left(1 - \frac{z}{z_j}\right).$$

Since $|z/z_j| \leq 1 - \delta$ by assumption, the Taylor series for $\log(1 - z/z_j)$ is absolutely convergent and for $j \in [n]$,

$$\left|\log\left(1 - \frac{z}{z_j}\right) - \sum_{i=1}^{m} \frac{(-1)^i (x/z_j)^i}{i}\right| \leq \sum_{i=m+1}^{\infty} \frac{(1-\delta)^i}{i} \leq \frac{(1-\delta)^m}{(m+1)\delta}.$$

Summing this inequality over j completes the proof. $\square$

This lemma shows that in order to compute an ϵ-relative approximation to $p(z)$ for z such that $|z| \leq (1-\delta)R$, it suffices to compute the first $O_\delta(\log(n/\epsilon))$ terms in the Taylor series of $\log p$. Therefore, if one can compute these early terms of the Taylor series of $\log p$ efficiently, one can approximate p efficiently. It is useful to observe that these coefficients can be recovered from the coefficients of p itself in an efficient manner. Indeed, we have $p' = (\log p)' p$ and so differentiating this product $k - 1$ times, we obtain

$$p^{(k)}(0) = \sum_{j=0}^{k-1} \binom{k-1}{j} (\log p)^{(k-j)}(0) \cdot p^{(j)}(0). \tag{5.1}$$

This gives a triangular system of linear equations which allows us to compute the coefficients $(\log p)^{(k)}(0)$ for $k = 1, \ldots, n$ from the coefficients $p^{(k)}(0)$ for $k = 0, \ldots, n$ in $O(n^2)$ time.

Example 5.4 (Approximating the hard-core model partition function) Let G be a graph of maximum degree Δ on n vertices. Let $\epsilon > 0$. Our goal is to efficiently compute an ϵ-relative approximation to $Z_G(\lambda)$. In this example, we restrict our attention to λ satisfying $|\lambda| < (1-\delta)\lambda_s(\Delta)$ where $\delta \in (0,1)$ is fixed and $\lambda_s(\Delta) := \frac{(\Delta-1)^{\Delta-1}}{\Delta^\Delta}$ is as in Theorem 2.7.

By Theorem 2.7, we know that $Z_G(\lambda)$ is zero-free in the disc $D(0; R)$ with $R = \lambda_s(\Delta) = \frac{(\Delta-1)^{\Delta-1}}{\Delta^\Delta}$. Moreover we may write $Z_G(\lambda) = \sum_{i=0}^{n} a_i \lambda^i$, a polynomial of degree at most n. By Lemma 5.3, in order to compute an ϵ-relative approximation of $Z_G(\lambda)$ it suffices to compute the first $k = O_\delta(\log(n/\epsilon))$ Taylor coefficients of $\log Z_G(\lambda)$ or equivalently, by (5.1), the first $k+1$ coefficients $a_0, \ldots, a_k$ of $Z_G(\lambda)$. To compute a_j, we may crudely go through all vertex subsets of G of size j and check whether or not each set is independent. This takes time $O(n^j)$ and so we can compute $a_0, \ldots, a_k$ in time $O(kn^{k+1}) = n^{O_\delta(\log(n/\epsilon))}$, a quasipolynomial time algorithm.

In a seminal work, Patel and Regts [54] showed that the kth Taylor coefficient of $\log Z_G(\lambda)$, can in fact be computed exactly in time $n \cdot \text{poly}(k) \cdot \Delta^{O(k)}$. Their results also apply to a wide class of partition functions that includes the hard-core model as a special case. This immediately reduces the run-time of the algorithm in Example 5.4 from quasipolynomial to polynomial in n and $1/\epsilon$. In other words it gives an FPTAS for $Z_G(\lambda)$ for λ in the range of Example 5.4.

The method of Patel and Regts [54] is closely related to the cluster expansion which can also be used to show that the Taylor coefficients of $\log Z_G(\lambda)$ can be computed efficiently. We discuss the algorithmic significance of the cluster expansion in the next section.

At first sight, it may seem that Lemma 5.3 is only useful when we have a polynomial with a zero-free *disc*. However, this is not the case. The full power of Barvinok's method is realised by using well-chosen polynomial maps that transform a given zero-free region into a disc. To demonstrate this approach we begin with the following remarkable theorem of Peters and Regts [58].

Theorem 5.5 (Peters and Regts [58]) *Let $\Delta \geq 3$. There exists a complex domain $D_\Delta \subseteq \mathbb{C}$ containing the real interval $[0, \lambda_c(\Delta))$ such that if G is a graph of maximum degree Δ and $\lambda \in D_\Delta$, then $Z_G(\lambda) \neq 0$.*

We highlight that D_Δ is not a disc, indeed the example of the tree $T_{\Delta,n}$, which has zeros approaching $-\lambda_s(\Delta)$ where $0 < \lambda_s(\Delta) < \lambda_c(\Delta)$, shows that it cannot be.

Let G be a graph of max degree Δ. In the next example we sketch how Theorem 5.5 can be used to approximate the $Z_G(\lambda)$ past $\lambda_s(\Delta)$ and all the way up to the tree uniqueness threshold $\lambda_c(\Delta) = \frac{(\Delta-1)^{\Delta-1}}{(\Delta-2)^\Delta}$, thereby giving an alternative proof of Theorem 5.1.

Example 5.6 (Example 5.4 revisited) Let G be as in Example 5.4. Fix $\delta > 0$. Our aim is to approximate Z_G evaluated at $x = (1-\delta)\lambda_c(\Delta)$. By Theorem 5.5, Z_G is zero-free in a complex neighbourhood $U = U_{\delta, \Delta}$ of the real interval $[0, x]$. One can construct a polynomial $\phi = \phi_{\delta, \Delta} : \mathbb{C} \to \mathbb{C}$ such that $\phi(0) = 0$, $\phi(x) = x$ and such that ϕ maps the disc $D(0; (1+\beta)x)$ into U for some $\beta = \beta(\delta, \Delta) > 0$ (see [3, Lemma 2.2.3] for an explicit construction). We then consider the composition $h(z) = Z_G(\phi(z))$, a polynomial of degree at most $n \cdot \deg \phi$ where $\deg \phi$ depends only on δ and Δ. Moreover h has no zeros in the disc $D(0; (1+\beta)x)$ and $h(0) = Z_G(0)$, $h(x) = Z_G(x)$. By Lemma 5.3, we can approximate $\log h(x) = \log Z_G(x)$ up to additive error ϵ by computing the kth order Taylor series of $\log h$ at $z = 0$ where $k = O_{\Delta, \delta}(\log(n/\epsilon))$. By (5.1) it suffices to compute the kth order Taylor series of h. For this, one can compute the kth order Taylor series $Z_{G,k}, \phi_k$ of Z_G and ϕ respectively, compute the composition $Z_{G,k}(\phi_k(z))$, and discard all monomials of degree higher than k.

For more on the Interpolation Method we refer the reader to the excellent survey [56].

5.2 Algorithmic Applications of the Cluster Expansion

When applied to the hard-core model, the algorithmic methods mentioned so far typically work when λ is *small* and typical independent sets are disordered. Moreover, Theorem 5.2 demonstrates that for general graphs there is a genuine barrier to algorithms for λ large. In this section we will see how the cluster expansion can be used to design algorithms that do work at large λ for certain classes of graphs. The intuition behind these algorithms is the same as that of Section 4: an ordered model can often be viewed as a perturbation of a ground state where the perturbations are disordered and therefore algorithmically tractable. In the language of spin systems (of which the hard-core model in one example), the large λ regime is referred to as 'low temperature'.

Low temperature algorithms first appeared in the breakthrough work of Helmuth, Perkins and Regts [32] where they designed algorithms for the hard-core model and the Potts model (a generalisation of the Ising model) on finite subregions of the integer lattice $\mathbb{Z}^d$. In the case of the hard-core model on $\mathbb{Z}^d$, the 'ground states' are those independent sets that are fully contained in either the even or odd sublattice of $\mathbb{Z}^d$. The hard-core model at low temperature (large λ) on a finite subregion of $\mathbb{Z}^d$, can then be understood as a perturbation of a ground state.

Theorem 5.7 (Helmuth, Perkins, Regts [32]) *For $d \geq 2$ there exists $\lambda^* = \lambda^*(d)$ such that for all $\lambda > \lambda^*$, there is an efficient sampling scheme and an FPTAS for the hard-core model on any finite subregion Λ of $\mathbb{Z}^d$ with even or odd padded boundary conditions.*

Even/odd padded boundary conditions means that we consider the hard-core model on Λ where we condition on even/odd vertices in Λ close to the complement of Λ being occupied. The proof uses Pirogov-Sinai theory and contour models along with their associated cluster expansions. Contour models are more complex than polymer models, and they capture the geometric nature of perturbations from a ground state in low temperature models on $\mathbb{Z}^d$.

Using the simpler theory polymer models and their associated cluster expansion from Section 3.2, the author, Keevash and Perkins [36] showed that low-temperature algorithms are available for the hard-core model and Potts model on bipartite *expanders* of bounded degree. We say that a bipartite graph $G = (\mathcal{O}, \mathcal{E}, E)$ is a *bipartite α-expander* if $|N(S)| \geq (1 + \alpha)|S|$ for all $S \subseteq \mathcal{O}$ with $|S| \leq |\mathcal{O}|/2$ and all $S \subseteq \mathcal{E}$ with $|S| \leq |\mathcal{E}|/2$.

Theorem 5.8 (J., Keevash, Perkins [36]) *There exists an absolute constant C such that for every $\alpha > 0$, $\Delta \geq 3$, and any $\lambda > C\Delta^{4/\alpha}$, there exists an FPTAS and a polynomial-time sampling scheme for the hard-core model at activity λ on bipartite α-expander graphs of maximum degree Δ.*

In [40] the dependence of λ on Δ was improved to $\lambda > C_\alpha (\log \Delta)/\Delta^{1/4}$ for Δ-regular α-expanders, by combining the polymer model approach of [36] with the graph container method of Sapozhenko. In particular, by setting $\lambda = 1$ this gives an FPTAS for counting the total number of independent sets in Δ-regular α-expander graphs for large enough Δ.

Theorems 5.7 and 5.8 both count (weighted) independent sets in *bipartite graphs* and belong to a complexity class known as #BIS (for 'bipartite independent set'). The class #BIS is the class of problems polynomial-time equivalent to approximating the number of independent sets of a bipartite graph [17], and many interesting approximate counting and sampling problems have been shown to be #BIS-hard [29, 11, 24] (that is, at least as hard as approximating the number of independent sets in a bipartite graph). In particular, Cai, Galanis, Goldberg, Guo, Jerrum, Štefankovič, and Vigoda [6] showed that for all $\Delta \geq 3$ and all $\lambda > \lambda_c(\Delta)$, it is #BIS-hard to approximate the hard-core partition function at activity λ on a bipartite graph of maximum degree Δ. Resolving the complexity of #BIS is a major open problem in the field of approximate counting.

Question 5.9 *Is there an FPRAS for approximating the number of independent sets in a bipartite graph?*

For the remainder of this section we sketch a proof of Theorem 5.8. The proof uses the very same polymer model as those introduced in Section 4.1 to count independent sets in the hypercube. Suppose $\alpha > 0$ and let G be a bipartite α-expander of maximum degree Δ with bipartition $(\mathcal{O}, \mathcal{E})$. As in Section 4.1, we define the 'odd' polymer model as follows: polymers are 2-linked subsets of $\mathcal{O}$ of size at most $|\mathcal{O}|/2$; the weight $w(\gamma)$ of a polymer γ is $\lambda^{|\gamma|}(1+\lambda)^{-|N(\gamma)|}$; and the two polymers γ, γ' are compatible if and only if $N(\gamma), N(\gamma')$ are disjoint. This polymer model comes with the associated partition function $\Xi_{\mathcal{O}}$. We define define the 'even' polymer model with partition function $\Xi_{\mathcal{E}}$ analogously. The first step in the proof of 5.8 is to show that $Z_G(\lambda)$ is well approximated by the linear combination

$$(1+\lambda)^{|\mathcal{E}|}\Xi_{\mathcal{O}} + (1+\lambda)^{\mathcal{O}}\Xi_{\mathcal{E}}. \tag{5.2}$$

The task then becomes to approximate the partition functions $\Xi_{\mathcal{O}}, \Xi_{\mathcal{E}}$ efficiently. The strategy, closely related to Barvinok's interpolation method, is to take an appropriate truncation of the cluster expansions of $\log \Xi_{\mathcal{O}}, \log \Xi_{\mathcal{E}}$. The accuracy of these truncations is guaranteed by Theorem 3.5. Let $\mathcal{C}_{\mathcal{O}}$ denote the set of all clusters in the odd polymer model, and suppose that we have verified (3.7) with $g(\gamma) = f(\gamma) = |\gamma|$, the number of vertices in γ. Define the truncated cluster expansion

$$T_m = \sum_{\substack{\Gamma \in \mathcal{C}_{\mathcal{O}}: \\ |\Gamma| < m}} \phi(\Gamma) \prod_{\gamma \in \Gamma} w(\gamma),$$

where $|\Gamma| = \sum_{\gamma \in \Gamma} |\gamma|$. By taking (3.8) with γ a polymer consisting of a single vertex $v \in \mathcal{O}$ and summing over all such v we see that

$$|\log \Xi_{\mathcal{O}} - T_m| \le ne^{-m},$$

where n is the number of vertices in the graph G. We conclude that if we take $m \ge \log(n/\epsilon)$, then e^{T_m} is an ϵ-relative approximation to $\Xi_{\mathcal{O}}$. The fact that we can compute T_m efficiently hinges mainly on two facts: (i) we can enumerate all clusters Γ with $|\Gamma| < m$ efficiently and (ii) we can calculate the Ursell function $\phi(\Gamma)$ efficiently. (i) is possible because clusters are *connected* objects, and so we may use an algorithm of Patel and Regts [55] for enumerating connected subgraphs in a bounded-degree graph. (ii) follows from the fact that $\phi(\Gamma)$ is an evaluation of the Tutte polynomial of the incompatibility graph H_Γ, which can be exactly evaluated in time $O\left(e^{|V(H_\Gamma)|}\right) = O(e^m) = O(n/\epsilon)$ by an algorithm of Björklund, Husfeldt, Kaski, and Koivisto [4].

The cluster expansion has by now been used to design low temperature algorithms for a number of different models and settings. We refer the reader to [7, 14, 9, 10, 23, 46, 8, 12, 31, 5] for further examples.

Acknowledgements

MJ is supported by a UKRI Future Leaders Fellowship MR/W007320/1.

References

[1] József Balogh, Ramon I Garcia, and Lina Li, *Independent sets in the middle two layers of boolean lattice*, Journal of Combinatorial Theory, Series A **178** (2021), 105341.

[2] Alexander Barvinok, *Computing the permanent of (some) complex matrices*, Foundations of Computational Mathematics **16** (2016), no. 2, 329–342.

[3] _____, *Combinatorics and complexity of partition functions*, Algorithms and Combinatorics, vol. 30, Springer, 2017.

[4] Andreas Björklund, Thore Husfeldt, Petteri Kaski, and Mikko Koivisto, *Computing the Tutte polynomial in vertex-exponential time*, Proceedings of the Forty-ninth Annual Symposium on Foundations of Computer Science, FOCS 2008, IEEE, 2008, pp. 677–686.

[5] Christian Borgs, Jennifer Chayes, Tyler Helmuth, Will Perkins, and Prasad Tetali, *Efficient sampling and counting algorithms for the potts model on $\mathbb{Z}^d$ at all temperatures*, Proceedings of the 52nd Annual ACM SIGACT Symposium on Theory of Computing, 2020, pp. 738–751.

[6] Jin-Yi Cai, Andreas Galanis, Leslie Ann Goldberg, Heng Guo, Mark Jerrum, Daniel Štefankovič, and Eric Vigoda, *# BIS-hardness for 2-spin systems on bipartite bounded degree graphs in the tree non-uniqueness region*, Journal of Computer and System Sciences **82** (2016), no. 5, 690–711.

[7] Jin-Yi Cai and Tianyu Liu, *An fptas for the square lattice six-vertex and eight-vertex models at low temperatures*, Proceedings of the 2021 ACM-SIAM Symposium on Discrete Algorithms (SODA), SIAM, 2021, pp. 1520–1534.

[8] Sarah Cannon and Will Perkins, *Counting independent sets in unbalanced bipartite graphs*, Proceedings of the Fourteenth Annual ACM-SIAM Symposium on Discrete Algorithms, SIAM, 2020, pp. 1456–1466.

[9] Charles Carlson, Ewan Davies, and Alexandra Kolla, *Efficient algorithms for the potts model on small-set expanders*, arXiv preprint arXiv:2003.01154 (2020).

[10] Charlie Carlson, Ewan Davies, Nicolas Fraiman, Alexandra Kolla, Aditya Potukuchi, and Corrine Yap, *Algorithms for the ferromagnetic potts model on expanders*, 2022 IEEE 63rd Annual Symposium on Foundations of Computer Science (FOCS), IEEE, 2022, pp. 344–355.

[11] Prasad Chebolu, Leslie Ann Goldberg, and Russell Martin, *The complexity of approximately counting stable matchings*, Theoretical Computer Science **437** (2012), 35–68.

[12] Zongchen Chen, Andreas Galanis, Leslie A Goldberg, Will Perkins, James Stewart, and Eric Vigoda, *Fast algorithms at low temperatures via markov chains*, Random Structures & Algorithms **58** (2021), no. 2, 294–321.

[13] Ewan Davies, Matthew Jenssen, and Will Perkins, *A proof of the upper matching conjecture for large graphs*, Journal of Combinatorial Theory, Series B **151** (2021), 393–416.

[14] David de Boer, Pjotr Buys, Han Peters, and Guus Regts, *On boundedness of zeros of the independence polynomial of tor*, arXiv preprint arXiv:2306.12934 (2023).

[15] RL Dobrushin, *Perturbation methods of the theory of gibbsian fields*, Lectures on Probability Theory and Statistics: Ecole d'Eté de Probabilités de Saint-Flour XXIV—1994 (1996), 1–66.

[16] Roland L Dobrushin, *Estimates of semi-invariants for the ising model at low temperatures*, Translations of the American Mathematical Society-Series 2 **177** (1996), 59–82.

[17] Martin Dyer, Leslie Ann Goldberg, Catherine Greenhill, and Mark Jerrum, *The relative complexity of approximate counting problems*, Algorithmica **38** (2004), no. 3, 471–500.

[18] John Engbers and David Galvin, *H-coloring tori*, Journal of Combinatorial Theory, Series B **102** (2012), no. 5, 1110–1133.

[19] Paul Erdős and László Lovász, *Problems and results on 3-chromatic hypergraphs and some related questions*, Infinite and finite sets **10** (1975), no. 2, 609–627.

[20] P. Erdös, D.J. Kleitman, and B. Rothschild, *Asymptotic enumeration of K_n-free graphs*, Colloquio Internazionale sulle Teorie Combinatorie (Rome, 1973) (1973), no. 17, 19–27.

[21] Paul Erdős and Alfréd Rényi, *On the evolution of random graphs*, Publ. Math. Inst. Hung. Acad. Sci **5** (1960), no. 1, 17–60.

[22] Andreas Galanis, Qi Ge, Daniel Štefankovič, Eric Vigoda, and Linji Yang, *Improved inapproximability results for counting independent sets in the hardcore model*, Approximation, Randomization, and Combinatorial Optimization. Algorithms and Techniques, Springer, 2011, pp. 567–578.

[23] Andreas Galanis, Leslie Ann Goldberg, and James Stewart, *Fast algorithms for general spin systems on bipartite expanders*, ACM Transactions on Computation Theory (TOCT) **13** (2021), no. 4, 1–18.

[24] Andreas Galanis, Daniel Stefankovic, Eric Vigoda, and Linji Yang, *Ferromagnetic Potts model: Refined #-BIS-hardness and related results*, SIAM Journal on Computing **45** (2016), no. 6, 2004–2065.

[25] David Galvin, *On homomorphisms from the hamming cube to $\mathbb{Z}$*, Israel Journal of Mathematics **138** (2003), no. 1, 189–213.

[26] _____, *A threshold phenomenon for random independent sets in the discrete hypercube*, Combinatorics, probability and computing **20** (2011), no. 1, 27–51.

[27] _____, *The independent set sequence of regular bipartite graphs*, Discrete Mathematics **312** (2012), no. 19, 2881–2892.

[28] _____, *Independent sets in the discrete hypercube*, arXiv preprint arXiv:1901.01991 (2019).

[29] Leslie Ann Goldberg and Mark Jerrum, *Approximating the partition function of the ferromagnetic Potts model*, Journal of the ACM (JACM) **59** (2012), no. 5, 25.

[30] J Groeneveld, *Two theorems on classical many-particle systems*, Phys. Letters **3** (1962).

[31] Tyler Helmuth, Matthew Jenssen, and Will Perkins, *Finite-size scaling, phase coexistence, and algorithms for the random cluster model on random graphs*, Annales de l'Institut Henri Poincare (B) Probabilites et statistiques **59** (2023), no. 2, 817–848.

[32] Tyler Helmuth, Will Perkins, and Guus Regts, *Algorithmic Pirogov-Sinai theory*, Probability Theory and Related Fields **176** (2020), 851–895.

[33] S. Janson, T. Łuczak, and A. Ruciński, *An exponential bound for the probability of nonexistence of a specified subgraph in a random graph*, Random graphs' 87, Proceedings, Poznán, 1987, eds. M. Karonński, J. Jaworski and A.Ruciński, 73–87.

[34] Svante Janson, Tomasz Luczak, and Andrzej Rucinski, *An exponential bound for the probability of nonexistence of a specified subgraph in a random graph*, Institute for Mathematics and its Applications (USA), 1988.

[35] Matthew Jenssen and Peter Keevash, *Homomorphisms from the torus*, arXiv preprint arXiv:2009.08315 (2020).

[36] Matthew Jenssen, Peter Keevash, and Will Perkins, *Algorithms for #BIS-hard problems on expander graphs*, SIAM Journal on Computing **49** (2020), no. 4, 681–710.

[37] Matthew Jenssen and Will Perkins, *Independent sets in the hypercube revisited*, Journal of the London Mathematical Society **102** (2020), no. 2, 645–669.

[38] Matthew Jenssen, Will Perkins, and Aditya Potukuchi, *On the evolution of structure in triangle-free graphs*, In preparation.

[39] _____, *Independent sets of a given size and structure in the hypercube*, Combinatorics, Probability and Computing **31** (2022), no. 4, 702–720.

[40] _____, *Approximately counting independent sets in bipartite graphs via graph containers*, Random Structures & Algorithms (2023).

[41] Jeff Kahn, *An entropy approach to the hard-core model on bipartite graphs*, Combinatorics, Probability and Computing **10** (2001), no. 3, 219–237.

[42] Jeff Kahn and Jinyoung Park, *The number of 4-colorings of the hamming cube*, Israel Journal of Mathematics **236** (2020), no. 2, 629–649.

[43] AD Korshunov and AA Sapozhenko, *The number of binary codes with distance 2*, Problemy Kibernet **40** (1983), 111–130.

[44] Roman Kotecký and David Preiss, *Cluster expansion for abstract polymer models*, Communications in Mathematical Physics **103** (1986), no. 3, 491–498.

[45] Lina Li, Gweneth McKinley, and Jinyoung Park, *The number of colorings of the middle layers of the hamming cube*, arXiv preprint arXiv:2304.03203 (2023).

[46] Chao Liao, Jiabao Lin, Pinyan Lu, and Zhenyu Mao, *Counting independent sets and colorings on random regular bipartite graphs*, arXiv preprint arXiv:1903.07531 (2019).

[47] Tomasz Łuczak, *On triangle-free random graphs*, Random Structures & Algorithms **16** (2000), no. 3, 260–276.

[48] Willem Mantel, *Problem 28*, Wiskundige Opgaven **10** (1907), no. 60-61, 320.

[49] Joseph E Mayer, *The statistical mechanics of condensing systems. i*, The Journal of Chemical Physics **5** (1937), no. 1, 67–73.

[50] Joseph E Mayer and Elliott Montroll, *Molecular distribution*, The Journal of Chemical Physics **9** (1941), no. 1, 2–16.

[51] Joseph Edward Mayer and Maria Goeppert Mayer, *Statistical mechanics*, vol. 28, John Wiley & Sons New York, 1940.

[52] Frank Mousset, Andreas Noever, Konstantinos Panagiotou, and Wojciech Samotij, *On the probability of nonexistence in binomial subsets*, The Annals of Probability **48** (2020), no. 1, 493–525.

[53] Deryk Osthus, Hans Jürgen Prömel, and Anusch Taraz, *For which densities are random triangle-free graphs almost surely bipartite?*, Combinatorica **23** (2003), no. 1, 105–150.

[54] Viresh Patel and Guus Regts, *Deterministic polynomial-time approximation algorithms for partition functions and graph polynomials*, SIAM Journal on Computing **46** (2017), no. 6, 1893–1919.

[55] _____, *Computing the number of induced copies of a fixed graph in a bounded degree graph*, Algorithmica **81** (2019), 1844–1858.

[56] _____, *Approximate counting using taylor's theorem: a survey*, arXiv preprint arXiv:2212.08143 (2022).

[57] Oliver Penrose, *Convergence of fugacity expansions for fluids and lattice gases*, Journal of Mathematical Physics **4** (1963), no. 10, 1312–1320.

[58] Han Peters and Guus Regts, *On a conjecture of Sokal concerning roots of the independence polynomial*, Michigan Mathematical Journal **68** (2019), no. 1, 33–55.

[59] Hans Jürgen Prömel and Angelika Steger, *Counting H-free graphs*, Discrete Mathematics **154** (1996), no. 1-3, 311–315.

[60] David Ruelle, *Correlation functions of classical gases*, Annals of Physics **25** (1963), no. 1, 109–120.

[61] AA Sapozhenko, *On the number of connected subsets with given cardinality of the boundary in bipartite graphs*, Metody Diskret Analiz **45** (1987), 42–70.

[62] Alexander D Scott and Alan D Sokal, *The repulsive lattice gas, the independent-set polynomial, and the lovász local lemma*, Journal of Statistical Physics **118** (2005), 1151–1261.

[63] _____, *On dependency graphs and the lattice gas*, Combinatorics, Probability and Computing **15** (2006), no. 1-2, 253–279.

[64] James B. Shearer, *On a problem of spencer*, Combinatorica **5** (1985), 241–245.

[65] Allan Sly, *Computational transition at the uniqueness threshold*, Proceedings of the Fifty-first Annual IEEE Symposium on Foundations of Computer Science, FOCS 2010, IEEE, 2010, pp. 287–296.

[66] Allan Sly and Nike Sun, *Counting in two-spin models on d-regular graphs*, The Annals of Probability **42** (2014), no. 6, 2383–2416.

[67] Frank Spitzer, *Markov random fields on an infinite tree*, The Annals of Probability **3** (1975), no. 3, 387–398.

[68] Dudley Stark and Nick Wormald, *The probability of non-existence of a subgraph in a moderately sparse random graph*, Combinatorics, Probability and Computing **27** (2018), no. 4, 672–715.

[69] Dror Weitz, *Counting independent sets up to the tree threshold*, Proceedings of the Thirty-Eighth Annual ACM Symposium on Theory of Computing, STOC 2006, ACM, 2006, pp. 140–149.

[70] Nicholas C Wormald, *The perturbation method and triangle-free random graphs*, Random Structures & Algorithms **9** (1996), no. 1-2, 253–270.

[71] Chen-Ning Yang and Tsung-Dao Lee, *Statistical theory of equations of state and phase transitions. i. theory of condensation*, Physical Review **87** (1952), no. 3, 404.

Department of Mathematics
King's College London
Strand Building
London WC2R 2LS, UK
matthew.jenssen@kcl.ac.uk

Sublinear Expanders and Their Applications

Shoham Letzter

Abstract

In this survey we aim to give a comprehensive overview of results using sublinear expanders. The term *sublinear expanders* refers to a variety of definitions of expanders, which typically are defined to be graphs G such that every not-too-small and not-too-large set of vertices U has neighbourhood of size at least $\alpha |U|$, where α is a function of n and $|U|$. This is in contrast with *linear expanders*, where α is typically a constant. We will briefly describe proof ideas of some of the results mentioned here, as well as related open problems.

1 Introduction

Very informally speaking, expanders are graphs which have good connectivity properties, yet may be quite sparse. Since their introduction by Bassalygo and Pinsker [9] in the 1970s, expanders have been studied extensively, and have seen numerous applications in combinatorics and computer science (see the concise expository paper by Sarnak [77] and the surveys by Hoory, Linial, and Wigderson [43], Lubotzky [68], and Krivelevich [55]). There is quite a large variety of definitions of expanders, but here we think of expanders as graphs whose every not-too-small and not-too-large set of vertices has a large neighbourhood. A very simple definition of expanders is the following, where $N_G(U)$, the *neighbourhood* of U in G, is the set of vertices in G that are not in U but have a neighbour in U.

Definition 1.1 (Linear expanders) Let $\alpha > 0$. A graph G on n vertices is called an α-*expander* if every subset $U \subseteq V(G)$ of size at most $n/2$ satisfies $|N_G(U)| \geq \alpha |U|$.

When α is a constant, which is often the case in applications, this expansion property is *linear*, namely, every not-too-large set of vertices expands linearly. See Krivelevich [55] for an excellent survey which mentions various results about α-expanders and their variants, and illustrates various ways in which they can be used in applications.

Our focus will be on *sublinear expanders*. These may be defined similarly, but with α being a function of n that tends to 0 as n grows, and oftentimes the rate of expansion of a set of vertices depends not only on n but also on the size of the set. Such expanders were first defined by Komlós and Szemerédi [53, 54] in 1994, who used them to solve a problem about subdivisions. A somewhat different notion of sublinear expanders was introduced by Shapira and Sudakov [78], in their work about finding small minors. These two papers gave rise to a host of other results, particularly in the last three years or so.

In this survey we aim to give a comprehensive overview of results that were proved using sublinear expanders, grouped into sections according the topics they address. We will also briefly describe proof ideas of some of the results presented here. See the survey of Liu [61] for a deeper dive into some of the methods developed for tackling problems using Komlós and Szemerédi's sublinear expanders. We will also highlight some open problems related to the results we will present.

- In Section 2 we mention several results about average degree conditions implying the existence of a subdivision of a certain graph.

- In Section 3 we mention a similar type of problem, about average degree conditions implying a small minor or subdivision of a complete graph.

- Section 4 discusses some results about immersions of complete graphs and digraphs. Next, we consider recent progress about the 'odd cycle problem' of Erdős and Hajnal, and a related problem about 'balanced subdivisions' in Section 5.

- Section 6 presents several extremal results, about tight cycles, rainbow cycles and clique subdivisions, and cycles with many chords.

- In Section 7 we consider two problems about the global structure of a graph, namely about decomposing a graph into cycles and edges, and about separating the edges of a graph by paths.

- Section 8 mentions two results about the number of 'Hamiltonian sets' in a graph with given average degree.

- Finally, in Section 9 we briefly mention a few other results that did not naturally belong in one of the previous sections.

Notation

Throughout the paper, we use log to denote the base 2 logarithm, and ln to denote the base e logarithm. The various results mentioned in this survey use various different definitions of sublinear expanders. In order to easily distinguish the names of different expanders, we use the initials of those who introduced a specific definition of expanders; for example, the name KS-expanders refers to expanders as defined by Komlós and Szemerédi.

2 Subdivisions

Recall that, for a graph F, an F-*subdivision* is a graph obtained by replacing each edge uv in F by a path with ends u and v, such that the interiors of these paths are pairwise vertex-disjoint and vertex-disjoint of the original vertices of F (see Figure 1).

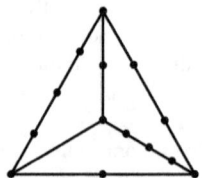

Figure 1: A subdivision of K_4

Sublinear expanders were first introduced by Komlós and Szemerédi [53] in 1994, who used them to solve an extremal problem about clique subdivisions. In this section we describe this first use, as well as several subsequent applications of sublinear expanders to similar problems about subdivisions.

2.1 Finding Clique Subdivisions

In the 1990s Komlós and Szemerédi [53, 54], as well as independently Bollobás and Thomason [11], proved the following result, estimating the average degree that guarantees the existence of a K_k-subdivision.

Theorem 2.1 (Komlós–Szemerédi [53, 54] and Bollobás–Thomason [11])
There exists $c > 0$ such that, for every integer $k \geq 1$, every graph with average degree at least ck^2 contains a K_k-subdivision.

This is tight up to the value of the constant c, as can be seen by considering a balanced complete bipartite graph on fewer than $2\binom{k/2}{2}$ vertices.

The following notion of expanders played a key part in Komlós and Szemerédi's proof.[1]

Definition 2.2 (Komlós–Szemerédi [53, 54]) For $\varepsilon, t > 0$, let $\rho(x) = \rho_{\varepsilon,t}(x)$ be the function defined (for $x \geq t/2$).

$$\rho(x) = \rho(x, \varepsilon, t) = \frac{\varepsilon}{(\log(15x/t))^2}.$$

An (ε, t)-*KS-expander* is a graph G in which every set of vertices U, with $t/2 \leq |U| \leq |G|/2$, satisfies $|N_G(U)| \geq \rho(|U|) \cdot |U|$.

As this is a somewhat cumbersome definition, let us digest it briefly. Writing $n = |G|$, we remark that, typically, t is much smaller than n; often we think of t as constant and n as large. Moreover, notice that every set of vertices of size $\Theta(t)$, and at least $t/2$, expands linearly, and sets of size $\Theta(n)$, and at most $n/2$, may expand at a rate as low as $O(\frac{1}{(\log n)^2})$. Nevertheless, we have no information at all about the expansion rate of sets of fewer than $t/2$ vertices.

At this point, it may not be at all clear why one may want to consider expanders as in Definition 2.2, when linear expanders, as in Definition 1.1, are much simpler to understand and work with. The basic reason is the fact that the sublinear (ε, t)-KS-expanders may be found in essentially any graph.

Theorem 2.3 (Komlós–Szemerédi [53, 54]) *Let $\varepsilon > 0$ be sufficiently small, and let $t > 0$. Then every graph G has a subgraph H which is an (ε, t)-KS-expander, and satisfies $d(H) \geq d(G)/2$ and $\delta(H) \geq d(H)/2$.*

Remark

- The bound $d(H) \geq d(G)/2$ could be replaced by $d(H) \geq (1-\delta)d(G)$, for any constant $\delta > 0$ using a slight variation of the proof (see, e.g., Theorem 2.14).

---
[1] Bollobás and Thomason's proof was quite different and influential in its own right: their main result was that every $22k$-connected graph contains a graph which is k-*linked*, namely for every sequence of distinct vertices $s_1, t_1, \ldots, s_k, t_k$, there is a collection of vertex-disjoint paths $P_1, \ldots, P_k$ such that P_i joins s_i with t_i.

- We note that the definition of $\rho(x)$ is somewhat arbitrary. For the proof of Theorem 2.3 from [54] to work, the following conditions need to hold: $\rho(x)$ is decreasing for $x \geq t/2$; $x\rho(x)$ is increasing for $x \geq t/2$; and $\int_{t/2}^{\infty} \frac{\rho(x)}{x} dx$ is finite. The exact choice of $\rho(x)$ was chosen for ease of presentation: the largest $\rho(x)$ can be while satisfying these conditions is $\rho(x) = \frac{1}{\log x (\log \log x)^{\Theta(1)}}$, which is close to best possible, as shown by Moshkovitz and Shapira [73].

- Additionally, notice that $|H|$ can be much smaller than $|G|$ (e.g. if G is a disjoint union of K_{d+1}'s, then $|H| \leq d+1$, while $|G|$ can be arbitrarily large); this is sometimes inconvenient in applications.

- Finally, note that there is some freedom in the choice of t. A natural choice is to take t to be a small constant times $d(G)$ (the average degree of G), because then the lower bound on the neighbourhood of any given vertex is large enough for this neighbourhood to be guaranteed to expand by Definition 2.2.

A useful property of (ε, t)-expanders is that any two not-too-small sets of vertices can be joined by a relatively short path, namely of length $O\big((\log n)^3\big)$.

Theorem 2.4 (Small diameter theorem; Komlós–Szemerédi [53, 54]) *Let G be an n-vertex (ε, t)-expander. Then for every $x \geq t/2$ and every three sets of vertices U_1, U_2, W, where $|U_1|, |U_2| \geq x$ and $|W| \leq \rho(x)x/4$, there is a path in $G - W$ between U_1 and U_2 of length at most $\frac{2}{\varepsilon}\big(\log(15n/t)\big)^3$.*

To see how KS-expanders can be used to prove Theorem 2.1, notice first that by Theorem 2.3, it suffices to show that every n-vertex (ε, t)-KS-expander H with average degree at least $c_1 k^2$ and with $t = c_2 k^2$, for some constants c_1, c_2, contains a K_k-subdivision.

To prove this, Komlós and Szemerédi proved it separately for dense graphs, namely when $t = \Theta(n)$, where $n = |H|$, using Szemerédi's regularity lemma [81]. If H is not dense, write diam $= \frac{2}{\varepsilon}(\log(15n/t))^3$, and notice that, roughly speaking, Theorem 2.4 shows that the diameter of H is close to diam. In this case, if there are at least k vertices with degree at least D, where $\binom{k}{2}$ diam $\leq \rho(D) \cdot D/4$, then it is very easy to find a K_k-subdivision: given a set K of k vertices with degree at least D, apply Theorem 2.4 for each pair of vertices in K (with U_1 and U_2 being the neighbourhoods of these vertices, and W the set of vertices used in previous connections) to find a path of length at most diam joining these vertices and avoiding previously used vertices, one by one.

Otherwise, the authors first apply Theorem 2.3 to the graph obtained by removing vertices of degree at least D, to find a new expander H' with maximum degree less than D. Then, they find a set K of $2k$ vertices, and associate a set $S(x)$ (called a 'stable neighbourhood') with each $x \in K$, which expands well around x, and moreover no vertex appears in too many sets $S(x)$ (this last point is where they use that H' has bounded degree). Now they again repeatedly apply Theorem 2.4 to join pairs of vertices in K through their stable neighbourhoods $S(\cdot)$, with a path of length at most diam avoiding previously defined paths, but whenever a set $S(x)$ becomes overused, they give up on the vertex x. A counting argument shows that at most k vertices are discarded, and so this yields a K_k-subdivision.

This idea of using sets $S(x)$ that are allowed to overlap somewhat, and starting with a bit more than the required k vertices, allowed the authors to prove the tight $O(k^2)$ bound in [54]; a similar argument where the sets $S(x)$ were required to be disjoint was used in their earlier paper [53], where a slightly weaker bound was proved. This idea, of allowing the $S(x)$ to overlap and then omitting x whenever $S(x)$ becomes overused will be used in many subsequent results, with $S(x)$ being replaced by a variety of structures.

While the bound ck^2 from Theorem 2.1 is tight up to a constant factor, it is still very interesting to find an asymptotically tight bound. Define $\text{sub}(k)$ to be the minimum d such that every graph with average degree at least d contains a K_k-subdivision. We have seen that $\text{sub}(k) = \Theta(k^2)$. The best known bound on $\text{sub}(k)$ are

$$\bigl(1+o(1)\bigr) \cdot \frac{9k^2}{64} \leq \text{sub}(k) \leq \bigl(1+o(1)\bigr) \cdot \frac{10k}{23}.$$

The lower bound is an observation of Łuczak, using bipartite random graphs, and the upper bound is due to Kühn and Osthus [57].

Question 2.5 *Determine* $\text{sub}(k)$ *asymptotically. Is* $\text{sub}(k) = \bigl(1+o(1)\bigr)\frac{9k^2}{64}$?

2.2 Clique Subdivisions in C_4-Free Graphs

While the requirement that $d(G) \geq ck^2$ is tight, up to a constant factor, for guaranteeing the existence of a K_k-subdivision, Mader [69] conjectured that the quadratic bound could be replaced by a linear one for C_4-free graphs. Namely, he conjectured that there is a constant $c > 0$ such that, if $d(G) \geq ck$ and G is C_4-free, then G has a K_k-subdivision.

In an early application of Komlós and Szemerédi's expanders, from 2004, Kühn and Osthus [56] proved a slightly weaker bound: they showed that there is a constant $c > 0$ such that if $d(G) \geq ck(\log k)^{12}$ and G is C_4-free, then G contains a K_k-subdivision. In the same paper, Kühn and Osthus also prove an analogous result for $K_{s,t}$-free graphs, for all $s, t \geq 2$.

In 2015, Balogh, Liu, and Sharifzadeh [8] proved Mader's conjecture under the additional assumption that G is C_6-free; namely, they showed that there is a constant $c > 0$ such that, if $d(G) \geq ck$ and G is $\{C_4, C_6\}$-free, then G has a K_k-subdivision.

Mader's conjecture was subsequently solved by Liu and Montgomery [62] in 2017.

Theorem 2.6 (Liu–Montgomery [62]) *There is a constant* $c > 0$ *such that, if* $d(G) \geq ck$ *and* G *is* C_4*-free, then* G *contains a* K_k*-subdivision.*

They also proved an analogous result for $K_{s,t}$-free graphs.

We describe some elements of their proof of Theorem 2.6 here, some of which are inspired by [8] and also a result of Montgomery [71] about small minors, that will be mentioned below.

Here, the authors apply the existence result Theorem 2.3 with $t = \Theta(k^2)$. This yields an n-vertex (ε, t)-expander H with minimum degree at least $ck/4$. By C_4-freeness, the second neighbourhood of every vertex has size at least $ck/4 \cdot (ck/4 - 1)$, so if t is taken to be a small factor of this number, then the expansion property of H guarantees that second neighbourhoods expand well.

The proof now splits into the following three cases, which are quite typical of a proof using $KS-expanders$.

- There are at least $2k$ vertices of large degree (at least $k \cdot m^{c_1}$, where $m = \log(n/t)$ and c_1 is a large constant; this bound is chosen so that a greedy algorithm using the small diameter result Theorem 2.4 would work).

- H has small maximum degree and is quite dense ($d(G) \geq (\log n)^{c_2}$ for some large constant c_2).

- H has small maximum degree and is sparse ($d(G) \leq (\log n)^{c_2}$).

The first case is resolved similarly to [54]: given a set K of $2k$ large degree vertices, join pairs of these vertices one by one, using the small diameter property from Theorem 2.4, and removing vertices from K whose neighbourhoods become overused. A counting argument shows that at least k vertices remain in K, yielding the desired K_k-subdivision.

If there are few vertices of large degree, they are removed from H, leaving a graph which is still an expander (with slightly worse parameters). The second case is resolved somewhat similarly to the first one, except that instead of large degree vertices, structures called 'units' are used; units are defined below, and inspired by a similar structure introduced by Montgomery [71] (see Figure 2). The small maximum degree helps to find units that do not overlap too much.

Definition 2.7 (Hub) An (h_1, h_2)-*hub* is a rooted tree of height 2, where the root has degree h_1 and its neighbours have degree h_2.

Definition 2.8 (Unit) An (h_0, h_1, h_2, h_3)-*unit* is a tree consisting of a core vertex v, h_0 pairwise vertex-disjoint (h_1, h_2)-hubs, and h_0 pairwise vertex-disjoint (except at v) paths of length at most h_3 joining the core vertex v with each of the roots of the hubs.

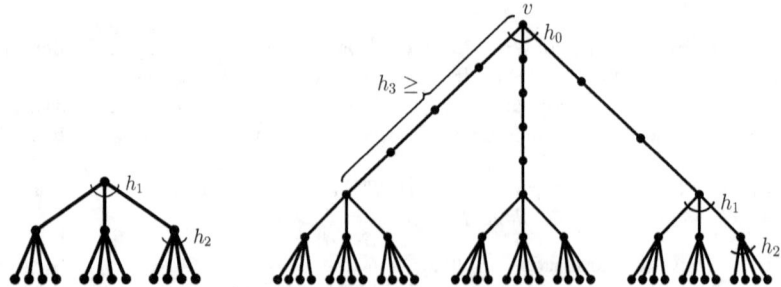

Figure 2: A $(3, 4)$-hub and a $(5, 3, 4)$-unit

The authors of [62] take $h_0, h_1 = \Theta(k)$ and $h_2, h_3 = m^{\Theta(1)}$, where $m = \log(n/t)$, and then find $2k$ many (h_0, h_1, h_2, h_3)-units whose non-leaf vertex sets are pairwise-disjoint. They construct them, one by one, by first constructing many disjoint hubs, and then joining their roots with short paths to find a unit. Notice that the requirement that $d(G)$ is relatively large is needed to guarantee the existence of many (h_1, h_2)-hubs, and the requirement that the maximum degree is not too large is used to show that the graph obtained by removing a not-too-large set of vertices still has large average degree, implying the existence of many hubs. With the $2k$ units at hand, one proceeds very similarly to the first case: connect, one by one, pairs of units, while avoiding vertices in any of the $2k$ units which are not leaves or their parents, and discard units whose sets of parents of leaves become overused.

Finally, in the third case, using the sparsity and bounded maximum degree, they find k vertices that are far from each other. They then connect pairs of vertices, one by one, by short paths that avoid previously chosen paths as well as the vicinity of other vertices.

2.3 Crux and Clique Subdivisions

Haslegrave, Hu, Kim, Liu, Luan, and Wang [40] introduced a graph parameter, which they called 'crux' and described as measuring the 'essential order' of a graph, defined as follows.

Definition 2.9 (Crux) For a constant $\alpha \in (0,1)$ and a graph G, a subgraph $H \subseteq G$ is an α-*crux* if $d(H) \geq \alpha \cdot d(G)$. The α-crux function of G, denoted $c_\alpha(G)$, is the minimum order of an α-crux in G, namely,

$$c_\alpha(G) = \min\{|H| : H \subseteq G \text{ and } d(H) \geq \alpha \cdot d(G)\}.$$

Thinking of α as a constant, notice that $c_\alpha(G) = \Omega(d(G))$. The authors of [40] proved various generalisations of results about cycles in graphs with large average degree, showing that in many cases the average degree can be replaced by crux. For example, they showed that every graph G contains a cycle of length $\Omega(c_\alpha(G))$, generalising (with a loss of a constant factor) Erdős and Gallai's theorem [29] asserting that every graph G contains a cycle of length at least $d(G)$.

Liu and Montgomery [62] suggested that a common generalisation of Theorem 2.6 and Theorem 2.1, that involves the average degree and crux[2] (yet yields slightly weaker bounds), might hold. Specifically, taking $\alpha = \frac{1}{100}$, say, they speculated that their methods could be used to show that every graph G contains a K_k-subdivision, where

$$k = \Omega\left(\min\left\{d(G), \sqrt{\frac{c_\alpha(G)}{\log c_\alpha(G)}}\right\}\right).$$

If true, this bound is tight: as noted in [62], the d-blow-up G of a d-vertex $O(1)$-regular expander satisfies $c_\alpha(G) = \Theta(d^2)$ and the largest clique subdivision has order $O(d \cdot (\log d)^{-1/2})$.

Im, Kim, Kim, and Liu [44] proved a slightly weaker bound.

Theorem 2.10 (Im–Kim–Kim–Liu [44]) *Let G be a graph, and write*

$$t = \min\left\{d(G), \sqrt{\frac{c_\alpha(G)}{\log c_\alpha(G)}}\right\}.$$

---
[2]They did not use the term 'crux' explicitly, as it was not introduced yet.

Then G contains a K_k-subdivision with $k = \Omega\bigl(t \cdot (\log \log t)^{-6}\bigr)$.

This paper is quite technical, and splits into several cases depending on density and value of t. In all cases the authors use vertex-disjoint stars, units whose non-leaf sets are disjoint, and webs, which are defined below, whose non-leaf sets are disjoint, to construct the desired subdivision. In some cases the crux is used to get an additional expansion property.

It is natural to ask if k can be taken to be $\Omega(t)$ in Theorem 2.10.

Question 2.11 *Is there a constant $c > 0$ such that for every graph G if $t = \min\left\{d(G), \sqrt{\frac{c_\alpha(G)}{\log c_\alpha(G)}}\right\}$ then G contains a K_k-subdivision with $k \geq c \cdot t$?*

2.4 Subdivisions of Sparse Graphs

So far we considered conditions guaranteeing a K_t-subdivision. Of course, one could instead look for F-subdivisions of other, sparser graphs F. In this direction, Haslegrave, Kim, and Liu [40] considered α-separable graphs: say that a graph F is α-separable if there is a set U of at most $\alpha|F|$ vertices, such that the components of $F - U$ have size at most $\alpha|F|$. They proved the following for α-separable graphs with bounded maximum degree.

Theorem 2.12 (Haslegrave–Kim–Liu [42]) *Let $\varepsilon > 0$ and $\Delta \geq 1$, and let α be sufficiently small and k sufficiently large. Then for every bipartite, α-separable graph F with $|F| \leq (1 - \varepsilon)k$, every graph G with $d(G) \geq k$ contains an F-subdivision.*

Notice that the assumption that F is bipartite is crucial: if F is a disjoint union of $1/\alpha$ complete graphs on $\alpha k/2$ vertices, then F is α-separable, $|F| = k/2$, yet any complete bipartite graph $K_{t,t}$ with $t < \binom{\alpha k/4}{2}$ does not contain a subdivision of F. The bound $|F| \leq (1 - \varepsilon)k$ is clearly optimal up to the error term εk (consider $G = K_{k+1}$), but perhaps the error term could be decreased or even removed. This result has various implications regarding subdivisions of various sparse graphs, such as the grid, planar graphs, and minor-closed families.

As part of their proof, the authors considered a 'robust' notion of the expanders in Definition 2.2.

Definition 2.13 (Robust KS-expanders) A graph G is called an (ε, t)-robust-KS-expander if for every vertex set U with $t/2 \leq |U| \leq |G|/2$, and every subgraph $F \subseteq G$ with $e(F) \leq d(G) \cdot \rho(|U|) \cdot |U|$, the following holds.

$$|N_{G-F}(U)| \geq \rho(|U|) \cdot |U|.$$

They also proved an existence result, analogous to, and in fact slightly stronger than, Theorem 2.3. (Here the notation $a \ll b$ means that there is a decreasing function f such that the statement holds for $a \leq f(b)$.)

Theorem 2.14 *Let $0 < \nu \ll \varepsilon_1, \varepsilon_2 \ll \delta$. Then every graph G with average degree d has a subgraph H which satisfies: H is an $(\varepsilon_1, \varepsilon_2 d)$-robust-$KS$-expander; $d(H) \geq (1 - \delta)d$; $\delta(H) \geq d(H)/2$; and H is νd-connected.*

This means, as usual, that it suffices to prove that every robust (ε, t)-KS-expander H, with $t = \Theta(k)$, $d(H) \geq k$, and $\delta(H) \geq k/2$, contains an F-subdivision, for F as in Theorem 2.12. And again, the proof differs depending on the density. The following three cases are considered.

- dense ($k = \Omega(n)$),
- medium ($k = o(n)$ and $k = \Omega((\log n)^c)$ for some constant c),
- sparse ($k \leq (\log n)^{O(1)}$).

In the dense case, the regularity lemma is used. In the medium case, when there are k vertices of large enough degree (at least $k \cdot m^{c_1}$ for some constant c_1, where $m = \log(n/t)$), a greedy algorithm suffices. Otherwise, the authors show that the graph obtained by removing the large degree vertices still has large average degree, using the robust expansion of H and the fact that H is medium. They then proceed somewhat similarly to Liu and Montgomery [62], finding 'webs', which are reminiscent of 'units' (see the next definition and Figure 3).

Definition 2.15 An (h_0, h_1, h_2, h_3)-*web* is a rooted tree, obtained by subdividing the height 3 tree T_{h_0, h_1, h_2} — whose root has degree h_0, its neighbours have h_1 children, and these children in turn have h_2 children — where the edges not touching leaves are subdivided at most h_3 times and edges touching leaves are not subdivided. The *centre* of a web is the set of vertices on the paths which subdivide the edges touching the root in T_{h_0, h_1, h_2}.

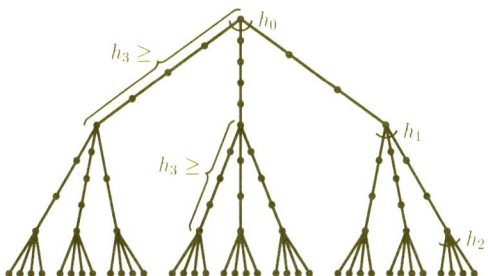

Figure 3: A $(3, 3, 4, 5)$-web

They then finds $2k$ webs, where h_0, h_1, h_3 are constant powers of $m = \log(n/t)$, and $h_0 h_1 h_2 \geq k \cdot m^{c_2}$ for some constant $c_2 > 0$, whose sets of non-leaf vertices are pairwise vertex-disjoint. They then connect them greedily, as usual, with connecting paths avoiding the centres of other webs, and omitting webs whose non-leaf sets become overused.

Finally, in the sparse case, the authors use 'nakjis'[3] (see Figure 4).

Definition 2.16 A (t, s, r, τ)-*nakji* in a graph G is a subgraph H, consisting of vertex-disjoint sets M and D_i, $i \in [t]$, and paths P_i, $i \in [t]$, such that the P_i's are

---
[3] *Nakji* means a 'long arm octopus' in Korean.

pairwise internally vertex-disjoint and their interiors are also disjoint of M and the D_j's; P_i joins M and D_i; $|D_i|, |M| \leq s$; D_i has diameter at most r; M is t-connected; the sets D_i and M are at pairwise distance at least τ.

We refer to M as the *head* of the nakji and to the D_i's as the *legs*[4].

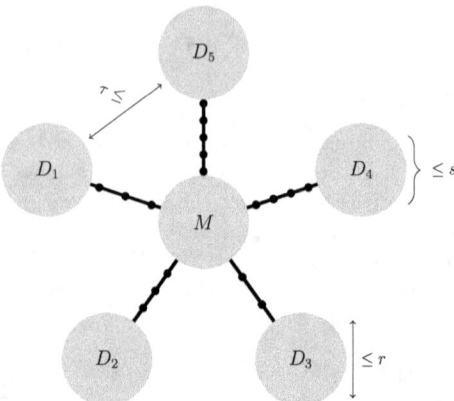

Figure 4: A nakji with $t = 5$. Its *head* is M and its *legs* are $D_1, \ldots, D_5$

In this case, the authors show that the graph obtained by removing large degree vertices still has average degree about as large as the original graph (otherwise, there is a copy of F as a subgraph). Now, in this remaining graph, if there is a large, almost regular subexpander, then in it one can find enough vertices that are far from each other (using sparsity and almost regularity), and join them in a greedy fashion. If, instead, there is a subexpander which is medium or dense, then the arguments from the previous cases can be invoked. Otherwise, there are many small sparse subexpanders that are far from each other. They use these to find k vertex-disjoint nakjis (with $t = \Delta$, i.e. the number of legs of each nakji corresponds to the maximum degree of F), where the head and legs of each nakji are distinct such subexpanders, which are joined together using the connectivity of the underlying expander H. The nakjis are then connected, vertex-disjointly and through their legs, using the expansion properties of H. The subdivision of H is then finalised using the connectivity of the heads of the nakji to find appropriate star subdivisions.

3 Small Minors and Subdivisions

Say that a graph G is an H-*minor* if H can be obtained from G by removing vertices and edges and contracting edges. Equivalently, G is an H-minor if there is a collection of vertex-disjoint connected subgraphs $(G_v)_{v \in V(H)}$, one for each vertex in H, such that if uv is an edge in H then there is an edge between G_u and G_v.

Recall that $\mathrm{sub}(k)$ is the minimum d such that every graph with average degree at least d contains a K_k-subdivision. Define, analogously, $\mathrm{minor}(k)$ to be the minimum d such that every graph with average degree at least d contains a K_k-minor. We

---

[4]Perhaps feet or hands would be more appropriate?

have seen that $\text{sub}(k) = \Theta(k^2)$ (see Theorem 2.1). Noting that a K_k-subdivision is also a K_k-minor, this implies that $\text{minor}(k) = O(k^2)$. In fact, $\text{minor}(k)$ is quite a lot smaller, and, moreover, its value is known quite precisely: Thomason [82] proved that $\text{minor}(k) = (\alpha + o(1)) \cdot k\sqrt{\ln k}$, for an explicit constant α.

Fiorini, Joret, Theis, and Wood [34] asked for the minimum d guaranteeing an H-minor on *few* vertices. Specifically, they asked if an average degree of at least $\text{minor}(k) + \varepsilon$ in an n-vertex graph guarantees an H-minor on $O_\varepsilon(\log n)$ vertices. The size estimate would be tight, as can be seen by considering random graphs, and the average degree condition is tight by definition of $\text{minor}(k)$.

3.1 Almost Logarithmically Small Clique Minors

Addressing this question, in 2015 Shapira and Sudakov [78] proved the following.

Theorem 3.1 (Shapira–Sudakov [78]) *For every $\varepsilon > 0$ and integer $k \geq 1$, there is a constant $\alpha = \alpha(\varepsilon, k)$ such that every n-vertex graph with average degree at least $\text{minor}(k) + \varepsilon$ has a K_k-minor on at most $\alpha \cdot \log n \cdot \log \log n$ vertices.*

In their paper, they introduced a new notion of sublinear expanders.

Definition 3.2 (Shapira–Sudakov [78]) *An n-vertex graph G is a δ-SS-expander if for every integer d with $0 \leq d \leq \log \log n - 1$ and subset $S \subseteq V(G)$ of size at most $n/2^{2^d}$, we have*

$$|N(S)| \geq \frac{\delta \cdot 2^d}{\log n \cdot (\log \log n)^2} \cdot |S|.$$

To better understand this expression, for a set S take d to be largest such that $|S| \leq n/2^{2^d}$, so $2^d \approx \log(n/|S|)$, yielding that $|N(S)| \gtrsim \frac{\delta \log(n/|S|)}{\log n \, (\log \log n)^2} \cdot |S|$. This shows that sets of size at most $n^{1-\Omega(1)}$ expand at rate $\Omega(\frac{1}{(\log \log n)^2})$, and sets of size $\Theta(n)$ expand at rate $\Omega(\frac{1}{\log n \, (\log \log n)^2})$.

As with KS-expanders, every graph contains a δ-SS-expander, as long as δ is sufficiently small.

Theorem 3.3 (Shapira–Sudakov [78]) *Let $\delta > 0$ be sufficiently small. Then every graph G contains a δ-SS-expander H with $d(H) \geq (1 - \delta)d(G)$.*

An important reason why KS-expanders are not appropriate here is that the small diameter result for them (Theorem 2.4) only yields a diameter of $\Theta((\log n)^3)$, which is way too large for Theorem 3.1. Indeed, SS-expanders do better in this respect. We state a special case of it informally (see Claim 3.3 in [78] for a precise statement): in a δ-SS-expander on n vertices, for every two sets U_1, U_2 of at least $(\log n)^4$ vertices, and a set W of at most $(\log n)^2$ vertices, there is a path between U_1 and U_2 that avoids W and has length $O(\log n \cdot (\log \log n)^3)$. This illustrates why this expansion notion is more suitable for the particular problem where we are interested in minors that are *small*, as the diameter here is quite a lot smaller. However, using just this notion of expansion, Shapira and Sudakov get a slightly weaker result than the one stated above, showing that average degree $c(k) + \varepsilon$ implies a K_k-minor on $O(\log n \, (\log \log n)^3)$ vertices.

Indeed, this is essentially immediate if there are at least k vertices with degree at least $(\log n)^4$, using the 'small diameter' result above, and in fact a K_k-subdivision is found in this case. If there are few vertices of high degree, the authors show that there are k disjoint structures they call 'expanding balls', which are relatively large sets of small radius and good expansion properties. They connect them via the same diameter argument, and show how to use these connections to get a small K_k-minor.

To get the slightly stronger result, with a bound of $O(\log n \log \log n)$, they use yet another notion of expansion, and proved a corresponding existence result for this latter notion. We mention the definition and existence result here, as several future papers use a special case of the existence result, or variants of it.

Definition 3.4 An m-vertex graph H is said to be a (δ, n)-SS-expander if for every integer d with $0 \leq d \leq \log \log m - 1$ and $S \subseteq V(H)$ with $|S| \leq m/2^{2^d}$ we have $|N(S)| \geq \frac{\delta 2^d}{\log n} \cdot |S|$.

Theorem 3.5 *For every sufficiently small $\delta > 0$, every graph G contains a subgraph H such that $d(H) \geq (1 - 2\delta)d(G)$ and H is a (δ, n)-SS-expander.*

3.2 Logarithmically Small Clique Minors and Subdivisions

Shortly afterwards, Montgomery [71] improved upon the above result, proving the following tight result.

Theorem 3.6 (Montgomery [71]) *For every $\varepsilon > 0$ and integer $k \geq 1$, there is a constant $\alpha = \alpha(\varepsilon, k)$ such that every n-vertex graph G with $d(G) \geq \text{minor}(k) + \varepsilon$ has a K_k-minor on at most $\alpha \cdot \log n$ vertices.*

He uses yet another definition of expanders.

Definition 3.7 An m-vertex graph is a (λ, η)-M-expander if every vertex set S of size at most $m^{1-\eta}$ satisfies $|N(S)| \geq \lambda |S|$.

In his application, η is a small constant, and λ a function of η, m, and the number of vertices n in a given graph. Notice that this definition does not imply a small diameter result: sets of not-too-small size can be shown to expand quickly to size $m^{1-\eta}$, but might not expand fast beyond this size, making it hard to connect two given fixed vertices.

As usual, Montgomery's first step is to prove an existence result, showing that for given δ and η and appropriate λ, every n-vertex graph G with average degree d contains a subgraph H with average degree at least $d(1 - \delta)$, which is either small or a (λ, η)-M-expander. Applying this to G with average degree at least $\text{minor}(t) + \varepsilon$, we get a graph H with $d(H) \geq \text{minor}(t)$ which is either small or an expander. If H is small, then in particular $|H| = O(\log n)$, and then it suffices to find a K_t-minor in H (with no size restrictions), which is possible by the definition of $\text{minor}(t)$. So suppose that H is a (λ, η)-M-expander. Similarly to [78], Montgomery shows that such H contains many disjoint sets of size at least $m^{1/4}$ and small radius. Using the quick expansion of not-too-small sets to size $m^{1-\eta}$, he shows that there is a vertex v which is at distance $O(\log n)$ from at least a $m^{-\eta}$ fraction of the sets, thereby overcoming the lack of a 'small diameter' result. He uses the vertex v as a basis of one of the k

vertices in the K_k-minor, and repeats the same argument after appropriate cleaning to find the desired K_k-minor.

In the same paper, Montgomery also proved an analogous result for subdivisions.

Theorem 3.8 (Montgomery [71]) *For every $\varepsilon > 0$ and integer $k \geq 1$, there is a constant $\alpha = \alpha(\varepsilon, k)$ such that every n-vertex graph G with $d(G) \geq \text{sub}(k) + \varepsilon$ has a K_k-subdivision on at most $\alpha \cdot \log n$ vertices.*

For this, Montgomery used (λ, η)-M-expanders, whose every vertex set S of size at most $m^{1/3}$ also expands at a rate of $\Omega(\frac{1}{(\log \log |S|)^2})$. This additional property is necessary here, as to find subdivisions we intuitively need single vertices to expand well. The proof here is more involved and uses the notion of 'units', which in this context are a collection of t disjoint, relatively large sets of small radius, along with t paths from these sets to a common ('corner') vertex v, which are vertex-disjoint other than at v. This is somewhat similar to the notion of units from Definition 2.8, and in fact served as inspiration for this latter definition of units.

4 Immersions

Say that a graph G *immerses* a graph H if there is an injective function $f : V(H) \to V(G)$, along with a collection of edge-disjoint paths P_{uv}, for $uv \in E(H)$, such that P_{uv} has ends u and v. Equivalently, G immerses H if H can be obtained from G by sequentially removing vertices and edges, and by replacing paths uvw by the edge uw (see Figure 5).

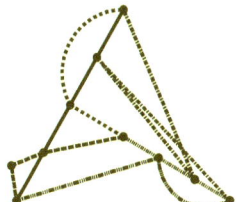

Figure 5: A K_4-immersion; each path P_{uv} is depicted with a different pattern.

Recall that $\text{sub}(k)$ is the minimum d such that every graph with average degree at least d contains a K_k-subdivision, and minor(k) is defined analogously for K_k-minors. We define $\text{imm}(k)$ similarly, as the minimum d such that every graph with average degree at least d immerses K_k. Since a K_k-subdivision is also an immersion of K_k, we immediately get $\text{imm}(k) \leq \text{sub}(k) = \Theta(k^2)$. Notice also that $\text{imm}(k) > k - 2$, because the existence of a K_k-immersion implies the existence of vertices of degree at least $k - 1$. This was proved to be tight, up to a constant factor, by DeVos, Dvořák, Fox, McDonald, Mohar, and Scheide [21], who proved $\text{imm}(k) \leq 400k$. This was improved to $\text{imm}(k) \leq 22k + 14$ by Dvořák and Yepremyan [26]. The answer might actually be k, which would imply the following conjecture and answer the subsequent question affirmatively. While neither [21] nor [26] use expanders, it is plausible that expanders can be used to make progress on the following conjecture and question.

Conjecture 4.1 (Lescure–Meyniel [58], Abu-Khzam–Langston [1]) *If $\chi(G) \geq k$ then G contains a K_k-immersion.*

Question 4.2 (Dvořák–Yepremyan [26]) *Does every graph with minimum degree k immerse K_k?*

We remark that Lescure and Meyniel [58] and, later and independently, DeVos, Kawarabayashi, Mohar, and Okamura [22] proved that minimum degree $k-1$ guarantees a K_k-immersion for $k \leq 7$, thereby proving the above conjecture and answering the question affirmatively for this range of k's, strengthening both conjecture and question slightly. Nevertheless, this is no longer true for large values of k (see [21, 19]).

Another natural question, given Theorem 3.6 and Theorem 3.8 above, asks whether an average degree slightly above imm(k) suffices to guarantee a small immersion.

Question 4.3 *Let $\varepsilon > 0$ and $k \geq 1$. Is there a constant $\alpha = \alpha(\varepsilon, k)$ such that every n-vertex graph G with $d(G) \geq \mathrm{imm}(k) + \varepsilon$ has a K_k-immersion on at most $\alpha \log n$ vertices?*

4.1 Clique Immersions in $K_{s,t}$-Free Graphs

Liu, Wang, and Yang [64] proved Conjecture 4.1 asymptotically for graphs which are $K_{s,t}$-free, for every $s, t \geq 2$.

Theorem 4.4 (Liu–Wang–Yang [64]) *For every $\varepsilon > 0$, integers $s, t \geq 2$, and large enough d, every $K_{s,t}$-free graph G with $d(G) \geq d$ contains a K_k-immersion with $k \geq (1-\varepsilon)d$.*

The proof uses many of the tools described before. First, by Theorem 2.14, it suffices to prove that every n-vertex $(\varepsilon_1, \varepsilon_2 k)$-KS-expander H, which is $K_{s,t}$-free, and has average degree at least $(1+\delta)k$, contains a K_k-immersion. Notice that k cannot be too close to n, by the $K_{s,t}$-freeness. Thus, there are two main cases.

- dense ($k \geq (\log n)^c$),
- sparse (otherwise).

In the dense case, the authors used a variant of 'hubs' and 'units' defined above: here an (h_1, h_2, h_3)-*unit* consists of h_1 vertex-disjoint stars of size h_2, that are joined by edge-disjoint paths of length at most h_3 to a 'core vertex'. Here edge-disjointness suffices because we are interested in immersions. They find k edge-disjoint (h_1, h_2, h_3)-units (with $h_1 = k$, $h_2 = \mathrm{diam}^c$, and $h_3 = 2\,\mathrm{diam}$, where diam is the expression from the small diameter result Theorem 2.4) with the centres of stars pairwise disjoint, and then connect them greedily.

The sparse case is somewhat more complex. If the maximum degree is at most $k \cdot (\log n)^c$, then one can find k vertices with degree a bit over k that are sufficiently far apart, and then join them, one by one, avoiding previously used edges and the vicinity of vertices not being currently joined. Also, as usual, if there are k vertices of degree at least $k(\log n)^c$ then a greedy strategy can join them edge-disjointly to form an immersion. Thus, we may assume that there are fewer than k vertices of degree at least $k(\log n)^c$. By $K_{s,t}$-freeness, the graph obtained by removing them still has large average degree. If the latter graph has a subexpander with average degree at bit above d and which is either dense or has small maximum degree, then previous arguments can be applied. If not, this means that there are many subexpanders with average degree a bit above k, that are far from each other and each have a large degree vertex. Take k such subexpanders and a largest degree vertex from each, and join these up as usual.

4.2 Immersions in Directed Graphs

Notice that the definition of immersions can be carried through to directed graphs. However, an analogue of imm(k) for digraphs does not exist: there are digraphs with arbitrarily large minimum in- and out-degree which do not immerse $\overrightarrow{K}_3$ (the complete digraph on 3 vertices); see Lochet [65]. Nevertheless, in the same paper Lochet showed that minimum out-degree k^3 guarantees an immersion of a transitive tournament on $\Omega(k)$ vertices, and it is plausible that minimum out-degree k already suffices for an immersion of a transitive tournament on $\Omega(k)$ vertices.

Question 4.5 *Is there a constant $c > 0$ such that, for every integer $k \geq 1$, every digraph with minimum out-degree at least ck contains an immersion of the transitive tournament on k vertices?*

DeVos, McDonald, Mohar, and Scheide [23, 24] showed that every Eulerian digraph with minimum in-degree k^2 immerses $\overrightarrow{K}_k$, and asked whether a linear bound would suffice. In [38], Girão and Letzter answered this question affirmatively.

Theorem 4.6 (Girão–Letzter [38]) *There exists $c > 0$ such that every Eulerian digraph with minimum in-degree at least ck immerses $\overrightarrow{K}_k$.*

The proof is a rare application of expanders in digraphs. We define an analogue of robust (ε, t)-KS-expanders (see Definition 2.13), where instead of lower bounding the size of the neighbourhood $N(U)$, we lower bound the sizes of the out- and in-neighbourhoods of U. To prove an existence result for such directed robust expanders, we use an undirected version guaranteeing the existence of an expander where every set of relevant size has large edge boundary (as opposed to the vertex boundary that was used above). Applying this to the underlying undirected graph obtained from the original Eulerian digraph G, which we additionally assume to be regular, we get a digraph $D' \subseteq G$ where every set of vertices U of relevant size has many outgoing edges (i.e. edges from U to $V(D') - U$) or many incoming edges (edges from $V(D') - U$ to U). Now, we observe that because G is Eulerian, it immerses an Eulerian *multidigraph* D that contains D' as a subgraph. By the property of D' mentioned above and by D being Eulerian, every set of vertices U in D of relevant size has many outgoing edges *and* many incoming edges, which readily implies that D is a directed robust (ε, t)-KS-expander (using that D is close to regular, which follows from regularity of G).

Additionally, we use the fact that imm(k) = $O(k)$, along with structural arguments, to show that every Eulerian multidigraph with minimum degree at least ck

and $O(k)$ vertices (with the technical condition that the underlying undirected graph has $\Omega(k^2)$ edges) immerses $\vec{K}_k$.

The last two paragraphs imply that it suffices to show that an n-vertex Eulerian expander D (which in our setting is a multidigraph), with average degree at least ck, immerses a (simple) digraph on $\Theta(k)$ vertices with $\Omega(k^2)$ edges. Conveniently, due to the nature of immersions, we may assume that D has maximum in- and out-degree $\Theta(k)$. As usual, we distinguish between dense ($k \geq (\log(n/k))^c$) and sparse expanders. In the sparse case we find $2k$ vertices which are far from each other, k of which have large out-degree (not counting multiplicities) and the other k having large in-degree. We then find edge-disjoint paths from each of the latter vertices to each of the former ones, thus forming an immersion of $\vec{K}_{k,k}$. In the dense case, we follow a similar strategy to that of Komlós and Szemerédi [54], finding $2k$ large out-stars and $2k$ large in-stars, with distinct centres and where leaves are not shared by too many stars, and connect as many of the out-stars to the in-stars with short edge-disjoint directed paths as possible, yielding an immersion of a dense subgraph of $\vec{K}_{2k,2k}$, as needed.

5 The Odd Cycle Problem and Balanced Subdivisions

In this section we mention recent developments about a conjecture of Erdős and Hajnal about cycle lengths in graphs with large chromatic number, known as the 'odd cycle problem', and about average degree conditions guaranteeing the existence of so-called balanced subdivisions. While seemingly unrelated at first glance, in both topics it is useful to be able to join two given vertices by a path of specific length.

5.1 Cycle Lengths in Graphs With Large Average Degree

For a graph G, let $\mathcal{C}(G)$ be the set of cycle lengths in a graph G. In 1966, Erdős and Hajnal [31] suggested to study the sum $\sum_{\ell \in \mathcal{C}(G)} \frac{1}{\ell}$ as a measure of the density of G's cycle lengths. In particular, they asked whether $\sum_{\ell \in \mathcal{C}(G)} \frac{1}{\ell}$ tends to infinity as $\chi(G) \to \infty$. In fact, Erdős later [27] suggested that the same should hold as $d(G) \to \infty$, and, moreover, he thought it likely that $\sum_{\ell \in \mathcal{C}(G)} \frac{1}{\ell} \geq (1/2 + o_d(1)) \ln d$, for every graph G with $d(G) \geq d$, which would be tight, as can be seen by considering $K_{d,d}$. In a slightly different direction, Erdős [28] asked if every graph G with sufficiently large average degree contains a cycle whose length is a power of 2.

In a breakthrough paper, Liu and Montgomery [63] answered both questions affirmatively.

Theorem 5.1 (Liu–Montgomery [63]) *For large enough d, and every graph G with $d(G) \geq d$ there exists $\ell \geq \frac{d}{10(\ln d)^{12}}$ such that $\mathcal{C}(G)$ contains every even number in $[(\ln \ell)^8, \ell]$.*

They use similar methods to prove a similar result about the odd cycle lengths in a graph G with large chromatic number, and, more generally, about cycle lengths of specific residues in graphs with large chromatic number. As an immediate corollary of their results, they solved a problem due to Erdős and Hajnal, known as the 'odd cycle problem': they asked whether $\mathcal{C}_{\text{odd}}(G)$, defined to be the sum of $1/\ell$ over all *odd* cycle lengths of G, tends to infinity as $\chi(G) \to \infty$. Liu and Montgomery answer this affirmatively, and, in fact, they show that $\mathcal{C}_{\text{odd}}(G) \geq (1/2 + o(1)) \ln \chi(G)$.

In a somewhat different direction, they considered balanced subdivisions of complete graphs. A *balanced subdivision* of a graph H is a subdivision of H where each edge is replaced by a path of the same length. Denote by $\text{TK}_k^{(\ell)}$ the balanced subdivision of K_k where each edge is replaced by a path of length ℓ (see Figure 6).

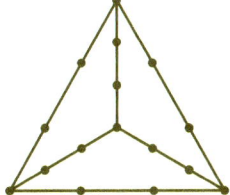

Figure 6: A balanced subdivision $\text{TK}_4^{(3)}$ of K_4

Liu and Montgomery proved that for any k, a large enough average degree guarantees the existence of a balanced K_k-subdivision, confirming a conjecture of Thomassen [83].

Theorem 5.2 (Liu–Montgomery [63]) *For every integer $k \geq 1$ and large enough d, every graph G with $d(G) \geq d$ contains a balanced subdivision of K_k.*

While the results about cycle lengths and balanced subdivisions sound quite different, a common theme is the need to join pairs of vertices by a path of specific length. This is in contrast with previously described results, where when joining two given vertices, or sets of vertices, we just wanted the path joining them to be short, but did not care about the exact length. As such, the following theorem is a key component in the proofs of the results in [63]. Here an (x,y)-*path* is a path with ends x and y.

Theorem 5.3 (Theorem 2.7 in Liu–Montgomery [63]) *Let $\varepsilon_1, \varepsilon_2 > 0$ be small and let d be large. Suppose that H is a $\text{TK}_{d/2}^{(2)}$-free bipartite n-vertex $(\varepsilon_1, \varepsilon_2 d)$-expander H with $\delta(H) \geq d$. Then for any two vertices x, y and $\ell \in [(\ln n)^7, n/(\ln n)^{12}]$ of the right parity[5], there is an (x,y)-path of length ℓ.*

It is easy to see that this implies the first result about even cycle lengths. For the other results, some more work is needed.

The main novelty in the proof of Theorem 5.3, is the introduction and use of 'adjusters'.

Definition 5.4 (Simple adjusters) *A simple (D, m)-adjuster is a subgraph consisting of a cycle C of length 2ℓ for some $\ell \leq 5m$, two vertices v_1, v_2 in C at distance $\ell - 1$ on C, and two pairwise vertex-disjoint graphs F_1, F_2, which are vertex-disjoint of $V(C) - \{v_1, v_2\}$, such that, for $i \in [2]$: $v_i \in V(F_i)$; $|F_i| = D$; and every vertex in F_i is at distance at most m in F_i.*

---

[5]By the right parity, we mean that if x, y are in the same part of the bipartition of H then ℓ is even, and otherwise it is odd.

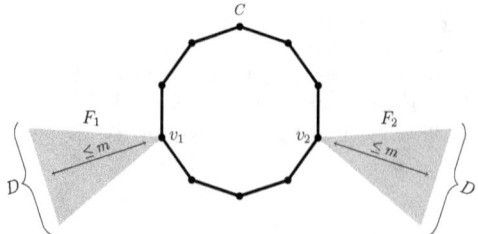

Figure 7: A simple adjuster

Assuming we have a long sequence of k simple (D, m)-adjusters (where $m = \Theta((\log n)^3)$, so that m is larger than the bound on the path length in Theorem 2.4, and D is polylogarithmic in n but significantly larger than m), the plan would be to join them up, one by one, via the sets F_2 and F_1 of consecutive adjusters, ensuring that the connecting paths are pairwise vertex-disjoint and relatively short. This yields the following structure, generalising simple adjusters, which gives (v_1, v_2)-paths of many different lengths.

Definition 5.5 (Adjusters) A (D, m, k)-adjuster is a subgraph consisting of vertices v_1, v_2 and graphs A, F_1, F_2, such that: A, F_1, F_2 are pairwise vertex-disjoint; $v_i \in V(F_i)$ and every vertex in F_i is at distance at most m from v_i in F_i; $|F_i| = D$ and $|A| \leq 10mk$; and for some $\ell_0 \leq 10mk$ and every $i \in \{0, 1, \ldots, k\}$ there is a (v_1, v_2)-path in $A \cup \{v_1, v_2\}$ of length $\ell_0 + 2i$.

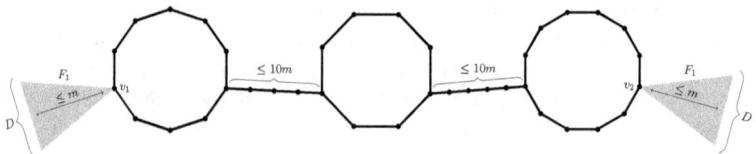

Figure 8: A $(D, m, 3)$-adjuster

Now, given two vertices x, y and a (D, m, k) adjuster, if we would like to find an (x, y)-path of length ℓ, it suffices to join x to F_1 and y to F_2 by paths of total length between $\ell - \ell_0 - 2k$ and $\ell - \ell_0 - 2m$, which can then be corrected to an (x, y)-path of length ℓ using the adjuster. These tasks, of joining the F_i's to form a (D, m, k)-adjuster, and of joining x and y to the adjuster by a path of length approximately ℓ, are both achievable via the diameter theorem Theorem 2.4, assuming that $k \geq m^c$ for large enough c.

As such, it suffices to show how to find many disjoint simple adjusters, which essentially amounts to finding a simple adjuster avoiding a forbidden set of vertices W of polylogarithmic size. Finding one simple adjuster in an expander (with no forbidden vertices) is quite simple: take a shortest cycle C, pick two vertices on it at appropriate distance in C, and expand from there, showing that C can be avoided due to it being a shortest cycle. To find an adjuster avoiding a set W, one can apply this reasoning to a subexpander of $G - W$. However, as we have no control over the order of such a subexpander, this might lead to an adjuster which is way too small. Thus, the authors take many disjoint such adjusters, and show that one of them can be expanded to be sufficiently large. This is a challenging task, which we do not elaborate on. We do briefly mention that a useful tool in overcoming this challenge is a lemma (Lemma 3.7 in [63]) about expanding a set A while avoiding forbidden sets with various ways of controlling the interaction between A and the forbidden sets.

5.2 Improved Bounds on Average Degree Implying Balanced Clique Subdivision

Notice that, through Theorem 5.2, Liu and Montgomery proved the existence of a function $\text{sb}(k)$ such that every graph with average degree at least $\text{sb}(k)$ contains a balanced subdivision of K_k. They do not calculate explicitly the upper bound on $\text{sb}(k)$ that their proof yields, but they suggested that $\text{sb}(k)$ might be $O(k^2)$. If true, this would be tight, and would generalise the aforementioned fact that $\text{sub}(k) = \Theta(k^2)$ (see Theorem 2.1).

The first progress towards estimating $\text{sb}(k)$ was made by Wang [91], who proved that $\text{sb}(k) \leq k^{2+o(1)}$. This was improved to the tight $\text{sb}(k) = O(k^2)$ by Luan, Tang, Wang, and Yang [67], and, independently, by Gil Fernandez, Hyde, Liu, Pikhurko, and Wu [35].

Theorem 5.6 (Luan–Tang–Wang–Yang [67] and Gil Fernandez–Hyde–Liu– Pikhurko–Wu [35]) *There is a constant $c > 0$ such that for every integer $k \geq 1$, every graph with average degree at least ck^2 contains a balanced subdivision of K_k.*

The former paper [91] also proves that there exists $c > 0$ such that average degree at least ck suffices to guarantee a balanced K_k-subdivision for C_4-free graphs, generalising Theorem 2.6.

As usual, it suffices to prove that every n-vertex robust $(\varepsilon_1, \varepsilon_2 d)$-KS-expander H contains the required balanced K_k-subdivision, with $k = \Omega(\sqrt{d})$. Again as usual, both proofs split into three cases, depending on density: sparse ($d = (\log n)^{O(1)}$); medium ($d \geq (\log n)^c$ for some constant c and $d = o(n)$); and dense ($d = \Omega(n)$).

The dense case follows from a result of Alon, Krivelevich, and Sudakov [5], that actually yields a $\text{TK}_k^{(2)}$. The sparse case was already addressed by Wang [91], and Wang addressed it as follows. Note that we may assume H is $\text{TK}_k^{(2)}$-free. The proof uses tools due to Liu and Montgomery [63] about joining two vertices by a path of specific length in a $\text{TK}_k^{(2)}$-free expander, and is otherwise quite routine. If there are enough vertices of large degree (at least $d(\log n)^c$ for some constant c) then they can be joined greedily. Otherwise, we may assume the maximum degree is small, allowing us to find many 'core' vertices (in the same part of the bipartition) that are far enough from each other. These can then be joined, one by one, by paths of specific length, while avoiding previously used vertices and the vicinity of other core vertices.

It thus remains to address the medium case. In [67], the authors find $\Theta(k)$ units (see Definition 2.8) with at least d^2m^c leaves (where $m = \log(n/d)$ and c is some large constant), such that the sets of non-leaves are pairwise vertex-disjoint. They then proceed to join pairs of units, one by one, removing units that become overused. To yield a balanced subdivisions, the connection here needs to be of a specific length, and for this the authors use (D, m, k)-adjusters (see Definition 5.5). Since here the adjusters need to be somewhat larger than in [63] (D here has size $dm^{\Theta(1)}$, to be able to use them in connections while avoiding $O(dm^{O(1)})$ vertices, as opposed to $m^{\Theta(1)}$ in [62]), their construction is somewhat different here. The authors of [35], follow a similar approach, using a variant of units, where a (h_0, h_1, h_2)-unit is defined to be a rooted tree of height $h_2 + 1$, such that the root has degree h_0, the vertices at level between 1 and $h_2 - 1$ have exactly one child, and the vertices at level h_2 have h_1 children. Units are also used in place of the graphs F_1, F_2 in the definition of adjusters.

6 Tight Cycles, Rainbow Subdivisions, and Cycles With Many Chords

In this section we describe several Turán type results, about tight cycles, rainbow subdivisions, and cycles with many chords. These results are connected to each other via the methods they use, which can be traced back to the paper of Shapira and Sudakov [78] about small minors. In particular, various notions of expansions are used here (though for the most part we will not give the exact definitions), but all are somewhat similar to the one in Definition 3.4.

6.1 Hypergraphs With No Tight Cycles

An r-uniform *tight cycle* is a hypergraph on vertices $\{v_1, \ldots, v_\ell\}$, for some $\ell \geq r+1$, with edges $\{v_i \ldots v_{i+r-1} : i \in [\ell]\}$ (addition of indices taken modulo r; see Figure 9).

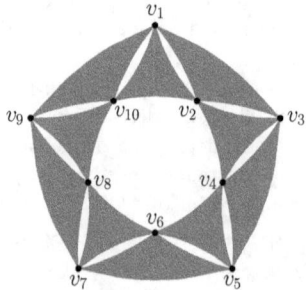

Figure 9: A 3-uniform tight cycle on 10 vertices

It is natural to ask: what is the maximum number of edges an n-vertex r-uniform hypergraph can have, if it has no tight cycles? Denoting this maximum by $\mathrm{ex}_r(n, \mathcal{C})$, notice that $\mathrm{ex}_r(n, \mathcal{C}) \geq \binom{n-1}{r-1}$ (take all edges containing a single vertex), which is best possible for $r = 2$. For larger values of r, this turns out not to be tight (disproving a conjecture due to Sós and, independently, Verstraëte; see [89, 74]). The best known lower bound is due to Janzer [45], who showed that $\mathrm{ex}_r(n, \mathcal{C}) = \Omega(n^{r-1} \cdot \frac{\log n}{\log \log n})$. Sudakov and Tomon [80] proved a close-to-matching upper bound.

Theorem 6.1 (Sudakov–Tomon [80]) *Let $r \geq 3$. Every n-vertex r-uniform hypergraph with no tight cycles has at most $n^{r-1} e^{O(\sqrt{\log n})}$ edges.*

In particular, this shows $\text{ex}_r(n, \mathcal{C}) = n^{r-1+o(1)}$.

They define r line graphs, which are graphs whose vertices are the edges of an r-partite r-uniform hypergraph, and whose edges correspond to two edges intersecting in $r - 1$ vertices, and define an appropriate notion of density for such graphs. They use a variant of a notion of expanders used by Shapira and Sudakov [78] (see Definition 3.4).

Definition 6.2 (Sudakov–Tomon [80]) A graph H is called a (λ, d)-*ST-expander* if it has minimum degree[6] at least d, and every vertex set $U \subseteq V(H)$ satisfies $|N(U)| \geq \lambda |U|$.

They then prove an existence result, showing that every r-line-graph G of density d (with an appropriate definition of density) contains a subgraph H which is a (λ, d')-ST-expander, with $\lambda = \Omega(1/\log n)$ and $d' = \Omega(d)$, and has density at least $d/2$. This proof is inspired by similar arguments made by Shapira and Sudakov [78] to prove an existence result for expanders as in Definition 3.4.

Notice that for this notion of expansion to be effective in expanding a single vertex, one would like to have $d = \Omega(\log n)$. This is indeed the case here; in fact, we have $d = e^{\Omega(\sqrt{\log n})}$. A key property of expanders in r-line-graphs that is proved here is that for every vertex e in an (λ, d)-ST-expander H (with appropriate parameters), which is an r-line-graph, one can reach almost every other vertex f via a short tight path in the hypergraph corresponding to H. Observe that, because of the nature of tight paths, this does not imply that every two vertices can be joined via a short tight path, analogously to Theorem 2.4, even if we insist that these vertices correspond to disjoint edges.

In order to find a tight cycle, the authors split the underlying vertex set of H into two sets, denoting the two corresponding r-line-graphs by H_1, H_2. If they could prove that each graph H_i maintains the expansion properties of H, then they would be able to find two tight paths, one from an r-set e to another r-set f in H_1, and one from f to e in H_2, using disjoint vertex sets (other than the vertices in e, f), thereby yielding a tight cycle, as required. The authors are unable to do so. Instead, they use the existence result from above to almost decompose each of H_1 and H_2 into expanders. If any of them are small, then we get a much denser expander than the original one, and the same argument is then repeated in this denser expander. Otherwise, the total number of expanders used in the decompositions is small, and then the expansion property can be used to essentially realise the strategy outlined above to get a tight cycle. Since the density cannot be increased indefinitely, at some iteration the latter option holds, yielding a tight cycle, as required.

In [60], the author improved Sudakov and Tomon's bound, proving the almost tight bound $\text{ex}_r(n, \mathcal{C}) = O(n^{r-1}(\log n)^5)$.

Theorem 6.3 (Letzter [60]) *Let $r \geq 3$. Every n-vertex r-uniform hypergraph with no tight cycles has at most $O(n^{r-1}(\log n)^5)$ edges.*

---

[6]Their definition of the minimum degree of an r-line-graph is somewhat different to the usual definition of the minimum degree. Denote by $\mathcal{H}$ the underlying r-partite r-graph and by $(U_1, \ldots, U_r)$ an appropriate partition of $V(\mathcal{H})$. The *minimum degree* of H is defined to be the minimum, over all $i \in [r]$ and vertices x in H, of the number of neighbours of x in H that agree with x on U_j for all $j \in [r] - \{i\}$ (plus 1).

Using the setup from [80], the improvement came from proving that every r-line-graph H, which is an ST-expander with appropriate parameters, and has sufficiently large density, has a tight cycle (so there is no need for a density increment argument as above). An important idea was to strengthen the expansion property above, showing that for every vertex e in an r-line-graph H as above, almost every vertex f in H can be reached from e via a short tight path, in such a way that no vertex of the underlying graph is used in too many of these paths (except for the vertices in e). Another trick is then needed to overcome the directed nature of the problem.

It would be interesting to prove an even tighter bound on $\mathrm{ex}_r(n,\mathcal{C})$. Recall that the best known lower bound [45] is $\mathrm{ex}_r(n,\mathcal{C}) = \Omega\bigl(n^{r-1} \cdot \frac{\log n}{\log \log n}\bigr)$.

Question 6.4 *Determine $\mathrm{ex}_r(n,\mathcal{C})$ asymptotically. Is $\mathrm{ex}_r(n,\mathcal{C}) = O\bigl(n^{r-1} \cdot \frac{\log n}{\log \log n}\bigr)$? A bit more crudely, is $\mathrm{ex}_r(n,\mathcal{C}) = n^{r-1} \cdot (\log n)^{1+o(1)}$?*

It is plausible that ideas described in subsequent subsections could yield a somewhat better than the one in Theorem 6.3, perhaps decreasing the exponent of $\log n$ (though likely they would require significant technical work), but it seems that to get a tight bound, new ideas will be needed.

6.2 Rainbow Clique Subdivisions

A similar notion of expansion was used by Jiang, Methuku, and Yepremyan [49] to tackle a rainbow Turán problem. A *rainbow* graph is an edge-coloured graph whose edges have distinct colours. Keevash, Mubayi, Sudakov, and Verstraëte [50] defined the *rainbow Turán number* of a graph H, denoted $\mathrm{ex}^*(n,H)$, as the maximum number of edges in an n-vertex properly edge-coloured graph which has no rainbow copies of H; a similar definition could be made for a collection of graphs $\mathcal{H}$.

A particularly interesting question here is to determine the rainbow Turán number of cycles, namely the maximum number of edges in an n-vertex properly edge-coloured graph which has no rainbow cycles, which we denote by $\mathrm{ex}^*(n,\mathcal{C})$. It is easy to see that the hypercube Q_d, where edges are coloured according to their direction, is a properly edge-coloured graph with no rainbow cycles, showing $\mathrm{ex}^*(n,\mathcal{C}) = \Omega(n \log n)$. The authors of [50] conjectured that the latter bound is tight, up to a constant factor.

Conjecture 6.5 (Keevash–Mubayi–Sudakov–Verstraëte [50]) *Every graph on n vertices whose edges are properly coloured so that there are no rainbow cycles has $O(n \log n)$ edges.*

Janzer [46] proved that $\mathrm{ex}^*(n,\mathcal{C}) = O\bigl(n(\log n)^4\bigr)$, getting quite close to proving the conjecture. His main tool was an upper bound on the number of homomorphic copies of an even cycle with some degeneracy, and he used this method to prove several other Turán type questions.

The problem Jiang, Methuku, and Yepremyan [49] considered was the following: for a constant k, what is the maximum number of edges in a properly edge-coloured n-vertex graph which has no rainbow subdivisions of K_k? Clearly, this number is at least $\mathrm{ex}^*(n,\mathcal{C}) = \Omega(n \log n)$. The authors of [49] proved the following bound, showing that the answer is $n^{1+o(1)}$.

Theorem 6.6 Let $k \geq 3$. If G is an n-vertex properly edge-coloured graph with no rainbow K_k-subdivision, then $e(G) \leq n \cdot e^{O(\sqrt{\log n})}$.

An attentive reader might notice the similarity between this bound and that of Sudakov and Tomon [80] in Theorem 6.1. This is not an accident: the outline of the proof here is very similar to that in [80]. The authors use a variant of expanders as in Definition 6.2, focusing on the edge boundary of vertex sets rather than the vertex boundary, and prove that in a properly-coloured expander, for every vertex u, almost every other vertex v can be reached from u via a short rainbow path. They then follow a density increment argument, similar to the one described above, to find a rainbow subdivision of K_k.

This bound was subsequently improved to a tighter $O(n(\log n)^{53})$ by Jiang, Letzter, Methuku, and Yepremyan [48].

Theorem 6.7 (Jiang–Letzter–Methuku–Yepremyan [48]) Let $k \geq 3$. If G is an n-vertex properly edge-coloured graph with no rainbow K_k-subdivision, then $e(G) = O(n(\log n)^{53})$.

Like the improvement of [60] regarding the Turán number of tight cycles, the key here was to prove that in a given properly-coloured expander there is a rainbow K_k-subdivision. To do so, the authors considered almost-regular expanders (meaning that the ratio between the minimum and maximum degree is at most polylogarithmic in n), and proved an appropriate existence result for almost-regular expanders. Next, the authors used the fact that random walks mix rapidly in expanders, as well as tools from [46] regarding homomorphism counts, to find the desired rainbow K_k-subdivision in a sufficiently dense almost-regular expander. The bound in Theorem 6.7 was improved further, first by Tomon [85] and then by Wang [92]; more on this soon.

6.3 Cycles With Many Chords

The connection between random walks and almost-regular expanders, described above, was used recently by Draganić, Methuku, Munhá-Correia, and Sudakov [25] to prove the following.

Theorem 6.8 (Draganić–Methuku–Munhá-Correia–Sudakov [25]) There is a constant $c > 0$ such that every n-vertex graph with at least $c \cdot n(\log n)^8$ edges contains a cycle C with at least $|C|$ chords.

This result is tight up to the polylogarithmic factor. It is plausible that a linear bound in n suffices.

Question 6.9 Is there $c > 0$ such that every n-vertex graph with at least cn edges contains a cycle C with at least $|C|$ chords?

The significantly lower exponent of $\log n$ here, when compared with Theorem 6.7, is at least in part due to the use of a sharper existence result, showing that in a graph G there is an 100-almost regular (namely, the ration between the maximum and minimum degrees is at most 100) expander H with $d(H) = \Omega(d(G)/\log n)$.[7]

---

[7] the notion of expansion here is similar to the one used in [48].

Using, among other things, the fact that random walks mix rapidly in expanders, the authors show that a random walk W of appropriate length t is, with positive probability: self-avoiding (i.e. a path); spans more than $2t$ edges among the middle $t/2$ vertices; and has an edge between the first and last $t/4$ vertices. This clearly yields a cycle of length at most t with at least t chords, as required.

6.4 A Sampling Trick

Theorem 6.7 was recently improved upon by Tomon [85], who showed that $n(\log n)^{6+\Omega(1)}$ edges suffice to guarantee the existence of a rainbow K_k-subdivision.

Theorem 6.10 (Tomon [85]) *Let $k \geq 3$. If G is an n-vertex properly edge-coloured graph which has no rainbow K_k-subdivisions, then $e(G) \leq n \cdot (\log n)^{6+o(1)}$.*

A key component in Tomon's proof is a clever sampling trick, which allowed him to prove the following lemma:

> ✺ In a properly edge-coloured expander H (with an appropriate definition of expanders[8]), if U and C are random sets of vertices and colours, obtained by including each vertex or colour with probability p, independently, then for every vertex v at least $\Omega(n)$ vertices in H can be reached from v via a short rainbow path whose colours are in C and interior vertices are in U, as long as $d(H)$ is large enough with respect to p.

Let us say a few words about the proof of ✺. Using the 'sprinkling' method, think of U and as the union of smaller disjoint random sets $U_1, \ldots, U_\ell$, and similarly think of C as the union of smaller disjoint random sets $C_1, \ldots, C_\ell$. Now define B_i to be the set of vertices reachable from v via a rainbow path of length at most i with interior in $U_1 \cup \ldots \cup U_i$ and colours in $C_1 \cup \ldots \cup C_i$; importantly, vertices in B_i need not be in $U_1 \cup \ldots \cup U_i$. Now it suffices to show that the sets $|B_i|$ grow sufficiently rapidly, until reaching size $\Omega(n)$. To prove this, Tomon distinguishes between the cases where the neighbourhood of B_i has many vertices with few neighbours in B_i, and vice versa.

Lemma ✺ implies that many pairs of vertices have many rainbow paths joining them, which have pairwise disjoint interiors and pairwise disjoint colour sets. It is then not hard to find a rainbow clique subdivision. Variants of this lemma allowed Tomon to improve the bound on the rainbow Turán number of cycles to $\text{ex}^*(n, \mathcal{C}) \leq n(\log n)^{2+o(1)}$, and to obtain interesting Turán type results about triangulations of the cylinder and Möbius strip in 3-uniform hypergraphs.

6.5 Further Improvements on Rainbow Cycles and Clique Subdivisions

Tomon's bound on the number of edges guaranteeing a rainbow K_t-subdivision was subsequently improved by Wang [92], who showed that $n(\log n)^{2+\Omega(1)}$ edges suffice.

---

[8]Tomon defines an α-*maximal* graph to be a graph G satisfying $\frac{d(G)}{|G|^\alpha} \geq \frac{d(H)}{|H|^\alpha}$ for every subgraph $H \subseteq G$, and shows that α-maximal graphs have strong expansion properties. In his paper α is taken to be either $\Theta(1/\log n)$ or a constant, depending on the context.

Theorem 6.11 *Let $k \geq 3$. If G is an n-vertex properly edge-coloured graph with no rainbow K_k-subdivisions, then $e(G) \leq n \cdot (\log n)^{2+o(1)}$.*

He did so by optimising Tomon's argument in the context of rainbow clique subdivisions, showing that in a properly edge-coloured expander (Wang used yet another notion which is quite similar to that in Definition 2.2), for every vertex v, if W is a small set of forbidden colours and C is a random colour set obtained by including each colour with probability $1/2$, independently, then, with high probability, more than half the vertices in the graph are reachable from v via a short rainbow path using colours in $C - W$. This readily implies that if W is a small set of forbidden colours, then every two vertices can be joined by a rainbow path. One can then construct a rainbow K_k-subdivision greedily.

Very recently, Tomon's bound on the rainbow Turán number of cycles was improved slightly, to $\text{ex}^*(n, \mathcal{C}) = O(n(\log n)^2)$, by Janzer and Sudakov [47] and, independently, by Kim, Lee, Liu, and Tran [51].

Theorem 6.12 (Janzer–Sudakov [47] and Kim–Lee–Liu–Tran [51]) *Let G be an n-vertex properly edge-coloured graph with no rainbow cycles. Then $e(G) = O(n \cdot (\log n)^2)$.*

Neither paper used expanders. Instead, they used inequalities regarding homomorphism counts, reminiscent of Janzer's methods from [46], with [47] considering a weighted count, and [51] using an unweighted count along with regularisation.

6.6 An Almost Tight Result Regarding Rainbow Cycles

Even more recently[9], Alon, Bucić, Sauermann, Zakharov, and Zamir [4] proved the following almost tight bound on the number of edges in a properly coloured graph with no rainbow cycles.

Theorem 6.13 (Alon–Bucić–Sauermann–Zakharov–Zamir [4]) *Let G be a properly edge-coloured n-vertex graph with no rainbow cycles. Then $e(G) \leq O(n \log n \log \log n)$.*

In their proof of Theorem 6.13 the authors use the following notion of expanders.

Definition 6.14 (Alon–Bucić–Sauermann–Zakharov–Zamir [4]) *An n-vertex graph G is an ABSZZ-expander if, for every ε with $0 \leq \varepsilon \leq 1$, every subset $U \subseteq V(G)$ with $1 \leq |U| \leq n^{1-\varepsilon}$, and every subset $F \subseteq E(G)$ with $|F| \leq (\varepsilon/3)d(G)|U|$, we have $|N_{G-F}(U)| \geq (\varepsilon/3)|U|$.*

Observe that in an n-vertex ABSZZ-expander, vertex sets U of size at most $n^{0.99}$, say, expand linearly (namely $|N(U)| \geq \frac{1}{100}|U|$, because we can take $\varepsilon = \frac{1}{100}$), and set of size at most $n/2$ expand at a rate of at least $\frac{1}{\log n}$ (by taking $\varepsilon = \frac{1}{\log n}$).

These expanders are relevant in the setting of Theorem 6.13 due to the following lemma.

---
[9]and after this survey was submitted for publication

Lemma 6.15 (Alon–Bucić–Sauermann–Zakharov–Zamir [4]) *Every graph G with at least one edge contains an ABSZZ-expander H with*

$$d(H) \geq \frac{1}{3} \cdot \frac{\log |H|}{\log |G|} \cdot d(G).$$

In particular, this lemma implies that if $d(G) \geq 3c \cdot \log |G| \log \log |G|$ then there is an ABSZZ-expander $H \subseteq G$ with $d(H) \geq c \cdot \log |H| \log \log |H|$. Thus, it suffices to show that, if H is a properly coloured n-vertex ABSZZ-expander with $d(H) \geq c \cdot \log n \log \log n$, where c is a large constant, then H has a rainbow cycle. Similarly to Tomon [85] (see ✽), the authors prove the following.

> ✽ Let H be an n-vertex properly coloured ABSZZ-expander with $d(H) \geq c \cdot \log n \log \log n$, where c is a large constant, and let v be a vertex in H. Let C be a random set of colours, obtained by including each colour in H with probability $1/2$, independently. Then, with probability larger than $1/2$, at least $\frac{n+1}{2}$ vertices in H can be reached from v via a rainbow path whose colours are in C.

Notice that ✽ readily implies Theorem 6.13. Indeed, as pointed out, it suffices to show that every expander H as in ✽ contains a rainbow cycle. Let $\{C_1, C_2\}$ be a random partition of the colour set of H. Then, by ✽, with positive probability the sets U_1, U_2, of vertices that are reachable from v through a rainbow path with colours in C_1, C_2, respectively, satisfy $|U_1|, |U_2| > n/2$. Fix such a partition $\{C_1, C_2\}$, and let $u \in U_1 \cap U_2$. Then there is a rainbow closed walk through v and u, which in turn contains a rainbow cycle, as required.

A key novelty in the proof of ✽ is another sampling trick. Define C_0 as above, namely to include each colour in H with probability $1/2$, independently. For appropriate t and p, define $C_1, \ldots, C_t$ so that C_i is obtained from C_{i-1} by removing each element of C_{i-1} with probability p, independently. Now define U_i to be the set of vertices reachable from v via a rainbow path with colours in C_i. Roughly speaking, the authors of [4] show that, in expectation,

$$|U_{i-1}| \geq (1 + \Omega(\varepsilon))|U_{i+1}|, \tag{6.1}$$

where ε satisfies $|U_i| = n^{1-\varepsilon}$. Assuming the U_i's indeed follow this behaviour (not just in expectation), then the parameters are such that $|U_0| > n/2$, as required. Turning this into an actual proof involves various probabilistic ideas that we do not mention here. Instead, we sketch how to prove that (6.1) holds in expectation.

Fix U_i, a corresponding ε (so that $|U_i| = n^{1-\varepsilon}$), and C_i. Then every colour in C_i is not in C_{i+1} with probability p, and one can check that every colour not in C_i is in C_{i-1} with probability $\Omega(p)$. The authors first prove a dichotomy, somewhat in spirit of ◆ below, according to which there are many edges E leaving U_i such that either E is coloured in C_i and no vertex in U_i is incident to many edges of E, or no edges in E are coloured in C_i and no vertex outside of U_i is incident to many edges in E.

Suppose the latter holds. Notice that every vertex outside U_i that is incident to an edge in E whose colour is in C_{i-1} is in U_{i-1}. Hence, using that E is large, no vertex outside of U_i is incident with many edges in E, and that the colour of every edge in E is in C_{i-1} with probability $\Omega(p)$, we get that, in expectation,

$|U_{i-1}| \geq (1+\Omega(\varepsilon))|U_i|$. Now consider the former case. Observe that if $u \in U_i$ is incident to an edge uw in E whose colour is not in C_{i+1} then $u \notin U_{i+1}$. Indeed, otherwise w would be in U_i (because there would be a rainbow path from v to u coloured in C_{i+1}, which can be extended by uw to a rainbow path from v to w coloured in C_i), a contradiction. Similar arguments to the previous case now show that, in expectation, $|U_i| \geq (1+\Omega(\varepsilon))|U_{i+1}|$. Either way, (✱) holds.

6.7 Cycles With All Diagonals

A *diagonal* in a cycle C is a chord joining two vertices at distance $\lfloor |C|/2 \rfloor$ in C. Erdős [27] asked the following question regarding cycles with all diagonals: what is the maximum number of edges in a graph on n vertices that has no cycle containing all diagonals? Notice that a cycle with all diagonals contains every cycle of length

Figure 10: A cycle with all diagonals

4. Hence, since there are graphs on n vertices with $\Omega(n^{3/2})$ edges and no 4-cycles, the answer is at least $\Omega(n^{3/2})$. Erdős observed also that $K_{3,3}$ is a 6-cycle with all diagonals, and so the answer is at most the Turán number $\text{ex}(n, K_{3,3})$ of $K_{3,3}$, which is $O(n^{5/3})$. Very recently, Bradač, Methuku, and Sudakov [13] gave a tight answer to the above question (up to a constant factor).

Theorem 6.16 (Bradač–Methuku–Sudakov [13]) *Every n-vertex graph with no cycles containing all diagonals has $O(n^{3/2})$ edges.*

While the proof of Theorem 6.16 uses expanders, it does not fall under the scope of this survey, as the expansion rate in their expanders is linear. Nevertheless, their proof uses the notion of expansion introduced by Tomon [85] in his proof of Theorem 6.10, and ideas of the author [60] used in her proof of Theorem 6.3, so we briefly sketch it.

An interesting new idea here is a way to find an almost spanning expander. They start with a bipartite n-vertex graph with at least $cn^{3/2}$ edges, for some (large) constant c. Within this graph, they find an expander (using Tomon's setup and an additional cleanup step) H, which has m vertices, at least $c'm^{3/2}$ edges for an appropriate constant c', and maximum degree $O(m^{1/2})$. They then define an auxiliary graph Γ, whose vertices are the edges of H, and where two edges xy and uv are joined if $(xyuv)$ is a 4-cycle in H. The main novelty in their proof is a way of finding an expander Γ' which is an almost spanning subgraph of Γ. This is achieved by showing that every set of vertices in Γ, which is not too large or too small, expands. They then use the expansion properties of Γ', and ideas from [60], to find an odd cycle $(e_1 \ldots e_\ell)$ in Γ, where the e_i's, viewed as edges of H, are pairwise vertex-disjoint.

This readily implies the existence of a 2ℓ-cycle in H that contains all diagonals (using that H is bipartite; see Figure 11).

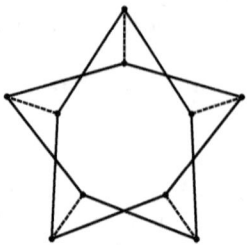

Figure 11: A 5-cycle in Γ whose vertices are pairwise vertex-disjoint edges in H (denoted by striped edges). The vertices of the 5-cycle are the diagonals in a 10-cycle in H.

7 Decompositions and Separation

All the results we have mentioned so far find certain substructures in graphs G with some properties (typically, with large average degree), which might be much smaller or sparse than G. In particular, these results do not aspire to say anything global about G. Often, they find an expander H within G, which might be much smaller than G, and focus just on H. In this section we will see how sublinear expanders can be used to prove more global results.

7.1 Decomposing Graphs Into Cycles and Edges

A *decomposition* of a graph G is a collection of subgraphs of G such that each edge in G appears in exactly one of these subgraphs. In the 1960's a lot of questions regarding decompositions of a given graph into simpler subgraphs were considered (see, e.g., [66, 39, 90]). One example of a conjecture along these lines is the following conjecture due to Erdős and Gallai.

Conjecture 7.1 (Erdős–Gallai (see [30])) *There is a constant $c > 0$ such that every n-vertex graph can be decomposed into at most cn cycles and edges.*

Notice that allowing edges in the decomposition is necessary, due to the potential existence of odd degree vertices. Moreover, at least $n - 1$ edges and cycles are necessary, as can be seen by considering a tree. A greedy algorithm that picks a longest cycle and removes it from the graph, one by one, yields a decomposition into $O(n \log n)$ cycles and edges. The first improvement on this was achieved in 2014 by Conlon, Fox, and Sudakov [20], who proved that $O(n \log \log n)$ cycles and edges suffice. This was recently improved upon significantly by Bucić and Montgomery [15].

Theorem 7.2 (Bucić–Montgomery [15]) *Every n-vertex graph can be decomposed into $O(n \log^* n)$ cycles and edges.*

Here $\log^* n$ is the *iterated logarithm*, namely the minimum number of times the (say) base 2 logarithm has to be applied, sequentially, to n, to reach a number which is at most 1.

To prove Theorem 7.2, the authors iterate the following statement:

❖ If G is an n-vertex graph with average degree d, then G can be decomposed into $O(n)$ cycles and $O\big(n(\log d)^{O(1)}\big)$ edges.

Notice that ❖ implies Theorem 7.2: iterating this statement $c\log^* n$ times, for large enough c, brings the average degree down from at most n to at most $O(1)$, using $O(n)$ edges at each step, and then the remaining edges can be covered by single edges, thereby using $O(n \log^* n)$ edges in total.

The high level sketch of the proof of ❖ is as follows.

- First, remove, one by one and as long as possible, cycles of length at least d.

- Next, decompose the remainder into expanders (on at most $O\big(d(\log d)^{O(1)}\big)$ vertices), which are almost vertex-disjoint.

- Finally, decompose each expander H into $O(|H|)$ cycles and $O\big(|H|(\log d)^{O(1)}\big)$ edges.

This outline is inspired by the proof of Conlon, Fox, and Sudakov [20], who used much denser expanders, to show that $O(n)$ cycles suffice to reduce the average degree from d to $O(d^{1-\Omega(1)})$. As such, the details are rather different, and quite a lot of new ideas were needed here.

The definition of expanders used in [15] is similar to the robust expanders defined in Definition 2.13.

Definition 7.3 (Bucić–Montgomery [15]) An n-vertex graph G is an (ε, s)-*BM-expander* if for every vertex set $U \subseteq V(G)$ and subgraph $F \subseteq G$, satisfying $|U| \leq 2n/3$ and $e(F) \leq s|U|$, the following holds: $|N_{G-F}(U)| \geq \frac{\varepsilon|U|}{(\log n)^2}$.

To realise the second step above, the authors prove a decomposition result.

Lemma 7.4 (Lemma 14 in [15]) *Let G be an n-vertex graph. Then it can be decomposed into (ε, s)-BM-expanders $G_1, \ldots, G_r$, with $\sum |G_i| = O(n)$, and $O(sn \log n)$ edges.*

To complete the second step, the authors apply the lemma with $s = 0$, and prove that every n-vertex $(\varepsilon, 0)$-expander contains a cycle of length $\Omega(n(\log n)^{-4})$, using a DFS algorithm. This shows that the expanders in the decomposition are rather small (they have at most $O\big(d(\log d)^{O(1)}\big)$ vertices), because the graph that remains after the first step has no cycles of length at least d.

The main challenge is in the proof of the final step. Applying the Lemma 7.4 again to each expander H, with $s = (\log n)^c$ for some constant c, it suffices to show that every n-vertex (ε, s)-expander can be decomposed into $O(n)$ cycles and at most $n(\log n)^{O(1)}$ edges.

A key result in this direction is the following (see Lemma 19 in [15]).

⋄ Given an n-vertex (ε, s)-expander H, if V is a random set of vertices that includes each vertex with probability $1/3$, independently, then, with high probability, for every vertex set U, and small subgraph F (of size at most $|U|/(\log n)^{27}$), more that $|V|/2$ vertices in V can be reached from U through a relatively short path with interior in V.

The proof of this uses the sampling trick of Tomon [85] described before (see Theorem 6.10 and ✿), as well as an expansion dichotomy for vertex sets U in (ε, s)-expanders (see Propositions 12 and 13 in [15]):

✯ If $|U| \leq 2n/3$ and $e(F) \leq s|U|/2$, then either the neighbourhood $N_{G-F}(U)$ is very large, or many vertices in the neighbourhood $N_{G-F}(U)$ have many neighbours in U.

The authors use this dichotomy to apply the union bound effectively. Using the key result ⋄, and a result about matchings in hypergraphs due to Aharoni and Haxell [2], they prove two results about joining pairs of vertices through a random vertex set, which they then use to complete the third step.

7.2 Separating the Edges of a Graph by Paths

A *separating path system* for a graph G is a collection $\mathcal{P}$ of paths such that for every two edges e and f there is a path $P \in \mathcal{P}$ that contains e but not f. This notion was introduced by Katona (2013). Writing $\text{sep}(G)$ for the size of a smallest separating path system for G, and $\text{sep}(n)$ for the maximum of $\text{sep}(G)$ over all n-vertex graphs, Katona asked to determine $\text{sep}(n)$. We claim that $\text{sep}(n) = O(n \log n)$. Indeed, given an n-vertex graph G, consider a collection $\mathcal{G}$ of $O(\log n)$ subgraphs of G that separates the edges of G, meaning that for every two edges e and f in G there is a subgraph in $\mathcal{G}$ that contains e but not f (that such a collection exists follows by noticing that every set of size m can be separated using $O(\log m)$ sets, an easy exercise). Using a result of Lovász [66], asserting that the edges of every n-vertex graph can be decomposed into at most n paths, for each $H \in \mathcal{G}$ there is a collection $\mathcal{P}_H$ of at most n paths decomposing the edges of H. It is easy to verify that $\bigcup_{H \in \mathcal{G}} \mathcal{P}_H$ is a separating path system for G, which has size $O(n \log n)$.

The first improvement on this initial bound was achieved by the author [59].

Theorem 7.5 (Letzter [59]) *Every n-vertex graph has a separating path system of size $O(n \log^\star n)$.*

The proof draws on many ideas from Bucić and Montgomery [15]. One difference is that here, unlike in [15], one needs to work with relatively sparse graphs, making the decomposition lemma (Lemma 7.4) too weak at times. To overcome this, we introduce a variant of BM-expanders.

Definition 7.6 (Letzter [59]) *An n-vertex graph G is an (ε, s, t)-L-expander if for every vertex set $U \subseteq V(G)$ and subgraph $F \subseteq G$, such that $1 \leq |U| \leq 2n/3$ and $|F| \leq s \cdot \min\{|U|, t\}$, we have $|N_{G-F}(U)| \geq \frac{\varepsilon |U|}{(\log |U|+1)^2}$.*

We also prove an analogue of Lemma 7.4.

Lemma 7.7 *If G is an n-vertex graph, then it can be decomposed into (ε, s, t)-L-expanders $G_1, \ldots, G_r$ with $\sum |G_i| = O(n)$ and $O(sn(\log t)^2)$ edges.*

The role of the parameter t is to allow for the expanders in the decomposition to have good expansion properties for small sets U, while not causing too many edges in G to remain uncovered.

To prove Theorem 7.5, a basic idea is that it is easy to find, given an n-vertex graph G, collections of paths $\mathcal{P}$ and matchings $\mathcal{M}$, of size $O(n)$, that separate the edges of G. As such, using the decomposition lemma, it suffices to be able to extend a given matching M to a path that avoids a small subgraph (corresponding to paths in $\mathcal{P}$ that intersect M). This is achieved for dense expanders using variants of tools from [15], and for sparse graphs using methods developed for KS-expanders, dealing separately with edges touching large degree vertices.

Shortly after [59] appeared, Bonamy, Botler, Dross, Naia, and Skokan [12] proved that $sep(n) \leq 19n$, using a simple inductive argument. This is tight up to the factor 19, confirming a conjecture from [7] and [33]. They raised the following question.

Question 7.8 (Bonamy–Botler–Dross–Naia–Skokan [12]) *Is there a constant $c > 0$ such that the edges of every properly edge-coloured n-vertex graph can be covered by $O(n)$ rainbow paths?*

This can be shown to imply the existence of a separating path system of size $O(n)$, via Lovász's result about decomposing a graph into paths.

8 Counting Hamiltonian Sets

An interesting different direction that has been addressed using sublinear expander involves the number of *Hamiltonian sets*. In a graph G, a set of vertices U is called *Hamiltonian* if $G[U]$ has a Hamiltonian cycle. Denote by $h(G)$ the number of Hamiltonian sets in G.

8.1 Maximising $h(G)$ Among Graphs With Given Average Degree

Komlós (see [86, 87, 88]) conjectured that among graphs G with minimum degree at least d, the complete graph K_{d+1} minimises $h(G)$; namely: if $\delta(G) \geq d$ then $h(G) \geq h(K_{d+1})$. In 2017, Kim, Liu, Sharifzadeh, and Staden [52] proved this conjecture (for large d).

Theorem 8.1 (Kim–Liu–Sharifzadeh–Staden [52]) *Let d be large, and suppose that G is a graph with $d(G) \geq d$. Then G has at least $h(K_{d+1}) = 2^{d+1} - \binom{d+1}{2} - (d+1) - 1$ Hamiltonian sets.*

In fact, they prove that if $d(G) \geq d$ and G is not isomorphic to K_{d+1} or to the union of two K_{d+1}'s that share a single vertex, then $h(G) \geq (2 + o(1))2^{d+1}$ (notice that $h(K_{d+1}) \approx 2^{d+1}$ and $h(H) \approx \frac{3}{2} \cdot 2^{d+1}$ for H the union of two K_{d+1}'s sharing a single vertex). They also sketch how their methods can be used to address a bipartite version of the same problem.

As a first step, the authors find an n-vertex $(\varepsilon_1, \varepsilon_2 d)$-expander H with $\delta(H) = \Omega(d)$ and $\Delta(H) = O(d)$, with $d = o(n)$. To do so, suppose there is no such expander. By analysis of "blocks" (maximal 2-connected subgraphs), assuming that G is a minimal counterexample to Theorem 8.1, and by a proof of Theorem 8.1 for dense graphs (using the regularity lemma), they may essentially assume that G is 2-connected and of order $\omega(d)$. Now, after removing the few large degree vertices, either there are two disjoint dense subexpanders H_1, H_2 with average degree a at least a bit below d, or there is a subexpander H with the required properties. Either way, this leads to a contradiction: either to the G being a counterexample, or to there being no expanders as above.

Now, given an almost-regular expander H as above, the authors again consider two cases: dense ($d \geq (\log n)^c$) and sparse (otherwise). In the sparse case, they find a set Z of $200d$ vertices that are far apart, and show that for every $U \subseteq Z$ of size $100d$ there is a cycle C with $V(C) \cap Z = U$, yielding $\binom{200d}{100d}$ Hamiltonian sets. This can be done in a routine way, by joining pairs of vertices in U by short paths avoiding the vicinity of other vertices in U. The dense case is a little more involved. They first find $200d$ many (h_0, h_1, h_2, h_3)-webs (see Definition 2.15; here $h_0, h_1, h_3 = (\log n)^{\Theta(1)}$ and $h_2 = \Theta(d)$), whose sets of non-leaf vertices are pairwise disjoint. Denoting the set of roots by Z, for every $U \subseteq Z$ of size $100d$, they find a cycle C that avoids $Z - U$ and contains at least $98d$ vertices of U (this yields at least the following number of different Hamiltonian sets $\binom{200d}{100d}/\binom{102d}{2d} \geq \Theta(d^{-1/2}2^{200d}/2^{102d}) \geq 2^{50d}$, with room to spare in the last inequality). To do so, they join vertices in U one by one through the webs, avoiding vertices with overused leaf sets.

8.2 Maximising $h(G)$ Among Graphs With Given Average Degree and Number of Vertices

Cambie, Gao, and Liu [17] considered a similar question: among n-vertex graphs G with $\delta(G) \geq d$, how small can $h(G)$ be? To describe their result, we need the notion of 'crux', given in Definition 2.9.

Theorem 8.2 (Cambie–Gao–Liu [17]) *Let G be a graph, $d = d(G)$, and $t = c_{1/5}(G)$. Then*
$$h(G) \geq n \cdot 2^{\Omega(t (\log t)^{-16})}.$$

At least in some regimes, this is tight up to the polylogarithmic factor, as can be seen by considering a disjoint union of $\lfloor n/(d+1) \rfloor$ copies of K_{d+1}.

The main technical result towards the proof of the above is the following.

✤ If $|G| = n$, $d(G) \geq d$, and $t = c_\alpha(G)$, then there is a vertex in at least $2^{\Omega(t(\log t)^{-c})}$ different Hamiltonian sets.

To prove ✤, they consider an $(\varepsilon_1, \varepsilon_2 d)$-KS-expander H on k vertices, find many disjoint cycles of length $(\log k)^{\Theta(1)}$, and join $k(\log k)^{-\Theta(1)}$ into a 'chain' by vertex-disjoint paths (see Figure 12). Then, every vertex in one of the connecting paths is in $2^{k(\log k)^{-\Theta(1)}}$ different Hamiltonian sets, because in forming a cycle through the chain, for each cycle there are two possible choices of a path through it to the next connecting path. By the definition of crux, we have $k \geq t$, and hence there is a vertex in at least the required number of different Hamiltonian sets.

Now, to get the main result, consider a graph G which is a minimal counterexample to the main result. First, remove, one by one, vertices which are in many different

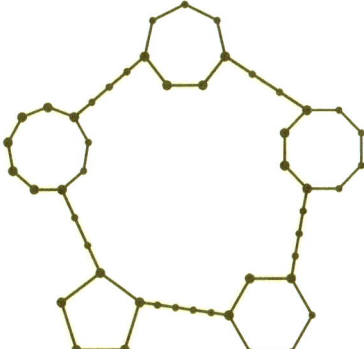

Figure 12: A 'chain' of cycles. Each of the vertices in the 'connecting paths' is in 32 cycles; one of them is marked in grey.

Hamiltonian sets, and denote the sets of removed vertices by S. Notice that $G - S$ has small average degree, because otherwise one could apply ✣ to a subexpander of $G - S$ with large enough average degree, contradicting the choice of S. Next, repeat a similar procedure in $G[S, V(G) - S]$, and move, one by one, a vertex from $V(G) - S$ to S if if is in many different Hamiltonian sets in $G[S, V(G) - S]$. If the resulting S is large, then $h(G)$ is large enough for a contradiction. Otherwise, it is small but has large average degree. This gives an upper bound on the crux of G, which together with the minimality assumption on G yields a contradiction.

Recall that Theorem 8.2 is close to tight, but unlikely to be tight. As such, it might be interesting to try to improve the bound.

Question 8.3 *Prove a tighter lower bound on $h(G)$ for an n-vertex graph G with average degree at least d. In particular, is it true that $h(G) \geq n \cdot 2^{\Omega(t)}$, where $t = c_{1/5}(G)$?*

9 Other Results

In this section we briefly mention a few other results, that used sublinear expanders in their proof. The first two are extremal results, showing that a large average degree implies the existence of a certain structure, the third is a result in Ramsey theory.

9.1 Nested Cycles

In 1975 Erdős [27] made two conjectures about the existence of nested cycles in graph with large average degree. The first conjecture asserts that there exists d_0 such that every graph G with $d(G) \geq d_0$ contains two edge-disjoint cycles C_1, C_2 such that $V(C_2) \subseteq V(C_1)$. The second is a strengthening of the former, additionally requiring that the cyclic ordering of C_2 respects that of C_1. More precisely, if $C_1 = (v_1 \ldots v_\ell)$ then there exist $i_1, \ldots, i_k$, such that $1 \leq i_1 < \ldots < i_k \leq \ell$ and $C_2 = (v_{i_1} \ldots v_{i_k})$. We refer to a sequence of nested cycles $C_1, \ldots, C_k$ (with $V(C_1) \supseteq \ldots \supseteq V(C_k)$), where the cyclic order of C_{i+1} respects that of C_i, as a sequence of *nested cycles without crossings* (see Figure 13).

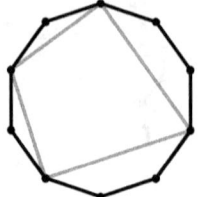 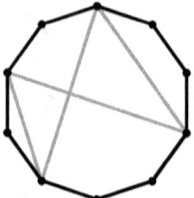

Figure 13: Two pairs of nested cycles (without and with crossings)

The former conjecture was proved by Bollobás [10] in 1978, and extended by Chen, Erdős, and Staton [18] in 1996 to longer sequences of nested cycles[10]. Gil Fernández, Kim, Kim, and Liu [36] recently proved the latter conjecture.

Theorem 9.1 (Gil Fernández–Kim–Kim–Liu [36]) *There exists d_0 such that every graph G with $d(G) \geq d_0$ has two nested cycles without crossings.*

The authors do not give an explicit value of d_0 for which the statement holds. Roughly speaking, their proof yields a value of d_0 which is large, but not regularity lemma large. I am not aware of any non-trivial lower bounds on d_0 for which the statement in Theorem 9.1 holds.

Very briefly, the idea here is to take a shortest cycle $C = (v_1 \ldots v_\ell)$ in an expander H, which is going to be the shorter cycle C_2, and then it suffices to find paths $P_1, \ldots, P_\ell$, whose interiors are pairwise vertex-disjoint and vertex-disjoint of C, such that P_i joins v_i with v_{i+1} (where $v_{\ell+1} := v_1$). To find these paths, they first join each vertex v_i to either two vertices of large degree, or to two large sets with small diameter (they call this structure a 'kraken'). Next, they join up these vertices of large degree, or the large sets with small diameter, to obtain the required connections between the vertices.

Theorem 9.1 raises the following interesting question.

Question 9.2 *Does there exist $d_0(k)$ such that, for every $k \geq 3$, every graph G with $d(G) \geq d_0(k)$ has a sequence $C_1, \ldots, C_k$ of nested cycles with no crossings?*

9.2 Pillars

In 1989 Thomassen [84] conjectured that large average degree guarantees the existence of a 'pillar', defined to consist of two vertex-disjoint cycles of the same length, denoted $C_1 = (v_1 \ldots v_\ell)$ and $C_2 = (u_1 \ldots u_\ell)$, and vertex-disjoint paths $P_1, \ldots, P_\ell$ of the same length such that P_i joins v_i with u_i (see Figure 14).

His conjecture was confirmed by Gil Fernández and Liu [37].

---

[10]Bollobás [10] proved that an average degree of 14 guarantees the existence of two edge-disjoint nested cycles, and this was improved by Chen, Erdős, and Staton [18] to 10. For general k they showed that an average degree of 6^k guarantees the existence of a sequence of k nested edge-disjoint cycles.

Sublinear Expanders and Their Applications

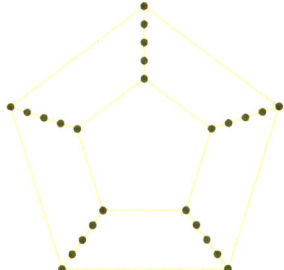

Figure 14: A pillar

Theorem 9.3 (Gil Fernández and Liu [37]) *There exists d_0 such that every graph G with $d(G) \geq d_0$ contains a pillar.*

In fact, the authors proved a more general result, about the existence of 'K_k-pillars' (involving k cycles, any two of which are joined as above).

To prove this, the authors work with an expander H which is Q_3-free (because the hypercube Q_3 contains a pillar). They then find many disjoint large 'krakens' (the structures used in [36]). Two of these would have the same cycle length, and then the kraken's 'legs' can be joined together by disjoint paths of the same length, using the approach of Liu and Montgomery [63].

9.3 Ramsey Goodness of Cycles

A final, quite different result we mention here involves Ramsey numbers. The *Ramsey number* of H_1 and H_2, denoted $r(H_1, H_2)$, is the minimum n such that every red-blue colouring of K_n contains a red copy of H_1 or a blue copy of H_2. Here we consider the case where H_1 is a cycle. It is well-known (and easy to see) that

$$r(C_n, H) \geq (n-1)(\chi(H) - 1) + \sigma(H), \tag{9.1}$$

where $\chi(H)$ is the chromatic number of H, and $\sigma(H)$ is the minimum possible size of a colour class in a proper $\chi(H)$-colouring of H. Burr [16] proved that, if n is sufficiently large in terms of H, then the bound in (9.1) is tight (in which case it is said that C_n is *H-good*). Allen, Brightwell, and Skokan [3] conjectured that the bound is tight already when $n \geq |H| \cdot \chi(H)$. Haslegrave, Hyde, Kim, and Liu [41] proved an even stronger statement, as follows.

Theorem 9.4 (Haslegrave–Hyde–Kim–Liu [41]) *There exists a constant $c > 0$ such that, if $n \geq c|H|(\log \chi(H))^4$, then $r(C_n, H) = (n-1)(\chi(H) - 1) + \sigma(H)$.*

Their approach uses ideas from Liu and Montgomery [63], specifically about the construction of adjusters (see Definition 5.5) in sublinear expanders, and is also inspired by a result of Pokrovskiy and Sudakov [75] who proved that (9.1) is tight when $n \geq 10^{60}|H|$ and $\sigma(H) \geq \chi(H)^{22}$.

The lower bound on n in Theorem 9.4 is close to optimal; the authors provide an example showing that n has to be taken to be at least $|H|(1 + o(1))$. It would be nice to get rid of the polylogarithmic factor in $\chi(H)$.

Question 9.5 (Haslegrave–Hyde–Kim–Liu [41]) *Is there a constant $c > 0$ such that, if $n \geq c|H|$, then $r(C_n, H) = (n-1)(\chi(H)-1) + \sigma(H)$?*

9.4 Transversals in Latin Squares

A *Latin square* is an $n \times n$ grid filled with n symbols so that each symbol appears exactly once in each row and column. A *transversal* in a Latin Square is a collection of cells that share no symbol, row, or column. A transversal in an $n \times n$ Latin square is said to be *full* if it has size n. (See Figure 15.)

0	1	**2**	3
1	2	3	0
2	**3**	0	1
3	0	1	2

0	1	2	3	**4**
1	2	3	4	0
2	**3**	4	0	1
3	4	**0**	1	2
4	0	1	**2**	3

Figure 15: A 4×4 Latin square with a transversal of size 3 (and with no full transversals), and a 5×5 Latin square with a full transversal

The study of Latin Squares dates back to the 18th century, during which Euler [32] considered Latin squares that can be decomposed into full transversals, and has been very prolific (see the surveys of Andersen [6], Wanless [93], and a forthcoming survey of Montgomery [70]). Ryser [76], Brualdi (see [14]), and Stein [79] made several related conjectures, which are now known in a combined form as the *Ryser–Brualdi–Stein conjecture*.

Conjecture 9.6 (Ryser–Brualdi–Stein conjecture [76, 14, 79]) *Every $n \times n$ Latin square has a transversal of size $n-1$, and a full transversal if n is odd.*

In a very recent breakthrough paper, Montgomery [72] proved the first part of this well-known conjecture, for large n.

Theorem 9.7 (Montgomery [72]) *For every large enough n, every $n \times n$ Latin square has a transversal of size $n-1$.*

The substantial proof of Theorem 9.7 contains many ideas and techniques. One of them is the use of sublinear expanders, similar to (δ, n)-SS-expanders (see Definition 3.4) and to (λ, d)-ST-expanders (see Definition 6.2). Due to the length of the paper [72], and the fact that this paper came out after the current survey was submitted for publication, we do not elaborate on the use of expanders in this proof. One can of course refer to [72] for more details, as well as to Montgomery's forthcoming survey [70].

Acknowledgements

My research is supported by the Royal Society. I would like to thank the anonymous referee for many useful and insightful comments. I would also like to thank Alp Müyesser for pointing out some typos.

References

[1] F. N. Abu-Khzam and M. A. Langston, *Graph coloring and the immersion order*, International Computing and Combinatorics Conference, Springer, 2003, pp. 394–403. 102

[2] R. Aharoni and P. Haxell, *Hall's theorem for hypergraphs*, J. Graph Theory **35** (2000), 83–88. 118

[3] P. Allen, G. Brightwell, and J. Skokan, *Ramsey-goodness–and otherwise*, Combinatorica **33** (2013), 125–160. 123

[4] N. Alon, M. Bucić, L. Sauermann, D. Zakharov, and O. Zamir, *Essentially tight bounds for rainbow cycles in proper edge-colourings*, arXiv:2309.04460 (2023). 113, 114

[5] N. Alon, M. Krivelevich, and B. Sudakov, *Turán numbers of bipartite graphs and related Ramsey-type questions*, Combin. Probab. Comput. **12** (2003), 477–494. 107

[6] L. D. Andersen, *The history of latin squares*, Combinatorics: Ancient & Modern (edited by R. Wilson and J.J. Watkins), 2013. 124

[7] J. Balogh, B. Csaba, R. R. Martin, and A. Pluhár, *On the path separation number of graphs*, Discr. Appl. Math. **213** (2016), 26–33. 119

[8] J. Balogh, H. Liu, and M. Sharifzadeh, *Subdivisions of a large clique in C_6-free graphs*, J. Combin. Theory Ser. B **112** (2015), 18–35. 93

[9] L. A. Bassalygo and M. S. Pinsker, *The complexity of an optimal non-blocking commutation scheme without reorganization*, Problemy Peredači Informacii **9** (1973), 84–87, Translated into English in Problems of Information Transmission, 9 (1974) 64-66. 89

[10] B. Bollobás, *Nested cycles in graphs*, Problémes combinatoires et théorie des graphes Colloq. Internat. CNRS, Univ. Orsay, Orsay, 1976), Colloq. Internat. CNRS, 260, 1978, pp. 49–50. 122

[11] B. Bollobás and A. Thomason, *Highly linked graphs*, Combinatorica **16** (1996). 313–320. 91

[12] M. Bonamy, F. Botler, F. Dross, T. Naia, and J. Skokan, *Separating the edges of a graph by a linear number of paths*, arXiv:2301.08707 (2023). 119

[13] D. Bradač, A. Methuku, and B. Sudakov, *The extremal number of cycles with all diagonals*, arXiv:2308.16163 (2023). 115

[14] R. A. Brualdi and H. J. Ryser, *Combinatorial matrix theory*, Cambridge University Press, 1991. 124

[15] M. Bucić and R. Montgomery, *Towards the Erdős-Gallai Cycle Decomposition Conjecture*, arXiv:2211.07689 (2022). 116, 117, 118, 119

[16] S. A. Burr, *Ramsey numbers involving graphs with long suspended paths*, J. London Math. Soc. **2** (1981), 405–413. 123

[17] S. Cambie, J. Gao, and H. Liu, *Many Hamiltonian subsets in large graphs with given density*, arXiv:2301.07467 (2023). 120

[18] G. Chen, P. Erdős, and W. Staton, *Proof of a conjecture of Bollobás on nested cycles*, J. Combin. Theory Ser. B **66** (1996), 38–43. 122

[19] K. L. Collins and M. E. Heenehan, *Constructing graphs with no immersion of large complete graphs*, J. Graph Theory **77** (2014), 1–18. 102

[20] D. Conlon, J. Fox, and B. Sudakov, *Cycle packing*, Random Structures Algorithms **45** (2014), 608–626. 116, 117

[21] M. DeVos, Z. Dvořák, J. Fox, J. McDonald, B. Mohar, and D. Scheide, *A minimum degree condition forcing complete graph immersion*, Combinatorica **34** (2014), 279–298. 101, 102

[22] M. DeVos, K.-I. Kawarabayashi, B. Mohar, and H. Okamura, *Immersing small complete graphs*, Ars Math. Contemp. **3** (2010), 139–146. 102

[23] M. DeVos, J. McDonald, B. Mohar, and D. Scheide, *Immersing complete digraphs*, Europ. J. Combin. **33** (2012), 1294–1302. 103

[24] _____, *A note on forbidding clique immersions*, Electron. J. Combin. **20** (2013), P55. 103

[25] N. Draganić, A. Methuku, D. Munhá-Correia, and B. Sudakov, *Cycles with many chords*, arXiv:2306.09157 (2023). 111

[26] Z. Dvořák and L. Yepremyan, *Complete graph immersions and minimum degree*, J. Graph Theory **88** (2018), 211–221. 101, 102

[27] P. Erdős, *Some recent progress on extremal problems in graph theory*, Congr. Numer. **14** (1975), 3–14. 104, 115, 121

[28] _____, *Some new and old problems on chromatic graphs*, Combinatorics and applications, 1984, pp. 118–126. 104

[29] P. Erdős and T. Gallai, *On maximal paths and circuits of graphs*, Acta Math. Acad. Sci. Hungar. **10** (1959), 337–356. 95

[30] P. Erdős, A. W. Goodman, and L. Pósa, *The representation of a graph by set intersections*, Canad. J. Math. **18** (1966), 106–112. 116

[31] P. Erdős and A. Hajnal, *On chromatic number of graphs and set-systems*, Acta Math. Acad. Sci. Hungar. **17** (1966), 61–99. 104

[32] L. Euler, *Recherches sur un nouvelle espéce de quarrés magiques*, Verhandelingen uitgegeven door het zeeuwsch Genootschap der Wetenschappen te Vlissingen (1782), 85–239. 124

[33] V. Falgas-Ravry, T. Kittipassorn, D. Korándi, S. Letzter, and B. Narayanan, *Separating path systems*, J. Combin. **5** (2014), 335–354. 119

[34] S. Fiorini, G. Joret, D. O. Theis, and D. R. Wood, *Small minors in dense graphs*, Europ. J. Combin. **33** (2012), 1226–1245. 99

[35] I. Gil Fernández, J. Hyde, H. Liu, O. Pikhurko, and Z. Wu, *Disjoint isomorphic balanced clique subdivisions*, J. Combin. Theory Ser. B **161** (2023), 417–436. 107, 108

[36] I. Gil Fernández, J. Kim, Y. Kim, and H. Liu, *Nested cycles with no geometric crossings*, Proc. Am. Math. Soc. Ser. B **9** (2022), 22–32. 122, 123

[37] I. Gil Fernández and H. Liu, *How to build a pillar: A proof of Thomassen's conjecture*, J. Combin. Theory Ser. B **162** (2023), 13–33. 122, 123

[38] A. Girão and S. Letzter, *Immersion of complete digraphs in Eulerian digraphs*, Israel J. Math. (2023). 103

[39] F. Harary, *Covering and packing in graphs, I.*, Annals of the New York Academy of Sciences **175** (1970), 198–205. 116

[40] J. Haslegrave, J. Hu, J. Kim, H. Liu, B. Luan, and G. Wang, *Crux and long cycles in graphs*, SIAM J. Discr. Math. **36** (2022), 2942–2958. 95, 96

[41] J. Haslegrave, J. Hyde, J. Kim, and H. Liu, *Ramsey numbers of cycles versus general graphs*, Forum Math. Sigma **11** (2023), e10. 123, 124

[42] J. Haslegrave, J. Kim, and H. Liu, *Extremal density for sparse minors and subdivisions*, Int. Math. Res. Not. **2022** (2022), 15505–15548. 96

[43] S. Hoory, N. Linial, and A. Wigderson, *Expander graphs and their applications*, Bull. Am. Math. Soc. **43** (2006), 439–561. 89

[44] S. Im, J. Kim, Y. Kim, and H. Liu, *Crux, space constraints and subdivisions*, arXiv:2207.06653 (2022). 95

[45] B. Janzer, *Large hypergraphs without tight cycles*, arXiv:2012.07726 (2020). 108, 110

[46] O. Janzer, *Rainbow Turán number of even cycles, repeated patterns and blow-ups of cycles*, Israel J. Math. **253** (2023), 813–840. 110, 111, 113

[47] O. Janzer and B. Sudakov, *On the Turán number of the hypercube*, arXiv:2211.02015 (2022). 113

[48] T. Jiang, S. Letzter, A. Methuku, and L. Yepremyan, *Rainbow clique subdivisions*, arXiv:2108.08814 (2021). 111

[49] T. Jiang, A. Methuku, and L. Yepremyan, *Rainbow turán number of clique subdivisions*, Europ. J. Combin. **110** (2023), 103675. 110

[50] P. Keevash, D. Mubayi, B. Sudakov, and J. Verstraëte, *Rainbow Turán problems*, Combin. Probab. Comput. **16** (2007), 109–126. 110

[51] J. Kim, J. Lee, H. Liu, and T. Tran, *Rainbow cycles in properly edge-colored graphs*, (2022), arXiv:2211.03291. 113

[52] J. Kim, H. Liu, M. Sharifzadeh, and K. Staden, *Proof of Komlós's conjecture on Hamiltonian subsets*, Proc. London Math. Soc. **115** (2017), 974–1013. 119

[53] J. Komlós and E. Szemerédi, *Topological cliques in graphs*, Combin. Probab. Comput. **3** (1994), 247–256. 89, 90, 91, 92, 93

[54] _____, *Topological cliques in graphs II*, Combin. Probab. Comput. **5** (1996), 79–90. 89, 91, 92, 93, 94, 104

[55] M. Krivelevich, *Expanders - how to find them, and what to find in them*, Surveys in Combinatorics 2019, 2019, pp. 115–142. 89

[56] D. Kühn and D. Osthus, *Large topological cliques in graphs without a 4-cycle*, Combin. Probab. Comput. **13** (2004), 93–102. 93

[57] _____, *Extremal connectivity for topological cliques in bipartite graphs*, J. Combin. Theory Ser. B **96** (2006), 73–99. 93

[58] F. Lescure and H. Meyniel, *On a problem upon configurations contained in graphs with given chromatic number*, Graph theory in memory of GA Dirac, Ann. Discrete Math., vol. 41, Elsevier, 1988, pp. 325–332. 102

[59] S. Letzter, *Separating paths systems of almost linear size*, arXiv:2211.07732 (2022). 118, 119

[60] _____, *Hypergraphs with no tight cycles*, Proc. Am. Math. Soc. **151** (2023), 455–462. 109, 111, 115

[61] H. Liu, *Robust sublinear expander*, (2020), https://homepages.warwick.ac.uk/staff/H.Liu.9/sublinear-expander-note-July16.pdf. 89

[62] H. Liu and R. Montgomery, *A proof of Mader's conjecture on large clique subdivisions in C_4-free graphs*, J. London Math. Soc. **95** (2017), 203–222. 93, 94, 95, 97, 108

[63] _____, *A solution to Erdős and Hajnal's odd cycle problem*, J. Am. Math. Soc. (2023). 104, 105, 107, 108, 123

[64] H. Liu, G. Wang, and D. Yang, *Clique immersion in graphs without a fixed bipartite graph*, J. Combin. Theory Ser. B **157** (2022), 346–365. 102

[65] W. Lochet, *Immersion of transitive tournaments in digraphs with large minimum outdegree*, J. Combin. Theory Ser. B **134** (2019), 350–353. 103

[66] L. Lovász, *On covering of graphs*, Theory of Graphs, Academic Press New York, 1968, pp. 231–236. 116, 118

[67] B. Luan, Y. Tang, G. Wang, and D. Yang, *Balanced subdivisions of cliques in graphs*, Combinatorica (2023), 1–23. 107

[68] A. Lubotzky, *Expander graphs in pure and applied mathematics*, Bull. Am. Math. Soc. **49** (2012), 113–162. 89

[69] W. Mader, *An extremal problem for subdivisions of K_5^-*, J. Graph Theory **30** (1999), 261–276. 93

[70] R. Montgomery, *Transversals in latin squares*, Surveys in Combinatorics. 124

[71] _____, *Logarithmically small minors and topological minors*, J. London Math. Soc. **91** (2015), 71–88. 93, 94, 100, 101

[72] _____, *A proof of the Ryser-Brualdi-Stein conjecture for large even n*, arXiv:2310.19779 (2023). 124

[73] G. Moshkovitz and A. Shapira, *Decomposing a graph into expanding subgraphs*, Random Structures Algorithms **52** (2018), 158–178. 92

[74] D. Mubayi, O. Pikhurko, and B. Sudakov, *Hypergraph Turán problem: Some open questions*, AIM workshop problem lists (2011), manuscript. 108

[75] A. Pokrovskiy and B. Sudakov, *Ramsey goodness of cycles*, SIAM J. Discr. Math. **34** (2020), 1884–1908. 123

[76] H. J. Ryser, *Neuere probleme der kombinatorik*, Vorträge über Kombinatorik, Oberwolfach **69** (1967), 91. 124

[77] P. C. Sarnak, *What is... an expander?*, Not. Am. Math. Soc. **51** (2004), 762–763. 89

[78] A. Shapira and B. Sudakov, *Small complete minors above the extremal edge density*, Combinatorica **35** (2015), 75–94. 89, 99, 100, 108, 109

[79] S. K. Stein, *Transversals of Latin squares and their generalizations*, Pacific J. Math. **59** (1975), 567–575. 124

[80] B. Sudakov and I. Tomon, *The extremal number of tight cycles*, Int. Math. Res. Not. **2022** (2022), 9663–9684. 108, 109, 110, 111

[81] E. Szemerédi, *Regular partitions of graphs*, Orsay, 1975. 92

[82] A. Thomason, *The extremal function for complete minors*, J. Combin. Theory Ser. B **81** (2001), 318–338. 99

[83] C. Thomassen, *Subdivisions of graphs with large minimum degree*, J. Graph Theory **8** (1984), 23–28. 105

[84] _____, *Configurations in graphs of large minimum degree, connectivity, or chromatic number*, Annals of the New York Academy of Sciences **555** (1989), 402–412. 122

[85] I. Tomon, *Robust (rainbow) subdivisions and simplicial cycles*, arXiv:2201.12309 (2022). 111, 112, 114, 115, 118

[86] Zs. Tuza, *Exponentially many distinguishable cycles in graphs*, J. Combin. Infor. Sys. Sci **15** (1990), 281–285. 119

[87] _____, *Unsolved combinatorial problems, Part I*, BRICS Lecture Series LS-01-1 (2001). 119

[88] _____, *Problems on cycles and colorings*, Discr. Math. **313** (2013), 2007–2013. 119

[89] J. Verstraëte, *Extremal problems for cycles in graphs*, Recent trends in combinatorics, Springer, 2016, pp. 83–116. 108

[90] V. G. Vizing, *On an estimate of the chromatic class of a p-graph*, Discret. Analiz. **3** (1964), 25–30. 116

[91] Y. Wang, *Balanced subdivisions of a large clique in graphs with high average degree*, arXiv:2107.06583 (2021). 107

[92] _____, *Rainbow clique subdivisions*, (2022), arXiv:2204.08804. 111, 112

[93] I. M. Wanless, *Transversals in latin squares: A survey*, (2011), 403–437. 124

Department of Mathematics
University College London
London WC1E 6BT, UK
s.letzter@ucl.ac.uk

Transversals in Latin Squares

Richard Montgomery

Abstract

A Latin square is an n by n grid filled with n symbols so that each symbol appears exactly once in each row and each column. A transversal in a Latin square is a collection of cells which do not share any row, column, or symbol. This survey will focus on results from the last decade which have continued the long history of the study of transversals in Latin squares.

1 Introduction

A *Latin square of order* n is an n by n grid filled with n symbols so that each symbol appears exactly once in each row and each column (see Figure 1). Important examples of Latin squares include the multiplication tables of finite groups, where the rows and columns of the Latin square of order n are both indexed by some group G with order n, and each entry is the product of its row with its column. In part due to their connection with magic squares, Latin squares have a long history dating back to early mathematics (for more on this we recommend the excellent historical survey by Andersen [10]). Euler initiated their more rigorous study in the 18th century, while, in modern mathematics, Latin squares have strong connections to design theory (as, for example, we will see in our discussion of Steiner triple systems) as well as links to 2-dimensional permutations, finite projective planes and error correcting codes (see [58, 104] for more on this).

A *partial transversal* in a Latin square is a collection of cells which share no row, column or symbol, while a *full transversal* is one which contains every symbol exactly once (sometimes referred to as simply a *transversal*). We will see that, when n is even, there are numerous Latin squares which may not contain a full transversal; for now, we recall only the canonical such example by remarking that it is easy to show that the Latin square corresponding to the addition table for $\mathbb{Z}_{2m}$ does not contain a full transversal for any $m \in \mathbb{N}$ (where a group is abelian we will use addition rather than multiplication). Indeed, suppose to the contrary that it did contain a full transversal, T. Then, the sum of the entries in T is the sum of the elements in $\mathbb{Z}_{2m}$, but also, by using the definition of the Latin square, the sum of the indices of the rows and the columns of the Latin square, and thus twice the sum of the elements of $\mathbb{Z}_{2m}$. That is, we have

$$\sum_{v \in \mathbb{Z}_{2m}} v = \sum_{x \in T} \text{entry}(x) = \sum_{x \in T} (\text{row}(x) + \text{column}(x)) = 2 \sum_{v \in \mathbb{Z}_{2m}} v.$$

As $\sum_{v \in \mathbb{Z}_{2m}} v = m$ in $\mathbb{Z}_{2m}$, this gives a contradiction, and thus the addition table for $\mathbb{Z}_{2m}$ contains no full transversal.

These examples already lead to two natural questions: Which Latin squares contain a full transversal? If a Latin square has no full transversal, then what is the size of a largest partial transversal? These are interesting and very challenging questions, and the progress on them will be the focus of this survey.

1	4	6	5	3	2
5	2	4	3	1	6
6	3	2	4	5	1
2	5	1	6	4	3
3	6	5	1	2	4
4	1	3	2	6	5

0	1	2	3	4	5
1	2	3	4	5	0
2	3	4	5	0	1
3	4	5	0	1	2
4	5	0	1	2	3
5	0	1	2	3	4

Figure 1: A Latin square of order 6 with a full transversal highlighted, and one with no full transversal (the addition table for $\mathbb{Z}_6$). For later illustration, an *intercalate*, or 2 by 2 Latin subsquare, is highlighted in the second Latin square.

In part due to the historical context, we will start by considering Latin squares with a much stronger property than simply the existence of a full transversal, those Latin squares which can be decomposed entirely into disjoint full transversals (see Section 2), while also discussing random Latin squares. Recently, Latin squares have been often studied using an equivalent formulation in edge-coloured complete bipartite graphs, and in Section 3 we will recall this formulation before using it to describe a large class of Latin squares with no full transversal. We will then discuss (in Section 4) Latin squares which are known to have a full transversal, before considering (in Section 5) large partial transversals in any Latin square and, in particular, the well-known Ryser-Brualdi-Stein conjecture. Finally, in Section 6, we will consider other problems related to the study of transversals in Latin squares.

Before continuing any further, we highlight two related surveys which, like the present survey, were written in connection with editions of the British Combinatorial Conference. Firstly, the 2011 survey by Wanless [104] also on 'Transversals in Latin squares'; while many of the questions and results covered by [104] are also covered here with their subsequent developments, other questions are also considered or are covered in more detail. Secondly, we discuss here the links to the study of rainbow subgraphs in edge-coloured graphs most closely related to transversals in Latin squares, but many further problems and more detailed discussion can be found in the 2022 survey by Pokrovskiy [90] on 'Rainbow subgraphs and their applications'.

2 Decompositions Into Full Transversals and Random Latin Squares

A Latin square decomposed into full transversals appears in a recreational mathematics problem which was popular from at least the 1720s when it appeared in the compendium 'Récréations mathématiques et physiques' by Jacques Ozanam [84]. Taking a standard deck of cards, can the aces, kings, queens and jacks be arranged into a 4 by 4 grid so that each column and each row contains an ace, king, queen and jack which are all from different suits? The answer is yes, and indeed a solution is given in Figure 2, displayed overleaf so as not to disappoint any enthusiastic readers. In such a solution, forgetting the suit of each card will give a Latin square of order 4, in which, furthermore, each suit marks out a full transversal. That is, it gives a Latin square of order 4 which can be decomposed into full transversals.

Euler [36] took this problem further around 1780, for Latin squares of order 6, in the form of his famous '36 officers problem'. In this problem, there are 36 officers of 6 different ranks from 6 different regiments, with one officer from each rank from each regiment, who are to stand in a 6 by 6 grid so that each row and column contains officers of different ranks from different regiments. To notate this, Euler used Latin letters to denote the regiments and Greek letters to denote the ranks; the Latin letters used thus form a Latin square, giving rise to the terminology 'Latin square'. While each rank (or Greek letter) would also mark out a full transversal and thus give a decomposition of a Latin square of order 6 into full transversals, the grid of ranks (or Greek letters) would also form a Latin square. We would say that the two Latin squares at play here are *orthogonal*: two Latin squares of order n are said to be *orthogonal* (sometimes *mutually orthogonal*) if the pairs of entries which appear in corresponding cells are distinct (so that all possible n^2 pairs appear). Finding two orthogonal Latin squares of order n is equivalent to finding one Latin square of order n that can be decomposed into full transversals.

Euler wrote that he could find no solution to his 36 officers problem, though could not prove none exists, and conjectured that, for each $n \equiv 2 \bmod 4$, there is no Latin square of order n which can be decomposed into disjoint full transversals [36]. This can immediately be seen to be true for $n = 2$, but it was not until 1900 that Tarry [100] confirmed this for $n = 6$, thus showing that Euler's 36 officers problem has no solution. More generally, however, Euler's conjecture is false! This was first shown by Bose and Shrikhande [17] in 1959 who constructed counterexamples for $n = 22$ and $n = 50$, before extending this with Parker [18] to show that Euler's conjecture is false for any $n \equiv 2 \bmod 4$ with $n \geq 10$.

Given this, an interesting question is to ask how common counterexamples are, and, in particular, is a random Latin square of order n likely to be a counterexample to Euler's conjecture? For each $n \in \mathbb{N}$, let $\mathcal{L}(n)$ be the set of Latin squares of order n which use the symbols in $[n] = \{1, 2, \ldots, n\}$, and let $L_n \in \mathcal{L}(n)$ be a Latin square chosen from $\mathcal{L}(n)$ uniformly at random. In 1990, van Rees [102] conjectured that a vanishingly small proportion of Latin squares have a decomposition into full transversals, or, equivalently, that the random Latin square L_n has no decomposition into full transversals with high probability. Wanless and Webb [105], however, observed in 2006 that numerical observations suggest the contrary, and, considering the recent advances in techniques and the results shown, the following conjecture now appears likely.

Conjecture 2.1 *The random Latin square L_n can be decomposed into full transversals with high probability.*

Kwan [66] showed in 2020 that the random Latin square L_n at least has a full transversal with high probability. Here, the rigidity of Latin squares make it difficult to study almost any non-trivial properties of a uniformly random Latin square. Kwan [66] approached the random Latin square L_n by approximating it via a modified random triangle removal process, adapting breakthrough methods of Keevash [56, 57] on the existence of designs and the analysis of the random triangle removal process carried out by Bohman, Friecze and Lubetzky [16]. To find a full transversal, Kwan [66] used the absorbing method, as codified by Rödl, Ruciński and Szemerédi [95] in 2006 (following earlier related work by Erdős, Gyárfás and Pyber [35] and Krivelevich [65]). As used in many different settings since, the aim here is find a small partial transversal (functioning as an absorption structure)

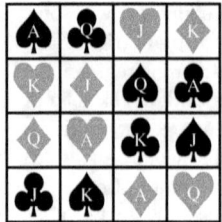

Figure 2: A Latin square of order 4 which can be decomposed into disjoint full transversals, as indicated by the card suits.

with some flexibility before extending it to an almost-full transversal. Finally, the flexibility of the absorption structure is used to make a relatively small adjustment to the overall structure to extend this to a full transversal, where each column, row and symbol must appear exactly once.

In this setting, it is typically not hard to extend some small partial transversal into an almost-full transversal in any Latin square. For some potential applications this can moreover be done so that the unused columns, rows and symbols look like a randomly chosen set, for example using the semi-random method (also known as the Rödl nibble) as introduced by Rödl [93]. This method was used, via a development by Pippenger and Spencer [88], in random Latin squares by Ferber and Kwan [42] (for the work discussed below), while it was implemented for generalised Latin squares (see Section 6) by Montgomery, Pokrovskiy and Sudakov [78] who showed that any generalised Latin square contains $(1 - o(1))n$ disjoint partial transversals of order $(1 - o(1))n$. (For more on nibble methods, see also the recent survey by Kang, Kelly, Kühn, Methuku and Osthus [55].)

The challenge then in using an absorption approach to find a full transversal in a random Latin square is to show that an appropriate absorption structure likely exists. To do this, Kwan [66] built up an absorption structure from much smaller flexible substructures. This uses distributive absorption, a specific style of absorption introduced by Montgomery [77] which has been applied in a range of settings to find large rigid substructures. More detail on distributive absorption is given in Section 5.

An alternative approach to studying full transversals in the random Latin square L_n was introduced by Gould and Kelly [47], who showed moreover that the random Latin square L_n has a *Hamilton* transversal with high probability. Here, a *Hamilton transversal* is a full transversal where the natural permutation of the indexing set of the rows/columns corresponding to the full transversal is a cyclic permutation. (For example, the full transversal highlighted in Figure 1 is not a Hamilton transversal as the corresponding natural permutation is (1)(2)(3465).) For their result, Gould and Kelly built on their work with Kühn and Osthus [48] (as described briefly in Section 6.3) by using 'switching methods' to study random Latin squares. Instead of working with random Latin squares, the idea is to consider random *Latin rectangles*, where a k by n Latin rectangle is a k by n array filled using n symbols, so that each symbol is used at most one in each row and column (see Figure 3). As Latin rectangles are less rigid than Latin squares, small alterations (known as 'switchings')

1	4	6	5	3	2
5	2	4	3	1	6
6	3	2	4	5	1

4	1	6	5	3	2
2	5	4	3	1	6
6	3	2	4	5	1

Figure 3: A 3 by 6 Latin rectangle on the left, where the comparative lack of rigidity allows many different small alterations to made while maintaining the Latin rectangles properties, with one such operation carried out on the highlighted cells to create the Latin rectangle on the right. Note that the same operation applied to the first Latin square in Figure 1 does not result in a Latin square.

can be made to a Latin rectangle in many different ways to reach another Latin rectangle. Gould and Kelly use the analysis of switching arguments to show that a uniformly random k by n Latin rectangle contains their desired small structures (for some $k = \Theta(n)$). These results can then be moved into a random Latin square by comparing its first k rows to a uniformly random k by n Latin rectangle, an approach pioneered by McKay and Wanless [75] which uses estimates on the permanent (more specifically an upper bound of Brègman [21] and a lower bound due to Egorychev [33] and Falikman [40]) to bound the number of extensions of a Latin rectangle to a Latin square.

Very recently, Eberhard, Manners and Mrazović [32] introduced yet another method for showing a random Latin square has a full transversal with high probability, giving a very different proof by using tools from analytic number theory. In particular, their methods are extremely powerful for counting the number of transversals that can be expected in a typical random Latin square. In [66], Kwan had shown that L_n has $\left((1-o(1))\frac{n}{e^2}\right)^n$ full transversals with high probability (and a similar counting result for Hamilton transversals was given by Gould and Kelly [47]). Here, the upper bound on the number of full transversals is due to Taranenko [99] (with a simpler proof subsequently given by Glebov and Luria [45]) and holds for any Latin square of order n. Eberhard, Manners and Mrazović [32] gave the remarkably strong bound that a random Latin square has $\left(e^{-1/2} + o(1)\right)\frac{(n!)^2}{n^n}$ full transversals with high probability. (See Section 4 for a further discussion of their methods and results.)

So far, the results mentioned sought one full transversal in a random Latin square (though finding many such single full transversals), some way from proving Conjecture 2.1! Further work in this direction has taken place in the closely related setting of perfect matchings in random Steiner triple systems. This is described in Section 6.2, but the results shown are likely to follow through with appropriate modification in Latin squares. That is, methods by Ferber and Kwan [42] on Steiner triple systems are likely with modification to show that the random Latin square L_n contains $(1-o(1))n$ disjoint full transversals with high probability (as noted in [42]). These methods build on those by Kwan in [66] by introducing sparse regularity techniques (as well as using a random partitioning argument of Ferber, Kronenberg and Long [41]). More specifically, they require a generalisation of the sparse regularity lemma of Kohayakawa and Rödl [61] to hypergraphs, to show a sparse version of a 'weak' hypergraph regularity lemma of Kohayakawa, Nagle, Rödl and Schacht [60], which is then used in conjunction with a generalisation to linear hypergraphs of the resolution of the KLR conjecture by Conlon, Gowers, Samotij and Schacht [29].

The results mentioned here on random Latin squares have all been shown within the last decade, with most of them more recent still. These new techniques open up the prospect of a proof of Conjecture 2.1 in the near future, though additional new ideas are no doubt needed. As highlighted by Ferber and Kwan [42], it is natural to take an absorption approach towards Conjecture 2.1. Most implementations of the absorption method in the literature correspond to finding a full transversal, but some more elaborate implementations have been used to decompose structures entirely. For example, Barber, Kühn, Lo and Osthus [12] introduced iterative absorption techniques to decompose dense graphs into small fixed subgraphs and Glock, Kühn, Montgomery and Osthus [46] decomposed large complete properly coloured graphs into rainbow spanning trees (see Section 3 for an exposition of the links of the current problem to rainbow subgraphs).

Finally, in this section, let us mention recent techniques to study the distribution of small structures within random Latin squares. The results mentioned above using distributive absorption (i.e., [42, 47, 66]) built their absorption structure from small structures found in Latin squares, and thus knowing how many such small structures are likely to appear is critical. One simple example of a small substructure in a Latin square is an *intercalate*, where two rows and two columns intersect on a 2 by 2 subsquare which is itself a Latin square (see Figure 1 for an example). Using switching arguments, Kwan and Sudakov [69] showed that a random Latin square L_n has at least $(1-o(1))\frac{n^2}{4}$ intercalates with high probability. This lower bound matched a conjecture of McKay and Wanless [75], whose proof was recently completed by Kwan, Sah and Sawhney [67] showing that a random Latin square L_n has at most $(1+o(1))\frac{n^2}{4}$ intercalates with high probability. If Conjecture 2.1 is to be proved by an absorption approach building on those in [47, 66], much of a random Latin square will need to be decomposed into small structures which are used to build absorbers. In recent years, related decompositions have been found by applying hypergraph matching results (see Ehard, Glock and Joos [34] for such a result) to some almost-regular auxiliary hypergraphs in an increasingly sophisticated way. For such an approach, upper bounds for the number of specific small substructures are needed as well as lower bounds, and thus the work of Kwan, Sah and Sawhney [67] (along with further developments by the same authors and Simkin [68]) is important here, where the upper bound is proved using the switching methods pioneered by McKay and Wanless [75] and the deletion method of Rödl and Ruciński [54, 94].

3 Latin Squares Without Full Transversals and Related Conjectures

It is convenient, and now commonplace, to translate transversals in Latin squares into a rainbow matching problem in optimally coloured complete bipartite balanced graphs, as follows (see also Figure 4). Given a Latin square L, create a vertex corresponding to each row and each column, and, for each row/column pair put an edge between their corresponding vertices coloured by the symbol in the cell indexed by that row and column. This gives a coloured bipartite graph $H(L)$, which is a copy of the complete bipartite n by n graph $K_{n,n}$ where the edges are coloured using the set of symbols of the Latin square. That each symbol appears only once in each row and in each column of the Latin square implies that every colour appears at each

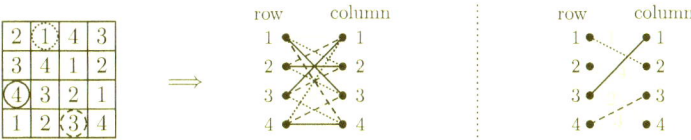

Figure 4: A Latin square of order 4 converted into an edge-coloured complete bipartite graph. A transversal in the Latin square is shown by highlighting each cell with the style used to represent the edges with colour corresponding to the cell entry. The corresponding perfect rainbow matching is then shown on the right.

vertex at most once, and therefore the colouring is *proper*. As the colouring uses the minimal number of colours for a proper colouring of $K_{n,n}$, we say it is *optimally coloured*. Given any optimal colouring of $K_{n,n}$, the reverse construction easily creates a corresponding Latin square of order n, and therefore the Latin squares of order n correspond exactly to the optimal colourings of the complete bipartite graph $K_{n,n}$.

Any transversal in a Latin square L corresponds to a subgraph in the corresponding graph $H(L)$ in which no edges share any vertices or any colours (see Figure 4). Subgraphs in edge-coloured graphs in which each colour appears at most once are known as *rainbow*, where rainbow subgraphs have in recent years seen a hive of activity (see, for example, the many results covered by the survey by Pokrovskiy [90]). A partial transversal then in L corresponds to a rainbow matching in $H(L)$, while a full transversal in L corresponds to a perfect rainbow matching in $H(L)$.

This translation allows us to use graph theory notation and techniques in the study of Latin squares. For example, for any abelian group G of order n, let $L(G)$ be the Latin square corresponding to its addition table and consider the graph, $H(G)$ say, corresponding to $L(G)$. If $H(G)$ contains a perfect rainbow matching, then, letting $c(e)$ be the colour of the edge e, we have

$$\sum_{v \in G} v = \sum_{e \in M} c(e) = \sum_{xy \in M} (x+y) = 2 \sum_{v \in G} v, \qquad (3.1)$$

so that we must have that $\sum_{v \in G} v = 0$, where 0 is the identity of G. Where this does not hold, then, $H(G)$ can have no perfect rainbow matching. Where $G = \mathbb{Z}_{2m}$, for any integer $m \geq 1$, this corresponds to the example given in Section 1, but already this gives us more examples of Latin squares without a full transversal, for example by taking $G = \mathbb{Z}_6 \times \mathbb{Z}_3$.

In fact, as, for simplicity, we have set G to be abelian, a characterisation of when the addition table for G has a full transversal has long been known due to Paige [85] in 1947 (using the language of complete mappings). That is, there is not a full transversal exactly when G has exactly one element of order 2, or, equivalently, when its only Sylow 2-subgroup is a cyclic group. In this case, this condition can be easily seen to be equivalent to $\sum_{v \in G} v \neq 0$.

In 1950, Paige [86] then considered whether $L(G)$ has a full transversal or not for finite non-abelian groups G, where $L(G)$ is the Latin square corresponding to the multiplication table of G. Paige proved that if $L(G)$ has a full transversal then there must be an ordering of the elements of G such that their product in that ordering is equal to the identity. Noting that this was shown to be a sufficient condition for finite abelian groups in [85], Paige conjectured that, likewise, this is a sufficient condition for a full transversal to occur in the multiplication table of a finite non-abelian group. As Paige noted, this is easy to show for finite non-abelian groups of odd order as the leading diagonal of the corresponding Latin squares forms a full transversal.

In 1955, Hall and Paige [51] considered an equivalent condition, that $\sum_{x \in G} x = 0$ in the abelianisation $G^{\mathrm{ab}} = G/G'$ of G, where G' is the commutator subgroup of G. This condition is now known as the Hall-Paige condition. Hall and Paige [51] conjectured that this is sufficient for a full transversal to occur in the multiplication table for any finite group, and also showed that this condition is equivalent to the condition that all the Sylow 2-subgroups of G are trivial or non-cyclic. The Hall-Paige conjecture was finally confirmed in 2009 through a combination of work by Wilcox, Evans and Bray [106, 38], as described in Section 4, and recorded below.

Conjecture 3.1 (The Hall-Paige conjecture, proved by Wilcox, Evans and Bray) *For any finite group G, $L(G)$ has a full transversal if and only if G satisfies the Hall-Paige condition.*

In the context of Section 2, we note that, for any finite group G, if $L(G)$ has a full transversal then it can be very easily seen to have a decomposition into full transversals. Indeed, if $f : G \to G$ is a bijection such that the cells indexed by $(g, f(g))$, $g \in G$, form a full transversal then, for each $h \in G$, the cells indexed by $(g, f(g) \circ h)$, $g \in G$, are also a full transversal, and thus gives a decomposition into full transversals.

Given a finite group G which does not satisfy the Hall-Paige condition, we can generate further Latin squares by using what, in modern parlance, is a 'blow-up' construction (see Figure 5 for a depiction with $G = \mathbb{Z}_2$). For convenience we will assume that G is an abelian group, of order n, and construct optimal colourings of complete bipartite graphs. Let $k \geq 1$ be an integer. Disjointly, for each $v \in G$, take sets A_v and B_v, each of k vertices, and a set C_v of k colours. Let $A = \cup_{v \in G} A_v$ and $B = \cup_{v \in G} B_v$ and, for each pair v and w, put all possible edges between A_v and B_w and properly colour them using the colours in C_{v+w}. This gives a graph H which is an optimally coloured copy of $K_{kn,kn}$ using the colours in $C = \cup_{v \in G} C_v$. If H contains a perfect rainbow matching M, then, for each edge $e \in M$, let $v_e, w_e \in G$ be such that e lies between A_{v_e} and B_{w_e}, noting that $c(e) \in C_{v_e + w_e}$. Then, the corresponding version of (3.1) is

$$k \cdot \sum_{u \in G} u = \sum_{u \in G} |C_u| \cdot u = \sum_{u \in G} \sum_{e \in M : c(e) \in C_u} u = \sum_{e \in M} (v_e + w_e) = 2k \cdot \sum_{u \in G} u.$$

Therefore, if $k \cdot \sum_{u \in G} u \neq 0$, then H can have no perfect rainbow matching, and thus the corresponding Latin square has no full transversal.

In the case $G = \mathbb{Z}_{2m}$ for any $m \geq 1$ and odd k, this example was given by Maillet [73] in 1894 (and later rediscovered by Parker [87]). As observed by Cavenagh and Wanless [27], this case already provides many Latin squares with no full transversals. For example, when $n = 2k$ and k is odd then the number of Latin squares with no full transversals is at least $\left(\left(\frac{1}{2}e^{-2} - o(1)\right)n\right)^{n^2}$. Of course, as shown

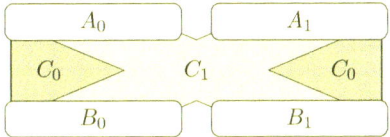

Figure 5: A blow-up construction based on the addition table for $\mathbb{Z}_2$. For some m, each labelled set has size m. The edges between A_0 and B_0, and between A_1 and B_1, are all present and properly coloured using the colours in C_0, while the edges between A_0 and B_1, and between A_1 and B_0, are all present and properly coloured using colours in C_1. When m is odd, there is no perfect rainbow matching.

by the result of Kwan [66] discussed in Section 2, this is a vanishingly small proportion of Latin squares of order n, as the number of Latin squares of order n is $|\mathcal{L}(n)| = ((e^{-2} + o(1))n)^{n^2}$ (see [101, Chapter 17]).

All the examples of Latin squares without a full transversal that we have covered here have even order. In 1967, Ryser [96] (see also [15]) conjectured that there are no Latin squares of odd order without a full transversal. Brualdi [24] later conjectured that every Latin square of order n has a partial transversal with $n - 1$ cells, while Stein [98] made some related conjectures in 1975 which are stronger than this (see Section 6). For at least the last decade, the following well-known combined conjecture has been known as the Ryser-Brualdi-Stein conjecture.

Conjecture 3.2 (The Ryser-Brualdi-Stein conjecture) *Every Latin square of order n has a partial transversal with at least $n - 1$ cells, and a full transversal if n is odd.*

In Section 4 we will discuss the proof of the Hall-Paige conjecture and other cases where Latin squares are known to have full transversals, before discussing progress towards the Ryser-Brualdi-Stein conjecture in Section 5. Before this, however, let us briefly note that we have seen no Latin squares of odd order n with any meaningful restriction on their transversal properties. However, there are Latin squares of order n which have no decomposition into full transversals for each odd $n \geq 5$. This was shown for every $n \geq 5$ with $n \equiv 1 \bmod 4$ by Mann [74], and for every $n \geq 7$ with $n \equiv 3 \bmod 4$ independently by Evans [37] and Wanless and Webb [105].

4 Full Transversals in Latin Squares

From its statement in 1955 until its proof in 2009, the Hall-Paige conjecture (Conjecture 3.1) was steadily shown to hold for various groups and saw much related work. For a detailed exposition of this, as well as a complete unified proof of the conjecture, we refer the reader to Part II of the book by Evans [39]. In their original paper, Hall and Paige [51] showed that, if G has a normal subgroup N for which both $L(N)$ and $L(G/N)$ has a full transversal, then $L(G)$ has a full transversal, and used this to prove Conjecture 3.1 for solvable groups. In a major breakthrough in 2009, also via an inductive argument, Wilcox [106] reduced Conjecture 3.1 to the case of simple groups. Thus, the classification of finite simple groups exactly characterised the remaining task. Hall and Paige [51] had already confirmed the conjecture for alternating groups. Wilcox [106] gave a unified proof for groups of Lie type, while Evans [38] combined Wilcox's methods with computer algebra to give a proof for the Tits group and all but one of the 26 sporadic groups. Bray then checked the final remaining case, which was the fourth Janko group (see [106], and [20] for the eventual publication), which completed the proof of the Hall-Paige conjecture.

More recently, Eberhard, Manners and Mrazović [31] gave a completely different proof of the Hall-Paige conjecture for large groups, using tools from analytic number theory. Remarkably, they were able to show that any group G satisfying the Hall-Paige condition has $\left(e^{-1/2}+o(1)\right)|G^{\mathrm{ab}}|\frac{(n!)^2}{n^n}$ full transversals, where, again, G^{ab} is the abelianisation of G, thus giving a precise asymptotic. (Indeed, they even determine the order of the $o(1)$ term, see [31, Theorem 1.4]!) The quantitative bounds in Eberhard, Manners and Mrazović's work are strong enough to already rule out many of the cases considered in the original proof of the Hall-Paige conjecture by Wilcox, Bray and Evans. This gives a proof of Conjecture 3.1 which avoids extensive case-checking for the sporadic groups or the Tits group, from these only requiring verification of the conjecture for the first two Mathieu groups (see [31]).

Even more recently, Eberhard, Manners and Mrazović [32] again used tools from analytic number theory (including a loose variant of the circle method) to find full transversals in Latin squares satisfying a quasirandomness condition. Here the quasirandomness condition is defined in terms of the spectral gap of an operator associated with the Latin square (see [32]). Where the Latin square coincides with the multiplication table of a group, their condition coincides with the definition of quasirandomness for subgroups by Gowers [49], and recovers the main result of their work in [31] for sufficiently quasirandom groups. Using recent results of Kwan, Sah, Sawhney and Simkin [68], Eberhard, Manners and Mrazović showed that a random Latin square of order n satisfies their quasirandomness condition with high probability, allowing them to improve Kwan's result (as discussed in Section 2) to show that a random Latin square of order n has $\left(e^{-1/2}+o(1)\right)\frac{(n!)^2}{n^n}$ full transversals with high probability.

A further independent proof of the Hall-Paige conjecture has been also recently been given, this time with combinatorial methods, by Müyesser and Pokrovskiy [83] as part of a generalised theorem with several applications resolving old problems in combinatorial group theory in the case of large groups (see [83]). To get an idea of this general theorem, first let G be a large group of order n and let $H(G)$ be the bipartite graph corresponding to the multiplication table of G (as described in Section 3) with vertex sets A and B and colour set C. Suppose we have sets $A' \subset A$, $B' \subset B$ and $C' \subset C$ of equal size. When can we expect a perfect rainbow matching between A' and B' in $H(G)$ using only the colours in C'? Similarly to the arguments we have covered, it is easy to see that a necessary condition must be that $\sum_{a \in A'} a + \sum_{b \in B'} b = \sum_{c \in C'} c$ in the abelianisation G^{ab} of G. Müyesser and Pokrovskiy [83] show that this is a sufficient condition for sets A', B' and C' which are sufficiently close to random subsets where each element is independently included with the same probability p, for any $p \geq n^{-1/10^{100}}$. Taking $p = 1$ in this result, then, recovers a proof of the Hall-Paige conjecture for large groups.

The main tool used by Müyesser and Pokrovskiy in [83] is the absorption method, as described loosely in Section 2. As we will discuss in detail an absorption approach to finding large rainbow matchings in the next section, we will outline how this can be used now. (This discussion is representative of the methods in [83], but slightly different to align with our subsequent discussion.) A natural approach for absorption here would be, for some $\ell_0, \ell_1 \leq n$, to look for a balanced set V^{abs} of $2\ell_0$ vertices and a set C^{abs} of $\ell_0 + \ell_1$ colours such that, for any balanced set $W \subset V(H) \setminus V^{\text{abs}}$ of $2\ell_1$ vertices, there is a rainbow matching in $H := H(G)$ with vertex set $V^{\text{abs}} \cup W$ and colour set C^{abs}. (A set is *balanced* if it has equally many vertices in each side of the bipartition.) In the terminology of absorption, we would say that $(V^{\text{abs}}, C^{\text{abs}})$ can absorb any balanced vertex set with $2\ell_1$ vertices. If this could be done, then the strategy would look for an initial rainbow matching with vertex set in $V(G) \setminus V^{\text{abs}}$ which uses exactly the colours not in C^{abs} (made easier than the original problem as there are $2\ell_1$ more vertices than have to be used in this matching) before applying the absorption property to the unused vertices in $V(H) \setminus V^{\text{abs}}$ to find a second matching which, added to the initial matching, completes a perfect rainbow matching in H.

However, there is an inherent complication here. If there is a rainbow matching in H with vertex set $V^{\text{abs}} \cup W$ and colour set C^{abs}, then, assuming from now for simplicity that G is abelian, we have

$$\sum_{v \in V^{\text{abs}} \cup W} v = \sum_{c \in C^{\text{abs}}} c, \tag{4.1}$$

and therefore the best we can hope for is that $(V^{\text{abs}}, C^{\text{abs}})$ can absorb any balanced set of $2\ell_1$ vertices for which $\sum_{v \in W} v = \sum_{c \in C^{\text{abs}}} c - \sum_{v \in V^{\text{abs}}} v$. We may as well say this sum is equal to 0, and hope to create an absorption structure $(V^{\text{abs}}, C^{\text{abs}})$ which can absorb any appropriate set of vertices with zero sum.

This is accomplished by Müyesser and Pokrovskiy in [83] with $\ell_0 = o(n)$ and $\ell_1 = n^{1-\varepsilon}$ for some small fixed $\varepsilon > 0$. (In fact, they prove something stronger that we have weakened to cohere with our subsequent discussion.) To build this absorber, they use distributive absorption (as mentioned in Section 2), an efficient method to build a global absorption property from small local absorbers. To avoid excessive detail here, we will comment no further than to say that the starting point is usually to robustly find small absorbers. For example, here, it would be useful to find a small absorber that can absorb either $\{x_1, y_1\}$ or $\{x_2, y_2\}$ (two balanced sets). The discussion around (4.1) implies that this is only possible if $x_1 + y_1 = x_2 + y_2$, or, equally, if $x_1 y_1$ and $x_2 y_2$ have the same colour in $H = H(G)$. (Note we are giving a minimal useful example as if we could absorb $\{x_1\}$ and $\{x_2\}$ then in fact $x_1 = x_2$.) However, subject only to the condition that $x_1 + y_1 = x_2 + y_2$ ($= c$, say) we can find such a small absorber in many ways (as depicted in Figure 6). Picking any two colours c_1 and c_2 (avoiding $O(1)$ bad options so the following is possible), label vertices w_1, z_1 and a colour c_3 such that $x_1 w_1 z_1 y_1$ is a path in H with colours c_1, c_2 and c_3 in that order, and c, c_1, c_2, c_3 are all distinct. Label vertices w_2, z_2 such that $x_2 w_2 z_2 y_2$ is a path in H with colours c_1, c_2 and c_3 in that order, noting that this is possible as $H = H(G)$ and $x_2 y_2$ is the same colour as $x_1 y_1$. Then, as long as we chose c_1, c_2 so that these paths are vertex-disjoint, it is easy to see that, setting $V^{\text{abs}} = \{w_1, z_1, w_2, z_2\}$ and $C^{\text{abs}} = \{c_1, c_2, c_3\}$, $(V^{\text{abs}}, C^{\text{abs}})$ can absorb either $\{x_1, y_1\}$ or $\{x_2, y_2\}$ (see Figure 6).

There are many complications dealt with in [83], not least of all of course when G is non-abelian, but we hope this gives some indication of the starting point from which a global absorber can be created. We return to this in Section 5 to discuss the challenges in attempting this more generally for colourings not generated from a group. Constructing absorbers for zero-sum sets in this manner is a natural extension in the development of absorption techniques, and appears also in the recent work by Bowtell and Keevash [19] on the famous n-queens problem.

In Section 2, we described work by Gould and Kelly [47] finding Hamilton transversals in a random Latin square with high probability. There, a Hamilton transversal corresponded to a full transversal where the natural permutation it defines is a cyclic permutation. More generally, we could ask for a full transversal in a random Latin square of order n where this permutation has any specific cycle type corresponding to a permutation of an n-element set. If there are few fixed points then it seems plausible this can be done with high probability. In the non-random case, Müyesser [82] has conjectured that if $k \geq 3$ is such that $k|(n-1)$, then for any group G of order n satisfying the Hall-Paige condition the Latin square $L(G)$ has a full transversal whose permutation corresponds to a single fixed point and otherwise has cycles of length k. (The case $k = n-1$ can be seen to correspond to a Hamilton transversal.) In related work, Müyesser [82] proved for large groups a conjecture of Friedlander, Gordon and Tannenbaum [43] from 1981 that, for $k \geq 2$ dividing $(n-1)$, if an *abelian* group G of order n satisfies the Hall-Paige condition then the Latin square formed like $L(G)$ but with entry $u - v$ in the cell indexed by row u and column v, for each $u, v \in G$, has a transversal whose permutation corresponds to a single fixed point and otherwise has cycles of length k.

5 Partial Transversals in Latin Squares

Shortly after the formation of the Ryser-Brualdi-Stein conjecture (or, more precisely, the portion due to Ryser [96]), Koksma [62] gave a simple argument that every Latin square of order n has a partial transversal with at least $2n/3$ cells. Drake [30] improved this bound to $3n/4$, before an approximate form of the conjecture was proved in 1978 independently by Brouwer, De Vries and Wieringa [23] and Woolbright [107], each showing that every Latin square of order n has a partial transversal with at least $n - \sqrt{n}$ cells. In 1982, Shor [97] gave a proof improving this

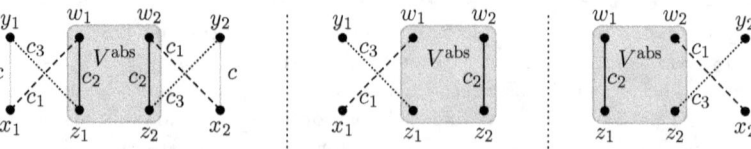

Figure 6: On the left a structure allowing $(V^{\text{abs}}, C^{\text{abs}})$ to absorb either $\{x_1, y_1\}$ or $\{x_2, y_2\}$, where $C^{\text{abs}} = \{c_1, c_2, c_3\}$). That is, there is both a matching with vertex set $V^{\text{abs}} \cup \{x_1, y_1\}$ and colour set C^{abs} (in the middle) and $V^{\text{abs}} \cup \{x_2, y_2\}$ and colour set C^{abs} (on the right). In a colouring arising from a group, this is possible as $x_1 y_1$ and $x_2 y_2$ have the same colour.

bound to $n - O(\log^2 n)$, though this paper contained an error that was only noticed and fixed in 2008 by Hatami and Shor [53], using effectively the original approach. Recently, Keevash, Pokrovskiy, Sudakov and Yepremyan [58] improved this bound, which had essentially stood for 40 years, to show any Latin square of order n has a partial transversal with $n - O(\log n / \log \log n)$ cells. Very recently, Montgomery [76] showed that every sufficiently large Latin square of order n has a partial transversal with $n - 1$ cells. While this confirms the Ryser-Brualdi-Stein for large even n, it seems very difficult to find a full transversal in a Latin square with large odd order using these methods and certainly further ideas would be needed.

To discuss the proof of the bounds in [58, 76], we will use the rainbow subgraph formulation of the problem. Suppose, then, that G is a bipartite copy of $K_{n,n}$ which is properly coloured with n colours. To find a find a large rainbow matching in G, Keevash, Pokrovskiy, Sudakov and Yepremyan [58] essentially began with a large random almost-perfect rainbow matching, before successively modifying it, each time getting a rainbow matching that is one edge larger, until it contains $n - O(\log n / \log \log n)$ edges. Suppose the initial large rainbow matching is M. Suppose further that there are vertices $x, y \in V(G) \setminus V(M)$ which are in different vertex classes in G, and an x,y-path P in G such that the even edges in this path are in M and the odd edges (starting with the edge containing x) all have different colours not used on M. Then, removing the even edges of P from M and adding the odd edges gives a rainbow matching in G with one more edge. In [58], the initial large rainbow rainbow matching M is found using the Rödl nibble (as discussed briefly in Section 2), and has $n - n^{1-\alpha}$ edges for some fixed small $\alpha > 0$. The randomness of the initial matching M implies (via some expansion properties) essentially that for every pair of vertices not in the matching and every set of $K \log n / \log \log n$ colours (for some fixed constant K) not on M such a path P can be found. Moreover, this property is sufficiently robust that it can be used to make iterative adjustments (with some additional care) until the rainbow matching uses all but $K \log n / \log \log n$ of the colours of G.

This brief outline belies the challenges overcome in [58], particularly in achieving a bound that seems likely to be optimal using these or related methods. Indeed, if $d = \log n / 2 \log \log n$ and C is a set of d colours, then however a matching M is chosen in G without using the colours in C, for any vertex $v \in V(G)$, there are at most $2d^d = o(\sqrt{n})$ paths from v which alternate between edges with colour in C and edges in M in the manner of the path P described above. Thus, such paths certainly cannot exist between an arbitrary pair of vertices. More broadly, the bound $n - O(\log n / \log \log n)$ is plausibly a natural barrier for any method that does not alter its approach based on the specific colour of G, and the underlying algebraic properties it might have (see, for example, the examples described in Section 3).

A significant part of the work in [76], then, is to find some way to identify and exploit (approximate) algebraic properties in the colouring of G. Following this, the approach taken is an implementation of the absorption method, where, as in the outline of the work in [82] by Müyesser and Pokrovskiy in Section 4, an absorption structure is found and set aside, before a large rainbow matching is found disjointly. Finding a large rainbow matching disjoint from the absorption structure in [76] is possible if a substantial set of random colours and vertices is reserved for this, following work by Montgomery, Pokrovskiy and Sudakov [79].

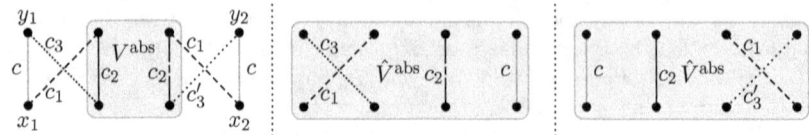

Figure 7: On the left, in contrast to Figure 6, with $C^{\mathrm{abs}} = \{c_1, c_2, c_3\}$, $(V^{\mathrm{abs}}, C^{\mathrm{abs}})$ can absorb $\{x_1, y_1\}$ but *not* $\{x_2, y_2\}$ if $c'_3 \neq c_3$. Instead, with $\hat{V}^{\mathrm{abs}} = V^{\mathrm{abs}} \cup \{x_1, y_1, x_2, y_2\}$ and $\hat{C}^{\mathrm{abs}} = \{c, c_1, c_2\}$, $(\hat{V}^{\mathrm{abs}}, \hat{C}^{\mathrm{abs}})$ can absorb either c_3 or c'_3, as depicted by the matchings in the middle and on the right, whose colour sets differ on $\{c_3, c'_3\}$.

The key, then, is in developing some absorption structure. We recall from Section 4 that we cannot hope to find an absorption structure that can absorb any small set of vertices. In contrast to the previous methods involving the absorption of 'zero-sum' sets (i.e., those in [19, 83]), here we do not have a structure with a well-determined, known, and exact algebraic structure. Some condition is thus needed to define the sets we will be able to absorb. In [76], a colour c_0 is chosen (fairly arbitrarily) to function as an 'identity colour' and the condition for absorption is that the vertex set must be the vertex set of a matching of colour-c_0 edges. (A few edges are deleted for this to be true, but here we gloss over this for simplicity.) More precisely, for some $\ell_0, \ell_1 \leq n$ we look for a balanced set V^{abs} of $2\ell_0$ vertices and a set C^{abs} of $\ell_0 + \ell_1$ colours such that, for any balanced set $W \subset V(H) \setminus V^{\mathrm{abs}}$ which is the vertex set of a matching of ℓ_1 edges with colour c_0, there is a rainbow matching in G with vertex set $V^{\mathrm{abs}} \cup W$ and colour set C^{abs}.

In [76], this absorption structure is found using distributive absorption, so, as in our discussion in Section 4, a good starting point would be to find an absorber capable of absorbing one of two sets of two vertices with 'zero sum' – here, instead, the 'zero sum' condition is replaced by the set of two vertices being the vertex set of an edge with the identity colour, c_0. However, trying the construction in Section 4 with $c = c_0$ (see Figure 6) to find an absorber here does not work as, without the algebraic structure conferred by a group, we may not have that the third edge of the x_2, y_2-path has colour c_3, but, perhaps, some other colour c'_3 (see Figure 7). Crucially, however, we can instead choose from what we have found a structure that can absorb either the colour c_3 or the colour c'_3 (as depicted in Figure 7). That is, using the labelling in Figures 6 and 7, taking $\hat{V}^{\mathrm{abs}}$ to be the set of all the vertices used and $\hat{C}^{\mathrm{abs}}$ to be the set of all the colours used except for c_3 and c'_3, we have that there is a rainbow matching with vertex set $\hat{V}^{\mathrm{abs}}$ and colour set $\hat{C}^{\mathrm{abs}} \cup \{c_3\}$ and a rainbow matching with vertex set $\hat{V}^{\mathrm{abs}}$ and colour set $\hat{C}^{\mathrm{abs}} \cup \{c'_3\}$ (see Figure 7). Note that if we knew G has many such small absorbers capable of absorbing either c_3 or c'_3, then if we tried the previous absorber construction for $\{x_1, y_1\}$ and $\{x_2, y_2\}$ and encountered the same problem we could now solve it. Indeed, if we take, using new colours and vertices, a small absorber that can absorb either c_3 or c'_3, then, in combination with the construction that failed because $c_3 \neq c'_3$ (adding both c_3 and c'_3), it is relatively easy to see that this can absorb either $\{x_1, y_1\}$ or $\{x_2, y_2\}$.

This is a small toe-hold into creating an absorption structure, but a crucial one, and it motivates how we might study the colouring of G to gather something representative of some of its (possible) algebraic structure. Looking at the pairs of colours, we consider how many different small absorbers can absorb either one (as in Figure 7). A good way to consider this to take a complete auxiliary graph K with the colour set of G as its vertex set where each edge is weighted by the number of absorbers in G with the construction in Figure 7 which can absorb either one of those two colours. If two colours c and d are connected by a short path of high weight edges in K, then, loosely speaking, we can chain disjoint absorbers together along this path to create an absorber that can absorb either c or d. Ideally, we want to know that, the majority of the time, when we attempt the construction in Figure 7, if it fails because $c'_3 \neq c_3$, then we will be able to find many small absorbers that can absorb c'_3 or c_3. To do this, in [76], most of the weight of K is essentially covered by well-connected subgraphs whose edges have roughly equal weight in such a way that no colour appears in too many subgraphs. This is done using techniques involving *sublinear expansion*, which allows well-connected graphs to be found where the length of the connections can be controlled (so that any absorbers created using such a connection in K are not too large). Sublinear expansion, as introduced by Komlós and Szemerédi [63, 64], is used here as the non-zero-weighted edges of K might be sparse, yet carry a meaningful weight. For more on sublinear expansion, and the many uses that have been found for it, we invite the reader to turn to the survey by Letzter [70] which also appears in this volume.

While our discussion so far gives some indication of the approach taken in [76] to study the algebraic properties of colourings, the actual implementation is somewhat more complicated. Aside from the omission here of many details around the distributive absorption approach, this is in part because the weight distribution on the auxiliary graph K can vary widely. For example, when the colouring of G arises directly from the addition table of an abelian group, then every weight will be 0. On the other hand, if the colouring of G is a randomly chosen optimal colouring, then we expect K to be roughly evenly weighted with total weight $\Theta(n^5)$, and, between these two examples, there are a wide variety of plausible auxiliary graphs K.

Suppose, however, that our absorption structure can be created. Note that the condition on the set to be absorbed (that it is the vertex set of a matching of identity colour edges) is quite restrictive – if a vertex is to be in the set to be absorbed then so must its neighbour across an identity-colour edge. To be able to consider an arbitrary balanced vertex set, an 'addition structure' is introduced in [76]. This consists of a set V^{add} of vertices and a set of colours C^{add} so that, given any sufficiently small set W of vertices not in V^{add}, G contains two vertex-disjoint matchings M_1 and M_2 and two remainder vertices r_1, r_2, such that these vertices altogether exactly form the set $V^{\text{add}} \cup W$, and such that M_2 is a rainbow matching using each colour in C^{add} exactly once and M_1 is a matching of identity colour edges. Thus, setting $\hat{W} = V(M_1)$, we have a set which can be absorbed by the absorption structure, while the matching M_2 is included in the rainbow matching that is found. The two remainder vertices are the two vertices not included in the final matching of $n - 1$ edges. The function of the addition structure here, effectively, is to take a general unstructured set W and transform it to a structured set $\hat{W}$ (with the loss of two remainder vertices) that is then suitable for absorption (this is rather abstractly represented in Figure 8). The construction of the addition structure is not particularly complex, though we include no further details here, but we note that it works in an iterative fashion where more

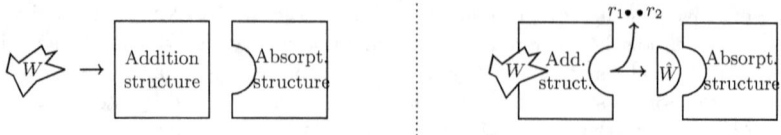

Figure 8: An addition structure and an absorption structure on the left, along with a small unstructured vertex set W. On the right, the addition structure transforms the unstructured set into two 'remainder vertices' and a structured vertex set $\hat{W}$ which is suitable for absorption.

vertices are added to the structure in pairs, all while a changing pair of 'remainder vertices' is considered along with two large matchings. Thinking of this as addition with remainder vertices is a very useful perspective when creating such a structure.

Finally, let us finish our discussion of the methods in [76] by noting how useful it is to have two vertices which are omitted from the final matching. Considering the case where the colouring arises from an abelian group, given an arbitrary balanced set of vertices, in order for the addition structure to function any group element must be representable as the sum of two vertices from the addition structure (i.e., any remainder can be represented as the sum of two remainder vertices). This would be true if, say, the addition structure contained a random set of $n^{2/3}$ vertices, while using only one remainder vertex is not possible with these methods. Thus, perhaps it is fair to say that finding a rainbow matching in G missing one edge is not as close to an n-edge rainbow matching as it might seem. In some sense, the degree of the discrepancy is not 1 but 3: we may choose two vertices and one colour to leave out of the matching. For the methods in [76] we need the flexibility to omit at least two of these — a brief discussion of a variant of the problem which is approachable using this is in Section 6.1. Finding a rainbow matching without any of this flexibility (i.e., proving the Ryser-Brualdi-Stein conjecture for large odd n) seems quite far beyond these methods, and would certainly need new ideas.

6 Related Problems

In the rainbow subgraph formulation, we have discussed the problem of finding large rainbow matchings in optimally coloured complete bipartite balanced graphs. To finish this survey, we will discuss some related work and conjectures. We consider what happens when the colouring is proper but non-optimal (Section 6.1), when the coloured graph is not bipartite but complete (Section 6.3), when the colouring is not required to be proper (Section 6.4) and when multigraphs are used (Section 6.5). Additionally, we discuss the closely related problem of large matchings in Steiner triple systems (Section 6.2). We cover these topics beginning with those where the methods discussed in Section 5 give some progress. Throughout this section, we touch only on the problems most closely connected to transversals in Latin squares, and, for additional background and a much wider variety of problems on rainbow subgraphs, we refer the reader to the excellent survey by Pokrovskiy [90].

6.1 Generalised Latin Squares

From the perspective of the rainbow subgraph formulation, it is natural to ask how large a rainbow matching can be found in any properly coloured $K_{n,n}$. This corresponds to the study of transversals in *generalised Latin squares*, also known as *Latin arrays*, of order n, where an n by n grid is filled with symbols so that no symbol appears more than once in any row or column. As non-optimal colourings have more colours than optimal colourings, it is generally expected that large, or even perfect, rainbow matchings should be easier to find in non-optimal colourings of $K_{n,n}$. For example, Montgomery, Pokrovskiy and Sudakov [78] showed that any proper colouring of $K_{n,n}$ where at most $(1-o(1))n$ colours appear more than $(1-o(1))n$ times has a perfect rainbow matching.

The techniques used in work towards the Ryser-Brualdi-Stein conjecture have often also been applied more generally to proper colourings of $K_{n,n}$. For example, Keevash, Pokrovskiy, Sudakov and Yepremyan (see [58, Theorem 1.6]) showed that any such colouring contains a rainbow matching with $n - O(\log n/\log\log n)$ edges. The results of [78] show that, for large n, we only need consider colourings which are, in some sense, close to an optimal colouring. For sufficiently large n, this allows the techniques described in Section 5 to show that any properly coloured $K_{n,n}$ has a rainbow matching with $n-1$ edges [76].

Beyond this, Best, Pula and Wanless [14] have conjectured that any properly coloured $K_{n,n}$ should have an $(n-1)$-edge rainbow matching despite the deletion of any chosen vertex, or, equivalently, that every proper colouring of $K_{n-1,n}$ has a rainbow matching with $n-1$ edges. As Georgakopoulos [44] observed, this is a special case of a conjecture of Haxell, Wilfong and Winkler (see also [5]). Though not proved in full in [76], the methods described in Section 5 are strong enough to confirm this conjecture for sufficiently large n. (Very roughly, that such a partial transversal would omit at least one colour and omit one vertex is enough flexibility to allow the methods described in Section 5 to work, though it is clearer from the sketches in Section 5 that the methods would work in a setting where two vertices and no colours are omitted, such as to find an $(n-1)$-edge rainbow matching in an optimal colouring of $K_{n,n}$ with the edges of any one colour removed.) This conjecture is a weak version of a very strong conjecture by Stein [98], which we cover in Section 6.5.

6.2 Steiner Triple Systems

In addition to rainbow matchings in properly coloured balanced bipartite complete graphs, transversals in Latin squares also have a natural expression as a matching in 3-uniform hypergraphs. Given a Latin square L of order n, form a hypergraph $\mathcal{H}(L)$ by first creating disjoint sets A and B of n vertices representing the rows and columns of L respectively and a set C of n vertices representing the symbols of L. Then, for each pair of vertices $a \in A$ and $b \in B$, take the symbol $c \in C$ in row a and column b of L and add abc to $\mathcal{H}(L)$. Note that a partial transversal in the Latin square L corresponds to a matching (a set of disjoint edges) in $\mathcal{H}(L)$, while a full transversal corresponds to a perfect matching.

Note further that any two vertices from different sets A, B and C are contained together in exactly one edge of $\mathcal{H}(L)$. Thus, a Latin square of order n can be represented as what is a tripartite version of a Steiner triple system. A *Steiner triple system (STS)* of order n is a 3-uniform hypergraph with n vertices in which every pair of vertices is contained in exactly one edge. Steiner triple systems are a type of design whose history dates back to Kirkman's famous schoolgirl problem [59] from 1850 (which asked for a particular type of Steiner triple system, called a *resolvable design*, with 15 vertices). Every vertex in an STS of order n must be in $\frac{n-1}{2}$ edges, and the number of vertex pairs, $\binom{n}{2}$, must be divisible by 3. Thus, Steiner triple systems of order n can only exist if $n \equiv 1$ or $3 \mod 6$. Kirkman [59] showed that this condition is sufficient for the existence of an STS of order n.

The corresponding version of the Ryser-Brualdi-Stein conjecture was made for Steiner triple systems in 1981 by Brouwer [22], as follows.

Conjecture 6.1 (Brouwer) *Every Steiner triple system of order n has a matching of at least $\frac{n-4}{3}$ edges.*

Conjecture 6.1 would be tight for infinitely many n as seen by constructions of Wilson (see [28]) and Bryant and Horsley [25, 26]. Improving on early bounds by Wang [103] and Lindner and Phelps [71], Brouwer [22] gave the first asymptotic version of Conjecture 6.1 by showing that any STS of order n has a matching with at least $n/3 - O(n^{2/3})$ edges. An asymptotic version of Conjecture 6.1 can now, however, be proved with a simple application of standard matching theorems for almost-regular hypergraphs proved using the nibble method (see, for example, [8]). Refining such an approach and using large deviation inequalities, Alon, Kim and Spencer [6] improved Brouwer's result to show that a matching with $n/3 - O(n^{1/2} \log^{3/2} n)$ edges exists in any STS of order n.

More recently, Keevash, Pokrovskiy, Sudakov and Yepremyan [58] made the following conversion of this problem to one on rainbow matchings, using it to apply their methods and make a very significant improvement towards Conjecture 6.1. Assuming, for a slight simplification, that $n = 3m$ for some integer m, take a Steiner triple system S of order n and partition $V(S)$ uniformly at random into sets A, B and C, each with size m. Form a bipartite graph G with vertex classes A and B, and, for each $a \in A$ and $b \in B$, if there is some $c \in C$ such that $abc \in S$, then put an edge between a and b in G with colour c. This gives a properly coloured bipartite graph, and, moreover, one in which a rainbow matching corresponds to a matching in the original Steiner triple system. The graph G is not a complete bipartite graph, but does look roughly like an optimal colouring of $K_{n,n}$ where edges have been deleted independently at random with probability $2/3$. This allowed Keevash, Pokrovskiy, Sudakov and Yepremyan [58] to hugely improve the previous bound to show that any STS of order n has a matching with $n/3 - O(\log n / \log \log n)$ edges. Applying the techniques described in Section 5 through this conversion introduces some complications, but in [76] it is shown that Conjecture 6.1 holds for sufficiently large n.

As mentioned in Section 2, we can also consider perfect matchings in random Steiner triple systems of order $n \equiv 3 \mod 6$. Kwan [66] showed that such a random STS of order $n \equiv 3 \mod 6$ does have a perfect matching with high probability, and, moreover contains $\left((1 - o(1))\frac{n}{2e^2}\right)^{n/3}$ perfect matchings with high probability. Morris [80] adapted Kwan's proof to show that a random STS of order $n \equiv 3 \mod 6$ has $\Omega(n)$ disjoint perfect matchings with high probability. Ferber and Kwan [42] were then able to show (see Section 2) that $(1 - o(1))n$ disjoint perfect matchings can be found with high probability, while making the corresponding conjecture to Conjecture 2.1, as follows.

Conjecture 6.2 (Ferber and Kwan) *A random Steiner triple system of order $n \equiv 3 \mod 6$ has a decomposition into perfect matchings with high probability.*

A Steiner triple system with such a decomposition as in Conjecture 6.2 is known as *resolvable*. Whether there existed any resolvable STSs for each $n \equiv 3 \mod 6$ was an old problem in combinatorics. Their existence was known for several infinite families of integers n (see [52]), but in general was only shown in famous work by Ray-Chaudhuri and Wilson [92] in 1971.

6.3 Complete Graphs and Rainbow Hamilton Cycles

What happens if we, instead, look for a large rainbow matching in a properly coloured n-vertex complete graph, K_n? The limit on the size of the matching imposed by the number of vertices ($\lfloor n/2 \rfloor$) is much stronger than that imposed by the number of colours needed in a proper colouring ($n - 1$ when n is even). This makes the problem much easier, and it is more interesting to ask for a rainbow subgraph with more edges, or even $n - 1$ edges, so that almost all the colours might be used. The most natural subgraph to consider is a Hamilton path. (For the consideration of trees more generally, we refer the reader to the survey by Pokrovskiy [90].) However, giving a counterexample to a conjecture by Hahn [50], in 1984 Maamoun and Meyniel [72] gave an optimal colouring of K_n which has no rainbow Hamilton path when $n \geq 4$ is any power of 2. To see this colouring, take the complete graph with the group $\mathbb{Z}_2^k$ (with $k \geq 2$) as its vertex set, where each edge xy is coloured $x + y$. Using that $2x = 0$ for every $x \in \mathbb{Z}_2^k$, this can easily be seen to be a proper colouring which never uses 0, so that the colouring with $\mathbb{Z}_2^k \setminus \{0\}$ is optimal. If this graph has a rainbow Hamilton path, then the corresponding equation to (3.1) shows that the sum of the two end vertices must be 0, a contradiction. Therefore, the following conjecture of Andersen [9] from 1989 would be best possible when $n \geq 4$ is a power of 2.

Conjecture 6.3 (Andersen) *All proper edge-colourings of K_n have a rainbow path with length $n - 2$.*

This question can also be asked for rainbow cycles, where we note that, when n is even, any optimal colouring of K_n does not have a rainbow Hamilton cycle simply as it does not have enough colours to support this. Akbari, Etesame, Mahini and Mahmoody [4] have asked whether all optimal colourings of K_n have a rainbow cycle with length $n - 2$, although the current author knows of no optimal colouring of K_n with no rainbow cycle with length $n - 1$. We also can note that the known optimal colourings of K_n with no rainbow Hamilton path are much rarer than the many examples of Latin squares with no full transversals given in Section 3 (and are only known for when $n \geq 4$ is a power of 2). It is, however, known that a random optimal colouring of K_n will have a rainbow Hamilton path with high probability, due to Gould, Kelly, Kühn and Osthus [48], who also showed that such a random colouring will have a rainbow cycle containing all of the colours with high probability.

To date, the progress towards Conjecture 6.3 has not needed to distinguish between the search for long rainbow paths or cycles. In contrast to rainbow matchings in the bipartite case, it turns out to be difficult even to find a rainbow path or cycle containing $(1-o(1))n$ vertices, and (after various improved bounds, as detailed in [7]) this was only done in 2017, by Alon, Pokrovskiy and Sudakov [7]. More precisely, in [7] it is shown that any proper colouring of K_n contains a rainbow cycle with $n - O(n^{3/4})$ vertices. Balogh and Molla [11] have used the techniques in [7] to show that in fact such a cycle with $n - O(\sqrt{n}\log n)$ vertices always exists.

Techniques used to work towards the Ryser-Brualdi-Stein conjecture do not seem to translate naturally to work on Conjecture 6.3, and, for example, Keevash, Pokrovskiy, Sudakov and Yepremyan [58] did not apply their techniques to this problem (see [90]). Similarly, the techniques described in Section 5 do not apply simply to work towards Conjecture 6.3. However, with significant development, it seems likely progress can be made by considering not only a single special 'identity colour', but a basis of colours that can be used to represent the full colour set, while carrying out a more complicated addition using these representations. This seems likely to show that, in any proper colouring of K_n, there is a rainbow cycle with $n - O(1)$ vertices and is the subject of forthcoming work by Benford, Bowtell and Montgomery [13].

6.4 Non-Proper Colourings

Interestingly, that the colourings we have considered are proper may be an unnecessarily strong condition if we want only to find a very large rainbow matching. In particular, in 1975 Stein [98] made a series of 7 conjectures on transversals in arrays where the restrictions of Latin arrays are replaced (at least in part) by a restriction only on the number of times a symbol appears in the entire array (or correspondingly only a restriction on the number of times a colour appears in our edge-coloured graphs). Granted, it turns out that these conjectures are mostly false, as shown by a counterexample by Pokrovskiy and Sudakov [91] and small variants of this (see [14]), but there remains the potential that very large partial transversals can still be found under the same conditions.

For example, Stein conjectured that every equi-n square should have a partial transversal of order $n-1$, where an *equi-n* square is an n by n square in which each one of n symbols appears exactly n times. Pokrovskiy and Sudakov [91] constructed an equi-n square with no partial transversal of order $n - \frac{1}{42}\log n$, but it remains an open question whether such squares contain a partial transversal of order $(1-o(1))n$. In this direction, the best bound known is by Aharoni, Berger, Kotlar, and Ziv [2], who used topological methods to show that every equi-n-square has a partial transversal of order at least $2n/3$.

Finally, here, let us note that one of Stein's conjectures in [98] does remain open (see [14]). In the language of coloured graphs, this states that any colouring of $K_{n,n-1}$ in which no vertex in the larger vertex class is adjacent to more than one edge of any colour (i.e., the colouring is proper on one side) should contain a rainbow matching with $n-1$ edges. This conjecture is much stronger than the special case conjectured by Best, Pula and Wanless [14] mentioned in Section 6.1, and, as such, remains widely open.

6.5 Multigraphs

Finally, let us turn to an interesting generalisation of the Ryser-Brualdi-Stein conjecture. Considering an optimal colouring of $K_{n,n}$, we have n monochromatic (disjoint) matchings of size n, and wish to form a matching with n or $n-1$ edges using at most one edge per matching. What if we no longer require these matchings to be disjoint, so that together they form a multigraph? This, in an even more general form, is considered by the following conjecture of Aharoni and Berger [1] (where n corresponds to $n-1$ in our discussion so far).

Conjecture 6.4 *If G is a bipartite multigraph formed of n matchings of size $n+1$, each of a different colour, then G contains a rainbow matching with n edges.*

After a sequence of improvements by various authors (see, for example, [90]), the best currently known bound towards Conjecture 6.4 that holds for all n is by Aharoni, Kotlar, and Ziv [3], who show that if the matchings in Conjecture 6.4 are made large, with at least $3n/2 + 1$ edges each, then there is always an n-edge rainbow matching. Conjecture 6.4 has also been proven asymptotically by Pokrovskiy [89], who showed that if there are $(1 + o(1))n$ edges in each matching then there is an n-edge rainbow matching. Though this asymptotic result now has a considerably easier proof, by Munhá Correia, Pokrovskiy and Sudakov [81], the full conjecture remains ambitious. The techniques described in Section 5 do not seem to be applicable and, thus, Conjecture 6.4 remains very open indeed.

Acknowledgements

The author would like to thank colleagues, in particular Ian Wanless and the anonymous referee, for their comments on this survey. The author would also like to acknowledge support by the European Research Council (ERC) under the European Union Horizon 2020 research and innovation programme (grant agreement No. 947978) and the Leverhulme trust.

References

[1] R. Aharoni and E. Berger, *Rainbow matchings in r-partite r-graphs*, The Electronic Journal of Combinatorics **16** (2009), R119.

[2] R. Aharoni, E. Berger, D. Kotlar, and R. Ziv, *On a conjecture of Stein*, In: Abhandlungen aus dem Mathematischen Seminar der Universität Hamburg, vol. 87, Springer, 2017, pp. 203–211.

[3] R. Aharoni, D. Kotlar, and R. Ziv, *Representation of large matchings in bipartite graphs*, SIAM Journal on Discrete Mathematics **31** (2017), no. 3, 1726–1731.

[4] S. Akbari, O. Etesami, H. Mahini, and M. Mahmoody, *On rainbow cycles in edge colored complete graphs*, Australasian Journal of Combinatorics **37** (2007), 33–42.

[5] N. Alon and V. Asodi, *Edge colouring with delays*, Combinatorics, Probability and Computing **16** (2007), no. 2, 173–191.

[6] N. Alon, J.-H. Kim, and J. Spencer, *Nearly perfect matchings in regular simple hypergraphs*, Israel Journal of Mathematics **100** (1997), no. 1, 171–187.

[7] N. Alon, A. Pokrovskiy, and B. Sudakov. *Random subgraphs of properly edge-coloured complete graphs and long rainbow cycles*, Israel Journal of Mathematics **222** (2017), no. 1, 317–331.

[8] N. Alon and J. H. Spencer, *The probabilistic method*, John Wiley & Sons, 2016.

[9] L. D. Andersen, *Hamilton circuits with many colours in properly edge-coloured complete graphs*, Mathematica Scandinavica **64** (1989), 5–14.

[10] _____, *The history of Latin squares*, Combinatorics: Ancient & Modern (edited by R. Wilson and J.J. Watkins), 2013.

[11] J. Balogh and T. Molla, *Long rainbow cycles and Hamiltonian cycles using many colors in properly edge-colored complete graphs*, European Journal of Combinatorics **79** (2019), 140–151.

[12] B. Barber, D. Kühn, A. Lo, and D. Osthus, *Edge-decompositions of graphs with high minimum degree*, Advances in Mathematics **288** (2016), 337–385.

[13] A. Benford, C. Bowtell, and R. Montgomery. *Long rainbow cycles in properly coloured graphs. In preparation*, 2023.

[14] D. Best, K. Pula, and I. M. Wanless, *Small Latin arrays have a near transversal*, Journal of Combinatorial Designs **29** (2021), no. 8, 511–527.

[15] D. Best and I. M. Wanless, *What did Ryser conjecture?*, arXiv preprint arXiv:1801.02893 (2018).

[16] T. Bohman, A. Frieze, and E. Lubetzky, *Random triangle removal*, Advances in Mathematics **280** (2015), 379–438.

[17] R. C. Bose and S. S. Shrikhande. *On the falsity of Euler's conjecture about the non-existence of two orthogonal Latin squares of order $4t + 2$*, Proceedings of the National Academy of Sciences **45** (1959), no. 5, 734–737.

[18] R. C. Bose, S. S. Shrikhande, and E. T. Parker, *Further results on the construction of mutually orthogonal Latin squares and the falsity of Euler's conjecture*, Canadian Journal of Mathematics **12** (1960), 189–203.

[19] C. Bowtell and P. Keevash, *The n-queens problem*, arXiv preprint arXiv:2109.08083 (2021).

[20] J. N. Bray, Q. Cai, P. J. Cameron, P. Spiga, and H. Zhang, *The Hall–Paige conjecture, and synchronization for affine and diagonal groups*, Journal of Algebra **545** (2020), 27–42.

[21] L. M. Brègman, *Some properties of nonnegative matrices and their permanents*, In: *Doklady Akademii Nauk*, vol. 211, Russian Academy of Sciences, 1973, pp. 27–30.

[22] A. E. Brouwer, *On the size of a maximum transversal in a Steiner triple system*, Canadian Journal of Mathematics **33** (1981), no. 5, 1202–1204.

[23] A. E. Brouwer, A. de Vries, and R. Wieringa, *A lower bound for the length of partial transversals in a Latin square*, Nieuw Archief Voor Wiskunde **26** (1978), no. 2, 330–332.

[24] R. A. Brualdi and H. J. Ryser, *Combinatorial matrix theory*, Cambridge University Press, 1991.

[25] D. Bryant and D. Horsley, *A second infinite family of Steiner triple systems without almost parallel classes*, Journal of Combinatorial Theory, Series A, **120** (2013), no. 7, 1851–1854.

[26] ———, *Steiner triple systems without parallel classes*, SIAM Journal on Discrete Mathematics **29** (2015), no. 1, 693–696.

[27] N. J. Cavenagh and I. M. Wanless, *Latin squares with no transversals*, The Electronic Journal of Combinatorics (2017), P2.45.

[28] C. J. Colbourn and A. Rosa, *Colorings of block designs*, In: *Contemporary design theory: A collection of surveys*, 1992, pp. 401–430

[29] D. Conlon, W. T. Gowers, W. Samotij, and M. Schacht, *On the KLR conjecture in random graphs*, Israel Journal of Mathematics **203** (2014), no. 1, 535–580.

[30] D. A. Drake, *Maximal sets of Latin squares and partial transversals*, Journal of Statistical Planning and Inference **1** (1977), no. 2, 143–149.

[31] S. Eberhard, F. Manners, and R. Mrazović, *An asymptotic for the Hall–Paige conjecture*, Advances in Mathematics **404** (2022), 108423.

[32] ———, *Transversals in quasirandom Latin squares*, Proceedings of the London Mathematical Society **127** (2023), no. 1, 84–115.

[33] G. P. Egorychev, *The solution of van der Waerden's problem for permanents*, Advances in Mathematics **42** (1981), no. 3, 299–305.

[34] S. Ehard, S. Glock, and F. Joos, *Pseudorandom hypergraph matchings*, Combinatorics, Probability and Computing **29** (2020), no. 6, 868–885.

[35] P. Erdős, A. Gyárfás, and L. Pyber, *Vertex coverings by monochromatic cycles and trees*, Journal of Combinatorial Theory, Series B **51** (1991), no. 1, 90–95.

[36] L. Euler, *Recherches sur un nouvelle espéce de quarrés magiques*, Verhandelingen uitgegeven door het zeeuwsch Genootschap der Wetenschappen te Vlissingen **9** (1782), 85–239.

[37] A. B. Evans, *Latin squares without orthogonal mates*, Designs, Codes and Cryptography **40** (2006), 121–130.

[38] _____, *The admissibility of sporadic simple groups*, Journal of Algebra **321** (2009), no. 1, 105–116.

[39] _____. *Orthogonal Latin squares based on groups*, Springer, 2018.

[40] D. I. Falikman, *Proof of the van der Waerden conjecture regarding the permanent of a doubly stochastic matrix*, Mathematical Notes of the Academy of Sciences of the USSR **29** (1981), 475–479.

[41] A. Ferber, G. Kronenberg, and E. Long, *Packing, counting and covering Hamilton cycles in random directed graphs*, Israel Journal of Mathematics **220** (2017), 57–87.

[42] A. Ferber and M. Kwan, *Almost all Steiner triple systems are almost resolvable*, Forum of Mathematics, Sigma **8** (2020), 39.

[43] R. J. Friedlander, B. Gordon, and M. D. Miller, *On a group sequencing problem of Ringel*, Congr. Numer **21** (1978), 307–321.

[44] A. Georgakopoulos, *Delay colourings of cubic graphs*, The Electronic Journal of Combinatorics (2013), P45.

[45] R. Glebov and Z. Luria, *On the maximum number of Latin transversals*, Journal of Combinatorial Theory, Series A **141** (2016), 136–146.

[46] S. Glock, D. Kühn, R. Montgomery, and D. Osthus, *Decompositions into isomorphic rainbow spanning trees*, Journal of Combinatorial Theory, Series B **146** (2021), 439–484.

[47] S. Gould and T. Kelly, *Hamilton transversals in random Latin squares*, Random Structures & Algorithms **62** (2023), no. 2, 450–478.

[48] S. Gould, T. Kelly, D. Kühn, and D. Osthus, *Almost all optimally coloured complete graphs contain a rainbow Hamilton path*, Journal of Combinatorial Theory, Series B **156** (2022), 57–100.

[49] W. T. Gowers, *Quasirandom groups*, Combinatorics, Probability and Computing **17** (2008), no. 3, 363–387.

[50] G. Hahn, *Un jeu de colouration*, In: Actes du Colloque de Cerisy, vol. 12, 1980, pp. 18–18.

[51] M. Hall and L. J. Paige, *Complete mappings of finite groups*, Pacific Journal of Mathematics **5** (1955), no. 4, 541–549.

[52] M. Hall Jr., *Combinatorial Theory*, Waltham, Mass., USA, 1967.

[53] P. Hatami and P. W. Shor, *A lower bound for the length of a partial transversal in a Latin square*, Journal of Combinatorial Theory, Series A **115** (2008), no. 7, 1103–1113.

[54] S. Janson and A. Ruciński, *The infamous upper tail*, Random Structures & Algorithms **20** (2002), no. 3, 317–342.

[55] D. Y. Kang, T. Kelly, D. Kühn, A. Methuku, and D. Osthus, *Graph and hypergraph colouring via nibble methods: A survey*, arXiv preprint arXiv:2106.13733 (2021).

[56] P. Keevash, *The existence of designs*, arXiv preprint arXiv:1401.3665 (2014).

[57] _____, *Counting designs*, Journal of the European Mathematical Society **20** (2018), no. 4, 903–927.

[58] P. Keevash, A. Pokrovskiy, B. Sudakov, and L. Yepremyan, *New bounds for Ryser's conjecture and related problems*, Transactions of the American Mathematical Society, Series B, **9** (2022), no. 8, 288–321.

[59] T. Kirkman, *On a problem in combinatorics*, Cambridge and Dublin Math J. **2** (1847), no. 1847, 191–204.

[60] Y. Kohayakawa, B. Nagle, V. Rödl, and M. Schacht, *Weak hypergraph regularity and linear hypergraphs*, Journal of Combinatorial Theory, Series B **100** (2010), no. 2, 151–160.

[61] Y. Kohayakawa and V. Rödl, *Szemerédi's regularity lemma and quasi-randomness*, In: *Recent advances in algorithms and combinatorics*, 2003, pp. 289–351.

[62] K. K. Koksma, *A lower bound for the order of a partial transversal in a Latin square*, Journal of Combinatorial Theory **7** (1969), no. 1, 94–95.

[63] J. Komlós and E. Szemerédi, *Topological cliques in graphs*, Combinatorics, Probability and Computing **3** (1994), no. 2, 247–256.

[64] _____, *Topological cliques in graphs II*, Combinatorics, Probability and Computing **5** (1996), no. 1, 79–90.

[65] M. Krivelevich, *Triangle factors in random graphs*, Combinatorics, Probability and Computing **6** (1997), no. 3, 337–347.

[66] M. Kwan, *Almost all Steiner triple systems have perfect matchings*, Proceedings of the London Mathematical Society **121** (2020), no. 6, 1468–1495.

[67] M. Kwan, A. Sah, and M. Sawhney, *Large deviations in random Latin squares*, Bulletin of the London Mathematical Society **54** (2022), no. 4, 1420–1438.

[68] M. Kwan, A. Sah, M. Sawhney, and M. Simkin, *Substructures in Latin squares*, arXiv preprint arXiv:2202.05088 (2022).

[69] M. Kwan and B. Sudakov, *Intercalates and discrepancy in random Latin squares*, Random Structures & Algorithms **52** (2018), no. 2, 181–196.

[70] S. Letzter, *Sublinear expanders and their applications*, In: *Surveys in Combinatorics 2024*, Cambridge University Press, 2024.

[71] C. C. Lindner and K. T. Phelps, *A note on partial parallel classes in Steiner systems*, Discrete Mathematics **24** (1978), no. 1, 109–112.

[72] M. Maamoun and H. Meyniel, *On a problem of G. Hahn about coloured Hamiltonian paths in K_{2^t}*, Discrete Mathematics **51** (1984), no. 2, 213–214.

[73] E. Maillet, *Sur les carrés latins d'Euler*, Assoc. Franc. Caen **23** (1894), 244–252.

[74] H. B. Mann, *On orthogonal Latin squares*, Bulletin of the American Mathematical Society **50** (1944), no. 4, 249–257.

[75] B. D. McKay and I. M. Wanless, *Most Latin squares have many subsquares*, Journal of Combinatorial Theory, Series A **86** (1999), no. 2, 323–347.

[76] R. Montgomery, *A proof of the Ryser-Brualdi-Stein conjecture for large even n*, In preparation.

[77] _____, *Spanning trees in random graphs*, Advances in Mathematics **356** (2019), 106793.

[78] R. Montgomery, A. Pokrovskiy, and B. Sudakov, *Decompositions into spanning rainbow structures*, Proceedings of the London Mathematical Society **119** (2019), no. 4, 899–959.

[79] _____, *A proof of Ringel's conjecture*, Geometric and Functional Analysis **31** (2021), no. 3, 663–720.

[80] P. Morris. Random Steiner triple systems. *Master's thesis, Freie Universität Berlin*, 2017.

[81] D. Munhá Correia, A. Pokrovskiy, and B. Sudakov, *Short proofs of rainbow matchings results*, International Mathematics Research Notices **2023** (2023), no. 14, 12441–12476.

[82] A. Müyesser, *Cycle type in Hall-Paige: A proof of the Friedlander-Gordon-Tannenbaum conjecture*, arXiv preprint arXiv:2303.16757 (2023).

[83] A. Müyesser and A. Pokrovskiy, *A random Hall-Paige conjecture*, arXiv preprint arXiv:2204.09666 (2022).

[84] J. Ozanam, *Récréations mathématiques et physiques*, vol. 1, 1723.

[85] L. J. Paige, *A note on finite abelian groups*, Bulletin of the American Mathematical Society **53** (1947), no. 6, 590–593.

[86] _____, *Complete mappings of finite groups*, Pacific J. Math. **1** (1951), no. 1, 111–116.

[87] E. T. Parker, *Pathological Latin squares*, Proc. Symbs. Pure Math **19** (1971), 177–181.

[88] N. Pippenger and J. Spencer, *Asymptotic behavior of the chromatic index for hypergraphs*, Journal of Combinatorial Theory, Series A **51** (1989), no. 1, 24–42.

[89] A. Pokrovskiy, *An approximate version of a conjecture of Aharoni and Berger*, Advances in Mathematics **333** (2018), 1197–1241.

[90] _____ , *Rainbow Subgraphs and their Applications*, In: *Surveys in Combinatorics 2022*, Cambridge University Press, 2022, pp. 191–214.

[91] A. Pokrovskiy and B. Sudakov, *A counterexample to Stein's equi-n-square conjecture*, Proceedings of the American Mathematical Society **147** (2019), no. 6, 2281–2287.

[92] D. K. Ray-Chaudhuri and R. M. Wilson, *Solution of Kirkman's schoolgirl problem*, In: *Proceedings of Symposia in Pure Mathematics*, vol. 19, 1971, pp. 187–203.

[93] V. Rödl, *On a packing and covering problem*, European Journal of Combinatorics **6** (1985), no. 1, 69–78.

[94] V. Rödl and A. Ruciński, *Threshold functions for Ramsey properties*, Journal of the American Mathematical Society **8** (1995), no. 4, 917–942.

[95] V. Rödl, A. Ruciński, and E. Szemerédi, *A Dirac-type theorem for 3-uniform hypergraphs*, Combinatorics, Probability and Computing **15** (2006), no. 1-2, 229–251.

[96] H. Ryser, *Neuere Probleme der Kombinatorik*, In: *Vorträge über Kombinatorik, Oberwolfach*, 1967, pp. 69–91.

[97] P. W. Shor, *A lower bound for the length of a partial transversal in a Latin square*, Journal of Combinatorial Theory, Series A **33** (1982), no. 1, 1–8.

[98] S. K. Stein, *Transversals of Latin squares and their generalizations*, Pacific J. Math. **59** (1975), 567–575.

[99] A. Taranenko, *Multidimensional permanents and an upper bound on the number of transversals in Latin squares*, Journal of Combinatorial Designs **23** (2015), no. 7, 305–320.

[100] G. Tarry, *Le problème des 36 officiers*, Secrétariat de l'Association française pour l'avancement des sciences, 1900.

[101] J. H. Van Lint and R. M. Wilson, *A course in combinatorics*, Cambridge University Press, 2001.

[102] G. van Rees, *Subsquares and transversals in Latin squares*, Ars Combinatoria **29** (1990), 193–204.

[103] S. P. Wang, *On self-orthogonal Latin squares and partial transversals of Latin squares*, PhD thesis, The Ohio State University, 1978.

[104] I. M. Wanless, Transversals in Latin squares: A survey, In: *Surveys in Combinatorics 2011*, Cambridge University Press, 2011, pp. 403–437.

[105] I. M. Wanless and B. S. Webb, *The existence of Latin squares without orthogonal mates*, Designs, Codes and Cryptography **40** (2006), 131–135.

[106] S. Wilcox, *Reduction of the Hall–Paige conjecture to sporadic simple groups*, Journal of Algebra **321** (2009), no. 5, 1407–1428.

[107] D. E. Woolbright, *An $n \times n$ Latin square has a transversal with at least $n - \sqrt{n}$ distinct symbols*, Journal of Combinatorial Theory, Series A **24** (1978), no. 2, 235–237.

Mathematics Institute
University of Warwick
Coventry CV4 7AL, UK
richard.montgomery@warwick.ac.uk

Finite Field Models in Arithmetic Combinatorics – Twenty Years On

Sarah Peluse

Abstract

About twenty years ago, Green wrote a survey article on the utility of looking at toy versions over finite fields of problems in additive combinatorics. This survey was extremely influential, and the rapid development of additive combinatorics necessitated a follow-up article ten years later, which was written by Wolf. Since the publication of Wolf's article, an immense amount of progress has been made on several central open problems in additive combinatorics in both the finite field model and integer settings. This survey covers some of the most significant results of the past ten years and suggests future directions.

The central problems in additive combinatorics typically concern subsets of the integers. For example, one of Erdős's most well-known conjectures, which dates back to the 1940s, is that any subset S of the natural numbers for which $\sum_{n \in S} \frac{1}{n} = \infty$ must contain arbitrarily long nontrivial arithmetic progressions. These foundational problems are all famously hard, and a nontrivial portion of that difficulty comes from the fact that the integers are not very convenient to work with–one disadvantage being that they have no nontrivial finite subgroups. Thus, in a survey article for the 2005 British Combinatorial Conference written almost twenty years ago, Green [42] suggested that a possible line of attack for some of these problems is to consider their analogue in high-dimensional vector spaces over finite fields.

The conjecture of Erdős mentioned above can be shown to be essentially equivalent to the statement that, for each natural number k, there exists a constant $c_k > 0$ such that any subset $A \subset \{1, \dots, N\}$ containing no nontrivial k-term arithmetic progressions, i.e., configurations of the form

$$x, x+y, \dots, x+(k-1)y$$

with $y \neq 0$, must satisfy a bound of the form $|A| = O\left(\frac{N}{(\log N)^{1+c_k}}\right)$. Thus, this conjecture can be addressed by answering the following question.

Question 0.1 *How large can a subset of the first N integers be if it contains no nontrivial k-term arithmetic progressions?*

This question also makes sense with $\{1, \dots, N\}$ replaced by $\mathbf{F}_p^n$, where we are interested in the regime with the prime $p \geq k$ fixed (so that the terms of a nontrivial arithmetic progression are distinct) and n tending to infinity.

Question 0.2 *Fix a prime $p \geq k$. How large can a subset of $\mathbf{F}_p^n$ be if it contains no nontrivial k-term arithmetic progressions?*

Most other problems in additive combinatorics solely involving addition or subtraction can be similarly formulated in $\mathbf{F}_p^n$, usually provided that the characteristic p is sufficiently large. For problems that also involve multiplication, some of which we will discuss later on in the survey, it can be similarly fruitful to consider analogues phrased in $\mathbf{F}_p$ in the regime where p tends to infinity, in $\mathbf{F}_{p^n}$ in the regime where p is fixed and n tends to infinity, or in the function field setting $\mathbf{F}_q[t]$.

Two very useful advantages of working in high-dimensional vector spaces versus working in $\{1,\ldots,N\}$ are that one has access to the tools of linear algebra in the former setting and high-dimensional vector spaces posses a huge number of subgroups. This means that many technical challenges that arise when working in the integer setting are trivial, or at least much easier to overcome, in vector spaces over finite fields. There are a few ways in which $\mathbf{F}_p^n$ is more complicated to work in than $\{1,\ldots,N\}$, since the number of generators of $\mathbf{F}_p^n$ grows as n grows. Overall, though, problems in additive combinatorics formulated in high-dimensional vector spaces over finite fields tend to be easier than the same problem formulated in the integers. Despite this, the $\mathbf{F}_p^n$-analogues of problems preserve most of the essential features of the problems they are based on, and there are several tools for translating sufficiently robust proofs from the high dimensional vector space setting to the integer setting. Similarly, the various finite field model analogues of problems involving multiplication tend to be easier than the same problem formulated in the integers for a variety of reasons (e.g., $\mathbf{F}_p$ is closed under addition and multiplication, while $\{1,\ldots,N\}$ is not), but still preserve the problem's essential features.

Even without the motivation of making progress on some famous problem concerning sets of integers, finite field analogues of problems are interesting in their own right, often have applications in theoretical computer science, and have led to the development of some beautiful mathematics.

Since Wolf's follow-up survey [125], there has been spectacular progress in the case $k = 3$ of Questions 0.1 and 0.2, and this survey will begin by covering these recent developments and some related problems, with a focus on the finite field model setting. We will then discuss progress on the inverse theorems for the Gowers uniformity norms in the setting of high-dimensional vector spaces over finite fields and the very recent resolution of the polynomial Freiman–Ruzsa conjecture, and then turn to the problem of proving a quantitative version of the polynomial Szemerédi theorem, where attacking the problem first in $\mathbf{F}_p$ led to the key developments needed to make progress in the integer setting. The last section will briefly cover a collection of other interesting topics: relations between different notions of rank of multilinear forms, the multidimensional Szemerédi theorem, and the notion of true complexity.

Notation and Conventions

Throughout this survey, we will use both the standard asymptotic notation O, Ω, and o and Vinogradov's notation $\ll, \gg$, and $\asymp$. So, for any two quantities X and Y, $X = O(Y)$, $Y = \Omega(X)$, $X \ll Y$, and $Y \gg X$ all mean that $|X| \leq C|Y|$ for some absolute constant $C > 0$, and the relation $X \asymp Y$ means that $X \ll Y \ll X$. We will also write $O(Y)$ to represent a quantity that is $\ll Y$ and $\Omega(X)$ to represent a quantity that is $\gg X$. The notation $\log_{(m)}$ means the m-fold iterated logarithm and $\exp^{(m)}$ means the m-fold iterated exponential.

For any subsets A and B of an abelian group $(G, +)$, we can form their *sumset* and *product set*,
$$A + B := \{a + b : a \in A \text{ and } b \in B\}$$

and
$$A - B := \{a - b : a \in A \text{ and } b \in B\}$$
respectively. We will denote the k-fold sumset of a set A with itself by
$$kA := \{a_1 + \cdots + a_k : a_1, \ldots, a_k \in A\}$$
for every natural number k. Note that this is not the dilation of the set A by k, which we will instead denote by $k \cdot A$, so that
$$k \cdot A := \{ka : a \in A\}.$$

We will denote the indicator function of a set S by 1_S, and set $e(z) := e^{2\pi i z}$ for all $z \in \mathbf{R}$ and $e_p(z) := e(z/p)$ for all $z \in \mathbf{F}_p$ and primes p. A complex-valued function is said to be *1-bounded* if its modulus is bounded by 1, so that indicator functions and complex exponentials $e(\cdot)$ and $e_p(\cdot)$ are always 1-bounded.

For any finite, nonempty set S and any complex-valued function f on S, we denote the average of f over S by $\mathbf{E}_{x \in S} f(x) := \frac{1}{|S|} \sum_{x \in S} f(x)$. The Fourier transform of $f : \mathbf{F}_p^n \to \mathbf{C}$ at the frequency $\xi \in \mathbf{F}_p^n$ is defined to be
$$\widehat{f}(\xi) := \mathbf{E}_{x \in \mathbf{F}_p^n} f(x) e_p(-\xi \cdot x).$$

We then have the Fourier inversion formula,
$$f(x) = \sum_{\xi \in \mathbf{F}_p^n} \widehat{f}(\xi) e_p(\xi \cdot x)$$
and, for any other function $g : \mathbf{F}_p^n \to \mathbf{C}$, Parseval's identity
$$\mathbf{E}_{x \in \mathbf{F}_p^n} f(x) \overline{g(x)} = \sum_{\xi \in \mathbf{F}_p^n} \widehat{f}(\xi) \overline{\widehat{g}(\xi)}.$$

Finally, we will use the standard notation $[N] := \{1, \ldots, N\}$ for the set of the first N integers.

1 Roth's Theorem, the Cap Set Problem, and the Polynomial Method

The first nontrivial case of Question 0.1 is the case $k = 3$, which has attracted by far the most attention and is one of the central problems in additive combinatorics.

Question 1.1 *How large can a subset of the first N integers be if it contains no nontrivial three-term arithmetic progressions?*

Roth [102] was the first to address this question, and also the first to show that if $A \subset [N]$ contains no nontrivial three-term arithmetic progressions, then $|A| = o(N)$. Using a Fourier-analytic argument, Roth proved the explicit bound $|A| \ll \frac{N}{\log \log N}$ for the size of such sets.

A subset of $\mathbf{F}_3^n$ having no nontrivial three-term arithmetic progressions is called a *cap set*, and the $k = 3$ case of Question 0.2 is known as the *cap set problem*.

Question 1.2 *How large can a subset of $\mathbf{F}_3^n$ be if it contains no nontrivial three-term arithmetic progressions?*

Brown and Buhler [19] were the first to prove that if $A \subset \mathbf{F}_3^n$ is a cap set, then $|A| = o(3^n)$. The first quantitative bound in the cap set problem was proven by Meshulam [84], who adapted Roth's proof to the setting of high-dimensional vector spaces over finite fields to show that if A is a cap set, then $|A| \ll \frac{3^n}{n}$. Note that the savings over the trivial bound $|\mathbf{F}_3^n|$ of $\log |\mathbf{F}_3^n|$ in Meshulam's theorem is exponentially stronger than the savings of $\log \log N$ in Roth's theorem–this turns out to be a direct consequence of the abundance of subspaces in $\mathbf{F}_3^n$.

Until very recently, all of the work on improving Roth's theorem focused on refining Roth's original Fourier-analytic proof to make it as efficient as in the finite field model setting. Work by Heath-Brown [58] and Szemerédi [116], Bourgain [16, 17], Sanders [107, 108], Bloom [11], and Schoen [111] improved the bound in Roth's theorem right up to the $O\left(\frac{N}{\log N}\right)$-barrier by following this approach, with Schoen's result giving a bound of $|A| \ll \frac{N(\log \log N)^{3+o(1)}}{\log N}$. By a very careful analysis of the set of large Fourier coefficients of cap sets, Bateman and Katz [5] proved that if $A \subset \mathbf{F}_3^n$ is a cap set, then $|A| \ll \frac{3^n}{n^{1+c}}$ for some absolute constant $c > 0$, thus breaking the logarithmic barrier in the cap set problem for the first time. Overcoming numerous difficult technical obstacles, Bloom and Sisask [13] managed to adapt the (already very complicated) argument of Bateman and Katz to the integer setting.

Theorem 1.3 *If $A \subset [N]$ contains no nontrivial three-term arithmetic progressions, then*

$$|A| \ll \frac{N}{(\log N)^{1+c}},$$

where $c > 0$ is some absolute constant.

This proves the first nontrivial case of the famous conjecture of Erdős mentioned in the introduction. In comparison, in 1946, Behrend [6] constructed a subset of $[N]$ lacking three-term arithmetic progressions of density $\gg \exp(-C\sqrt{\log N})$, where $C > 0$ is an absolute constant. This is still, essentially, the best known lower bound construction.

Earlier this year, there was another spectacular breakthrough in the study of sets lacking three-term arithmetic progressions. Kelley and Meka [64] proved almost optimal bounds in Roth's theorem using a beautiful argument that almost completely avoids the use of Fourier analysis and is significantly simpler than the arguments of Bateman–Katz and Bloom–Sisask, though, like all improvements to the best known bound in Roth's theorem since the work of Sanders, it uses a result of Croot and Sisask [21] on the almost-periodicity of convolutions.

Theorem 1.4 *If $A \subset [N]$ contains no nontrivial three-term arithmetic progressions, then*

$$|A| \ll \frac{N}{\exp(C(\log N)^{1/12})},$$

where $C > 0$ is some absolute constant.

Bloom and Sisask [14] have since shown that the exponent $\frac{1}{12}$ of $\log N$ appearing in the Kelley–Meka theorem can be improved to $\frac{1}{9}$ by optimizing the argument in [64].

In contrast to the integer setting, all known constructions of cap sets are exponentially small, and thus Frankl, Graham, and Rödl [27], and, independently, Alon and Dubiner [1], asked whether there exists a positive constant $c < 3$ such that $|A| \ll c^n$ whenever $A \subset \mathbf{F}_3^n$ is a cap set. In breakthrough work, Croot, Lev, and Pach [20] introduced a new variant of the polynomial method, which they used to show that any subset of $(\mathbf{Z}/4\mathbf{Z})^n$ containing no nontrivial three-term arithmetic progressions must have size $\ll 3.61^n$. Ellenberg and Gijswijt [23] then used the Croot–Lev–Pach polynomial method to obtain a similar power-saving bound in the cap set problem.

Theorem 1.5 *If $A \subset \mathbf{F}_3^n$ is a cap set, then*

$$|A| \ll 2.756^n.$$

Though it produces weaker bounds than those of Ellenberg and Gijswijt in the cap set problem, the argument of Kelley and Meka is still very interesting in the setting of high-dimensional vector spaces over finite fields. Answering a question of Schoen and Sisask [112], Kelley and Meka showed that if $A \subset \mathbf{F}_q^n$ has density α, then its three-fold sumset $A + A + A$ must contain an affine subspace of codimension $\ll \log(1/\alpha)^9$. Bloom and Sisask [15] showed that their argument adapts to prove an integer analogue of this result: if $A \subset [N]$ has density α, then $A + A + A$ must contain an arithmetic progression of length $\gg N^{\Omega(1/\log(2/\alpha)^9)} \exp(-O(\log(2/\alpha)^2))$, which improves substantially on the previous best lower bound of $\gg N^{\alpha^{1+o(1)}}$ due to Sanders [106]. It is likely that the new ideas of Kelley and Meka will have further applications in both the finite field model and integer settings. For more on the work of Kelley and Meka, the interested reader should consult their paper [64] or the very short and clear exposition of their argument by Bloom and Sisask [15], which gives an almost self-contained proof (minus a now standard almost-periodicity result and some basic facts about Bohr sets) of Theorem 1.4 in less than ten pages.

The remainder of this section will be devoted to the cap set problem and the Croot–Lev–Pach polynomial method.

1.1 The Slice Rank Method

In this subsection, we will present a full proof of a slightly weaker version of Theorem 1.5.

Theorem 1.6 *If $A \subset \mathbf{F}_3^n$ is a cap set, then*

$$|A| \ll 2.838^n.$$

The improved power-saving bound in Theorem 1.5 can be obtained using essentially the same argument, but with a bit more (tedious) work.

The proof of Theorem 1.6 uses the "slice rank method", which is a symmetric reformulation, due to Tao [120], of the proof of Ellenberg and Gijswijt [23]. The slice rank of a function is a measure of its complexity, and the basic idea of the argument is the following. If S is any finite set, then the indicator function $1_{\Delta(S)} : S \times S \times S \to \mathbf{F}_3$ of the diagonal $\Delta(S) := \{(s, s, s) : s \in S\}$ of $S \times S \times S$ must have slice rank at least

$|S|$ (in the same way that the $N \times N$ identity matrix has rank N). On the other hand, if $S \subset \mathbf{F}_3^n$ is a cap set, then $1_{\Delta(S)}$ can be seen to have very small slice rank. Combining these two facts produces an upper bound for $|S|$.

Now we will define slice rank, specializing to functions taking values in $\mathbf{F}_3$.

Definition 1.7 Let k be a natural number and $S \neq \emptyset$ be a finite set.

1. A function $f : S^k \to \mathbf{F}_3$ has *slice rank* 1 if there exist functions $g : S \to \mathbf{F}_3$ and $h : S^{k-1} \to \mathbf{F}$ and an index $i \in [k]$ such that
$$f(x_1, \ldots, x_k) = g(x_i)h(x_1, \ldots, x_{i-1}, x_{i+1}, \ldots, x_k)$$
for all $x_1, \ldots, x_k \in S$.

2. A function $f : S^k \to \mathbf{F}_3$ has *slice rank at most m* if there exist functions $f_1, \ldots, f_m : S^k \to \mathbf{F}_3$, all with slice rank 1, such that
$$f = \sum_{j=1}^m f_j.$$

3. The *slice rank* of a function $f : S^k \to \mathbf{F}_3$ is defined to be the smallest natural number m such that f has slice rank at most m.

Observe that, when $k = 2$, the slice rank of f is just the (usual) rank of the $|S| \times |S|$ matrix $(f(s, s'))_{s,s' \in S}$. This will be important in the proof of Lemma 1.8 below.

Lemma 1.8 *Let $S \neq \emptyset$ be a finite set. The slice rank of $1_{\Delta(S)}$ is $|S|$.*

Proof First, observe that the slice rank of $1_{\Delta(S)}$ is at most $|S|$, since it can be written in the following way as the sum of slice rank 1 functions from S^3 to $\mathbf{F}_3$:
$$1_{\Delta(S)}(x, y, z) = \sum_{s \in S} \delta_s(x)\delta_s(y)\delta_s(z).$$

So, the real content of this lemma is that $1_{\Delta(S)}$ cannot have slice rank strictly less than $|S|$, and the strategy is to reduce the problem of bounding the slice rank of $1_{\Delta(S)}$ from below to the problem of bounding the rank of a diagonal matrix from below, which is much easier to get our hands on. Suppose by way of contradiction that the slice rank of $1_{\Delta(S)}$ is less than $|S|$, so that there exists a natural number $n < |S|$, nonnegative integers n_1 and n_2, and functions $g_1, \ldots, g_n : S \to \mathbf{F}_3$ and $h_1, \ldots, h_n : S^2 \to \mathbf{F}_3$ for which
$$1_{\Delta(S)}(x, y, z) = \sum_{j=1}^{n_1} g_j(x)h_j(y, z) + \sum_{j=n_1+1}^{n_2} g_j(y)h_j(x, z) + \sum_{j=n_2+1}^{n} g_j(z)h_j(x, y)$$
where, we may assume without loss of generality that $0 \leq n_1 \leq n_2 < n$. Set $n_3 := n - n_2$, so that n_3 is positive.

Now, observe that if $r : S \to \mathbf{F}_3$ is any function, then the function $f_r : S^2 \to \mathbf{F}_3$ defined by
$$f_r(x, y) := \sum_{z \in S} 1_{\Delta(S)}(x, y, z)r(z)$$

Finite Field Models in Arithmetic Combinatorics   165

is supported on the diagonal of S^2, and takes the values $f_r(s,s) = r(s)$ there. Thus, the slice rank of f_r is exactly the size of the support of r. We will obtain a contradiction by finding a function $r : S \to \mathbf{F}$ with support of size greater than n_2 that is orthogonal to each of the functions $g_{n_2+1}, \ldots, g_n$, since then f_r will have slice rank greater than n_2 but will be expressible in the form

$$\sum_{j=1}^{n_1} g_j(x) \left(\sum_{z \in S} h_j(y,z) r(z) \right) + \sum_{j=n_1+1}^{n_2} g_j(y) \left(\sum_{z \in S} h_j(x,z) r(z) \right),$$

which is the sum of at most n_2 functions of slice rank 1, yielding a contradiction.

Showing that such an $r : S \to \mathbf{F}_3$ exists just requires some simple linear algebra. Let

$$V := \left\{ f : S \to \mathbf{F}_3 : \sum_{s \in S} f(s) g_j(s) = 0 \text{ for all } j = n_2+1, \ldots, n \right\}$$

be the vector space over $\mathbf{F}_3$ of functions on S that are orthogonal to each of $g_{n_2+1}, \ldots, g_n$, and suppose that $r \in V$ has maximal support size over all functions in V. Since $\operatorname{codim} V \leq n_3 \leq n < |S|$, V must contain some nonzero function, which means that r is not identically zero. Further, the subspace V' of V of functions vanishing on the support $\operatorname{supp} r$ of r has codimension at most $|\operatorname{supp} r| + n_3$. If $|\operatorname{supp} r| \leq n_2$, this means that $\operatorname{codim} V' \leq n_2 + n_3 = n < |S|$, and so there would exist a nonzero function $r' \in V'$, which necessarily has support disjoint from that of r. But then $r + r'$ would have support strictly larger than the support of r, which contradicts the maximality of $\operatorname{supp} r$. Thus, we must have $|\operatorname{supp} r| > n_2$, showing that a function with the properties desired above exists. We therefore conclude that $1_{\Delta(S)}$ has slice rank exactly $|S|$. □

Note that the above lemma required no information about S aside from the assumption that it is finite and nonempty. In contrast, the next lemma relies crucially on the assumption that A is a cap set.

Lemma 1.9 *If $A \subset \mathbf{F}_3^n$ is a cap set, then the slice rank of $1_{\Delta(A)}$ is bounded above by*

$$3 \cdot \# \left\{ \mathbf{a} \in \{0,1,2\}^n : \sum_{i=1}^n a_i \leq \frac{2n}{3} \right\}. \tag{1.1}$$

Proof Note that
$$1_{\Delta(A)}(x,y,z) = \delta_0(x+y+z)$$
since A is a cap set, and a cap set contains only trivial solutions to $x+y+z = 0$ (which, in characteristic 3, is the same as the equation $x + z = 2y$ characterizing three-term arithmetic progressions). Thus, we can express $1_{\Delta(A)}$ as

$$1_{\Delta(A)}(x,y,z) = \prod_{i=1}^n \left(1 - (x_i + y_i + z_i)^2\right)$$

since $\delta_0(w) = 1 - w^2$ for all $w \in \mathbf{F}_3$. The polynomial on the right-hand side above has degree $2n$, and every monomial appearing in it takes the form

$$\prod_{i=1}^n x_i^{a_i} y_i^{b_i} z_i^{c_i},$$

where $a_i, b_i, c_i \in \{0, 1, 2\}$ for each $i \in [n]$ and $\sum_{i=1}^n (a_i + b_i + c_i) \leq 2n$. It follows from this second fact that, for any such monomial, $\sum_{i=1}^n a_i$, $\sum_{i=1}^n b_i$, or $\sum_{i=1}^n c_i$ is at most $2n/3$ (for otherwise $\sum_{i=1}^n (a_i + b_i + c_i)$ would be greater than $2n$). This means that there exist functions $g_{\mathbf{a}}^{(1)}, g_{\mathbf{b}}^{(2)}, g_{\mathbf{c}}^{(3)} : A^2 \to \mathbf{F}_3$ such that

$$1_{\Delta(A)}(x, y, z) = \sum_{\substack{\mathbf{a} \in \{0,1,2\}^n \\ \sum_{i=1}^n a_i \leq 2n/3}} x_1^{a_1} \cdots x_n^{a_n} g_{\mathbf{a}}^{(1)}(y, z) + \sum_{\substack{\mathbf{b} \in \{0,1,2\}^n \\ \sum_{i=1}^n b_i \leq 2n/3}} y_1^{b_1} \cdots y_n^{b_n} g_{\mathbf{b}}^{(2)}(x, z)$$

$$+ \sum_{\substack{\mathbf{c} \in \{0,1,2\}^n \\ \sum_{i=1}^n c_i \leq 2n/3}} z_1^{c_1} \cdots z_n^{c_n} g_{\mathbf{c}}^{(3)}(x, y),$$

from which is follows that $1_{\Delta(A)}$ has slice rank at most

$$3 \cdot \# \left\{ \mathbf{a} \in \{0, 1, 2\}^n : \sum_{i=1}^n a_i \leq \frac{2n}{3} \right\}.$$

$\square$

Now we can prove Theorem 1.6.

Proof The above two lemmas say that the slice rank of $1_{\Delta(A)}$ is both equal to $|A|$ and bounded above by (1.1). Thus,

$$|A| \ll \# \left\{ \mathbf{a} \in \{0, 1, 2\}^n : \sum_{i=1}^n a_i \leq \frac{2n}{3} \right\},$$

and it remains to bound the right-hand side of the above. Note that this is exactly 3^{-n} times the probability that, if $X_1, \ldots, X_n$ is a sequence of independent uniform random variables taking values in $\{0, 1, 2\}$, the random variable $X_1 + \cdots + X_n$ is at most $2n/3$. Hoeffding's inequality says that this probability is $\ll e^{-n/18}$, so that

$$|A| \ll \left(\frac{3}{e^{1/18}} \right)^n \ll 2.838^n,$$

as desired. $\square$

To improve the upper bound to $\ll 2.756^n$ as in Theorem 1.5, one just needs to put a bit more effort into estimating the size of (1.1).

1.2 Further Results and Questions

The slice rank method has had numerous applications to a wide variety of problems in combinatorics since the work of Croot–Lev–Pach and Ellenberg–Gijswijt. We will mention a few that are relevant to the theme of this survey.

A *tricolored sum-free set* in $\mathbf{F}_p^n$ is a collection of triples $((x_i, y_i, z_i))_{i=1}^M$ of elements of $\mathbf{F}_p^n$ such that $x_i + y_j + z_k = 0$ if and only if $i = j = k$. Note that if $A \subset \mathbf{F}_3^n$ is a cap set, then the diagonal $\Delta(A)$ is a tricolored sum-free set in $\mathbf{F}_3^n$. Blasiak, Church, Cohn, Grochow, Naslund, Sawin, and Umans [10] and, independently, Alon, observed that the slice rank method can also be used to bound the size of tricolored sum-free sets in $\mathbf{F}_p^n$, thus generalizing Theorem 1.5. Kleinberg, Sawin, and Speyer [66] showed that the bound obtained in [10] is essentially optimal for each prime p, assuming a conjecture that was shortly after proved by Norin [91] and Pebody [92]. These results combined yield the following theorem.

Theorem 1.10 *For every prime p, there exists a constant $c_p \in (0,1)$ such that the following holds. Any tricolored sum-free set in $\mathbf{F}_p^n$ has size at most $p^{(1-c_p)n}$, and there exists a tricolored sum-free set in $\mathbf{F}_p^n$ of size at least $p^{(1-c_p)n-o(1)}$.*

The largest known cap sets in $\mathbf{F}_3^n$ have size on the order of 2.218^n [124], so, in contrast to the tricolored sum-free set problem, there is still an exponential gap between the best known bounds in the cap set problem.

Recall that the triangle removal lemma, a standard result in extremal graph theory, states that for every $\varepsilon > 0$ there exists a $\delta > 0$ such that any graph on n vertices containing δn^3 triangles (i.e., copies of the complete graph on three vertices) can be made triangle-free by removing at most εn^2 edges. The quickest way to prove the triangle removal lemma is by using Szemerédi's regularity lemma, which produces an upper bound for $\frac{1}{\delta}$ that is a tower of height polynomial in $\frac{1}{\varepsilon}$. The best known bounds for $\frac{1}{\delta}$ in the triangle removal lemma are due to Fox [24], who showed that one can take $\frac{1}{\delta}$ bounded by a tower of height $\ll \log(1/\varepsilon)$.

Now let A, B, and C be subsets of a finite abelian group $(G, +)$. A *triangle* in $A \times B \times C$ is a triple $(x, y, z) \in A \times B \times C$ satisfying $x+y+z=0$. Green [44] developed an arithmetic regularity lemma, and used it to prove the following "arithmetic triangle removal lemma": for every $\varepsilon > 0$, there exists a $\delta > 0$ such that if $A \times B \times C$ contains $\delta|G|^2$ triangles, then $A \times B \times C$ can be made triangle-free by removing at most $\varepsilon|G|$ elements from A, B, and C. Green's argument produced a bound for $\frac{1}{\delta}$ that is a tower of height polynomial in $\frac{1}{\varepsilon}$. Král, Serra, and Vena [67] observed that Green's arithmetic triangle removal lemma also follows from the triangle removal lemma for graphs, and used this observation to prove a generalization of Green's result for nonabelian groups. Indeed, consider the graph with vertex set consisting of three copies of G,

$$V := (G \times \{1\}) \cup (G \times \{2\}) \cup (G \times \{3\}),$$

and edge set

$$E := \{((g,1),(g+a,2)) : g \in G, a \in A\} \cup \{((g,2),(g+b,3)) : g \in G, b \in B\}$$
$$\cup \{((g,3),(g+c,1)) : g \in G, c \in C\}.$$

Then a triangle in the graph (V, E) is exactly a triple of the form

$$((g,1),(g+a,2)), ((g+a,2),(g+a+b,3)), ((g+a+b,3),(g+a+b+c,1)),$$

where $g \in G$, $(a,b,c) \in A \times B \times C$, and $a+b+c=0$. Thus, the number of triangles in the graph (V, E) equals $|G|$ (the number of possible choices for $g \in G$) times the number of triangles in $A \times B \times C$. The arithmetic triangle removal lemma now follows from the triangle removal lemma for graphs and the pigeonhole principle. Thus, Fox's improved bound in the triangle removal lemma leads to the same improved bound in the arithmetic triangle removal lemma.

Green asked in [44] whether a polynomial bound in the arithmetic triangle removal lemma could hold in the setting of high-dimensional vector spaces over finite fields. This was answered in the affirmative by Fox and L. M. Lovász [25], who obtained a massive improvement over the previous best known bound, and also show that their result is essentially tight.

Theorem 1.11 *For every prime p, there exists a constant $C_p \in (0,1)$ such that the following holds. If $\varepsilon > 0$ and $\delta = (\varepsilon/3)^{C_p}$, then whenever $A, B, C \subset \mathbf{F}_p^n$ are such that $A \times B \times C$ has at most δp^{2n} triangles, then $A \times B \times C$ can be made triangle-free by removing at most εp^n elements from A, B, and C. The best possible δ for which this result holds satisfies $\delta \leq \varepsilon^{C_p - o(1)}$.*

Fox and Lovász prove Theorem 1.11 by a clever argument using Theorem 1.10 as its key input. In [44], Green actually proved a more general arithmetic k-cycle removal lemma. A polynomial bound in this theorem was later shown by Fox, L. M. Lovász, and Sauermann [26].

We will now turn to the problem of finding nonlinear configurations, both in subsets of the integers and in various finite field model settings. Let $P \in \mathbf{Z}[y]$ be a polynomial with zero constant term and degree greater than 1. Proving a conjecture of L. Lovász (which was also confirmed independently by Furstenberg [29], though without quantitative bounds), Sárközy [110] showed in 1978 that if $A \subset [N]$ contains no nontrivial two-term polynomial progressions $x, x + P(y)$, then $|A| \ll \frac{N}{\log \log N}$. Analogously to Question 0.1, the following problem is of great interest, in particular for the simplest nonlinear case $P(y) = y^2$.

Question 1.12 *How large can a subset of the first N integers be if it contains no nontrivial progressions of the form $x, x + P(y)$?*

There is a huge gap between the best-known upper and lower bounds for Question 1.12. Bloom and Maynard [12] showed that any $A \subset [N]$ lacking the progression $x, x + y^2$ must satisfy $|A| \ll \frac{N}{(\log N)^{C \log \log \log N}}$ for some constant $C > 0$, improving on a bound of Balog, Pelikán, Pintz, and Szemerédi [4] that was almost thirty years old. Arala [3] extended the argument of Bloom and Maynard to prove bounds of the same shape for sets lacking the progression $x, x + P(y)$ for any P having zero constant term (and, even more generally, to any intersective P). The largest known subsets of $[N]$ lacking two-term polynomial progressions (which come from either a greedy construction or a construction of Ruzsa [103]) all have size on the order of N^{γ_P} for some $\gamma_P \in (0,1)$. I think it is more likely that these lower bounds are closer to the truth in Question 1.12, especially given Green's recent proof of power-saving bounds for subsets of $[N]$ whose difference sets avoid the set of shifted primes $\{p - 1 : p \text{ prime}\}$ [47].

There are a few natural settings involving finite fields in which one can formulate a version of Question 1.12. Using the Weil bound for additive character sums, it is not difficult to prove power-saving bounds for the size of subsets of $\mathbf{F}_p$ lacking any fixed polynomial progression $x, x + P(y)$. So, replacing $[N]$ by $\mathbf{F}_p$ in Question 1.12 leads to too simple a problem (though, as we will discuss later in this survey, the analogous question for longer polynomial progressions in subsets of finite fields is much more interesting).

Question 1.12 makes sense and is highly nontrivial in the *function field setting*, where $\mathbf{Z}$ is replaced by $\mathbf{F}_q[t]$ for a fixed prime power q and $\{1, \ldots, N\}$ is replaced by $\mathbf{F}_q[t]_{<n} := \{f \in \mathbf{F}_q[t] : \deg f < n\}$. Observe that $|\mathbf{F}_q[t]_{<n}| = q^n$.

Question 1.13 *Let $P \in \mathbf{F}_q[X]$ have zero constant term. How large can a subset of $\mathbf{F}_q[t]_{<n}$ be if it contains no nontrivial progressions of the form $f, f + P(g)$?*

This question was first studied by Le and Liu [74], who proved a bound of the form $|A| \ll \frac{q^n (\log n)^{O(1)}}{n}$ when $A \subset \mathbf{F}_q[t]_{<n}$ lacks the progression $f, f + g^2$.

Shortly after the proof of Theorem 1.5, Green [46] used the slice rank method to prove power-saving bounds in Question 1.13, provided that the polynomial P satisfies certain technical conditions. These conditions were removed in recent work of Li and Sauermann [76], who also used the slice rank method.

Theorem 1.14 *Fix a prime power q, and let $P \in \mathbf{F}_q[X]$ have zero constant term. If $A \subset \mathbf{F}_q[t]_{<n}$ has no nontrivial progressions*

$$f, f + P(g),$$

then

$$|A| \ll_{q, \deg P} (q^n)^{1 - \gamma_{q, \deg P}},$$

for some constant $1 > \gamma_{q, \deg P} > 0$.

This theorem implies power-saving bounds for Sárközy's theorem in $\mathbf{F}_{p^n}$ in the regime where the prime p is fixed and n tends to infinity.

Corollary 1.15 *Fix a prime p and let $P \in \mathbf{F}_{p^n}[x]$ have zero constant term. If $A \subset \mathbf{F}_{p^n}$ has no nontrivial progressions*

$$x, x + P(y),$$

then

$$|A| \ll_{p, \deg P} (p^n)^{1 - \gamma_{p, \deg P}}.$$

Perhaps the most important open question regarding the slice rank method is whether it, or some variant, can be used to prove power-saving bounds for subsets of $\mathbf{F}_5^n$ lacking four-term arithmetic progressions. The best-known bound for subsets of $\mathbf{F}_5^n$ free of four-term arithmetic progressions is due to Green and Tao [51] (see also the corrected arXiv version [54]).

Theorem 1.16 *If $A \subset \mathbf{F}_5^n$ contains no nontrivial four-term arithmetic progressions, then*

$$|A| \ll \frac{p^n}{n^c},$$

where $c > 0$ is a (very small) absolute constant.

Any substantial improvement to the bound in Theorem 1.16 would be of great interest.

It is also not known whether the slice rank method, or some variant of it, could apply to prove power-saving bounds for subsets of $(\mathbf{F}_2^n)^2$ lacking nontrivial *corners*,

$$(x,y), (x, y+z), (x+z, y) \qquad (z \neq 0).$$

The best-known bounds for corner-free sets comes from adapting a proof of Shkredov [113, 114] in the integer setting to the finite field model setting [42], with the largest exponent of $\log n$ coming from work of Lacey and McClain [73].

Theorem 1.17 *If $A \subset (\mathbf{F}_2^n)^2$ contains no nontrivial corners, then*

$$|A| \ll \frac{4^n \log \log n}{\log n}.$$

The problem of adapting the slice rank method to prove power-saving bounds for sets lacking four-term arithmetic progressions or for sets lacking corners has received a considerable amount of attention, and is likely very difficult. However, I am optimistic that some of the ideas from the recent breakthrough of Kelley and Meka [64] on Roth's theorem in the integer setting could be used to improve the bounds for corner-free sets, in both the integer and finite field model setting. In particular, I expect the following problem should be attackable.

Problem 1.18 *Show that if $A \subset (\mathbf{F}_2^n)^2$ contains no nontrivial corners, then*

$$|A| \ll \frac{4^n}{n^{c'}}$$

for some absolute constant $c' > 0$.

2 The Inverse Theorems for the Gowers Uniformity Norms

Confirming a conjecture of Erdős and Turán from 1936, Szemerédi [115] proved in 1975 that if $A \subset [N]$ contains no nontrivial k-term arithmetic progressions, then $|A| = o_k(N)$. Answering Question 0.1 is therefore equivalent to determining the best possible bounds for $|A|$ in Szemerédi's theorem. The bounds for $o_k(N)$ that can be extracted from Szemerédi's argument, in which he introduced his now-famous regularity lemma for graphs, are extremely weak–the savings over the trivial bound of N are of inverse-Ackermann type.

No reasonable bounds in Question 0.1, i.e., a savings over the trivial bound of N that grows at least as fast as a finite number of iterated logarithms of N, were known for any k larger than three until pioneering work of Gowers [31, 33] in the late 1990s and early 2000s, who initiated the study of "higher-order Fourier analysis" and used it to prove that

$$|A| \ll_k \frac{N}{(\log \log N)^{2^{-2^{k+9}}}}.$$

for any $k \geq 4$. Before we explain what higher-order Fourier analysis is, we will briefly illustrate why Fourier analysis is relevant to the study of three-term arithmetic progressions.

2.1 The Fourier-Analytic Approach to Roth's Theorem

Let $A \subset \mathbf{F}_3^n$ be a nonempty subset of density α. Note that if we construct a random subset A' of $\mathbf{F}_3^n$ by including each element independently and uniformly with probability α, then A' will almost always have density very close to α and contain very close to $\alpha^2 |\mathbf{F}_3^n|^2$ three-term arithmetic progressions. The distance between the number of three-term arithmetic progressions in A and the number $\alpha^2 |\mathbf{F}_3^n|^2$ expected in a random set of the same density can be controlled using Fourier analysis.

For any functions $f_0, f_1, f_2 : \mathbf{F}_3^n \to \mathbf{C}$, we define the trilinear average $\Lambda_3(f_0, f_1, f_2)$ by

$$\Lambda_3(f_0, f_1, f_2) := \mathbf{E}_{x,y \in \mathbf{F}_3^n} f_0(x) f_1(x+y) f_2(x+2y).$$

Thus, $\Lambda_3(1_A, 1_A, 1_A)$ equals the normalized count of the number of three-term arithmetic progressions in A. Observe that, by Fourier inversion, $\Lambda_3(1_A, 1_A, 1_A)$ also equals

$$\sum_{\xi_1, \xi_2, \xi_3 \in \mathbf{F}_3^n} \widehat{1_A}(\xi_1) \widehat{1_A}(\xi_2) \widehat{1_A}(\xi_3) \left(\mathbf{E}_{x,y \in \mathbf{F}_3^n} e_3([\xi_1 + \xi_2 + \xi_3]x + [\xi_2 + 2\xi_3]y) \right),$$

which, by orthogonality of characters and Parseval's identity, equals

$$\sum_{\xi \in \mathbf{F}_3^n} \widehat{1_A}(\xi)^2 \widehat{1_A}(-2\xi) = \alpha^3 + O\left(\max_{0 \neq \xi \in \mathbf{F}_3^n} \left| \widehat{1_A}(\xi) \right| \right).$$

Thus, if A has far from the $\alpha^3 |\mathbf{F}_3^n|^2$ three-term arithmetic progressions expected in a random set of density α (which happens, for example, when A is a cap set and n is large enough that the number of trivial three-term progressions is significantly smaller than $\alpha^3 |\mathbf{F}_3^n|^2$), then $\widehat{1_A}(\xi)$ must be large for some nonzero $\xi \in \mathbf{F}_3^n$. This gives us strong structural information about A, which can be used to continue the argument–full details can be found in the surveys of Green [42] and Wolf [125], or in my Bourbaki seminar article on quantitative bounds in Roth's theorem [96].

Fourier analysis is not sufficient for the study of four-term, and longer, arithmetic progressions, in the sense that a set can have no large nontrivial Fourier coefficients but still contain far from the number of four-term arithmetic progressions expected in a random set of the same size. Indeed, consider the set $S \subset \mathbf{F}_5^n$ defined by

$$S := \{ x \in \mathbf{F}_5^n : x \cdot x = 0 \}.$$

It is easy to verify that S has density $\frac{1}{5} + O(\frac{1}{\sqrt{5^n}})$ and all nontrivial Fourier coefficients satisfy $\left| \widehat{1_S}(\xi) \right| \ll \frac{1}{\sqrt{5^n}}$. However, since

$$x \cdot x - 3(x+y) \cdot (x+y) + 3(x+2y) \cdot (x+2y) - (x+3y) \cdot (x+3y) = 0$$

for all $x, y \in \mathbf{F}_5^n$, if the first three terms of a four-term arithmetic progression lie in S, then the last term is forced to as well. This means that S contains

$$\left(\frac{1}{125} + O(\frac{1}{\sqrt{5^n}}) \right) |\mathbf{F}_5^n|^2$$

four-term arithmetic progressions, many more than the roughly $\frac{1}{625} |\mathbf{F}_5^n|^2$ expected in a random set of density $\frac{1}{5}$. Thus, something beyond Fourier analysis is needed to get a handle on four-term arithmetic progressions.

2.2 Higher-Order Fourier Analysis

Let $(G, +)$ be a finite abelian group (which, for us, will be either $\mathbf{Z}/N\mathbf{Z}$ or $\mathbf{F}_p^n$) and s be a natural number. For any complex-valued function f on G and $h \in G$, the *multiplicative discrete derivative* $\Delta_h f : G \to \mathbf{C}$ is defined by

$$\Delta_h f(x) := f(x)\overline{f(x+h)}.$$

For $h_1, \ldots, h_s \in G$, we denote the s-fold multiplicative discrete derivative of f by

$$\Delta_{h_1,\ldots,h_s} f := \Delta_{h_1} \cdots \Delta_{h_s} f.$$

Observe that $\Delta_{h_1,\ldots,h_s} f = \Delta_{h_{\sigma(1)},\ldots,h_{\sigma(s)}} f$ for any permutation σ of $\{1,\ldots,s\}$, so $\Delta_{h_1,\ldots,h_s}$ depends only on the multiset of differences $\{h_1,\ldots,h_s\}$. We can similarly define the *additive discrete derivative* of a function $g : G \to G'$, where $(G', +)$ is also an abelian group, by

$$\partial_h g(x) = g(x) - g(x+h),$$

as well as the s-fold additive discrete derivative

$$\partial_{h_1,\ldots,h_s} g = \partial_{h_1} \cdots \partial_{h_s} g.$$

For any natural number s, the *Gowers U^s-norm* $\|f\|_{U^s}$ of f is defined by

$$\|f\|_{U^s} := (\mathbf{E}_{x,h_1,\ldots,h_s \in G} \Delta_{h_1,\ldots,h_s} f(x))^{1/2^s}.$$

(note that $\mathbf{E}_{x,h_1,\ldots,h_s \in G} \Delta_{h_1,\ldots,h_s} f(x)$ is always real-valued and nonnegative, and thus possesses a nonnegative 2^s-th root). For example, $\|f\|_{U^1} = |\mathbf{E}_{x \in G} f(x)|$ and

$$\|f\|_{U^2}^4 = \mathbf{E}_{x,h_1,h_2 \in G} f(x)\overline{f(x+h_1)f(x+h_2)}f(x+h_1+h_2).$$

We will now list some basic facts about these norms.

Lemma 2.1 *Let $f : G \to \mathbf{C}$.*

1. *$\|\cdot\|_{U^1}$ is a seminorm.*

2. *$\|\cdot\|_{U^s}$ is a norm when $s \geq 2$.*

3. *$\|f\|_{U^1} \leq \|f\|_{U^2} \leq \cdots \leq \|f\|_{U^s} \leq \|f\|_{U^{s+1}} \leq \cdots \leq \|f\|_{\ell^\infty}$.*

4. *$\|f\|_{U^{s+1}}^{2^{s+1}} = \mathbf{E}_{h \in G} \|\Delta_h f\|_{U^s}^{2^s}$ for all $s \geq 1$.*

5. *$\|f\|_{U^2} = \|\widehat{f}\|_{\ell^4}$.*

None of these statements are very hard to prove, but hints can be found in [45] and [119], for example. While the U^2-norm is simply the ℓ^4-norm of the Fourier transform of f, the study of the U^s-norms when $s \geq 3$ is called *higher-order Fourier analysis*.

Gowers observed (in the integer setting) that the U^s-norm controls the count of $(s+1)$-term arithmetic progressions in subsets of abelian groups. We will present a proof of this statement in the setting of vector spaces over finite fields.

Lemma 2.2 Let $p \geq k \geq 2$ and $f_0, \ldots, f_{k-1} : \mathbf{F}_p^n \to \mathbf{C}$ be 1-bounded functions. Then

$$|\mathbf{E}_{x,y \in G} f_0(x) f_1(x+y) \cdots f_{k-1}(x+(k-1)y)| \leq \|f_{k-1}\|_{U^{k-1}}.$$

With a bit more work, one can replace the right-hand side with the stronger bound $\min_{0 \leq i \leq k-1} \|f_i\|_{U^{k-1}}$.

Proof We proceed by induction on k, starting with the base case $k=2$. Observe, by making the change of variables $y \mapsto y - x$, that

$$\left|\mathbf{E}_{x,y \in \mathbf{F}_p^n} f_0(x) f_1(x+y)\right| = \left|\mathbf{E}_{x \in \mathbf{F}_p^n} f_0(x)\right| \left|\mathbf{E}_{y \in \mathbf{F}_p^n} f_1(y)\right| \leq \|f_1\|_{U^1}.$$

Now suppose that we have proven the result for a general $k \geq 2$. Writing

$$\left|\mathbf{E}_{x,y \in \mathbf{F}_p^n} f_0(x) f_1(x+y) \cdots f_k(x+ky)\right|^2 \tag{2.1}$$

as

$$\left|\mathbf{E}_{x \in \mathbf{F}_p^n} f_0(x) \left(\mathbf{E}_{y \in \mathbf{F}_p^n} f_1(x+y) \cdots f_k(x+ky)\right)\right|^2,$$

we have by that Cauchy–Schwarz inequality that (2.1) is bounded above by

$$\mathbf{E}_{x,y,h \in \mathbf{F}_p^n} \Delta_h f_1(x+y) \cdots \Delta_{kh} f_k(x+ky),$$

which equals

$$\mathbf{E}_{x,y,h \in \mathbf{F}_p^n} \Delta_h f_1(x) \cdots \Delta_{kh} f_k(x+(k-1)y)$$

by making the change of variables $x \mapsto x - y$. It now follows from the induction hypothesis and our assumption that $p \geq k+1$ that

$$\left|\mathbf{E}_{x,y \in \mathbf{F}_p^n} f_0(x) f_1(x+y) \cdots f_k(x+ky)\right|^2 \leq \mathbf{E}_{h \in \mathbf{F}_p^n} \|\Delta_{kh} f_k\|_{U^{k-1}} = \mathbf{E}_{h \in \mathbf{F}_p^n} \|\Delta_h f_k\|_{U^{k-1}}.$$

By Hölder's inequality,

$$\left|\mathbf{E}_{x,y \in \mathbf{F}_p^n} f_0(x) f_1(x+y) \cdots f_k(x+ky)\right|^{2^k} \leq \mathbf{E}_{h \in \mathbf{F}_p^n} \|\Delta_h f_k\|_{U^{k-1}}^{2^{k-1}} = \|f_k\|_{U^k},$$

completing the inductive step. □

An important consequence is that the U^s-norms are measures of pseudorandomness as far as counting k-term arithmetic progressions is concerned. Let $A \subset \mathbf{F}_p^n$ have density α in $\mathbf{F}_p^n$ and set $f_A := 1_A - \alpha$. Then, by Lemma 2.2, we have

$$\left|\frac{\#\{(x,y) \in (\mathbf{F}_p^n)^2 : x, x+y, \ldots, x+(k-1)y \in A\}}{|\mathbf{F}_p^n|^2} - \alpha^k\right| \ll_k \|f_A\|_{U^{k-1}}.$$

Thus, if A has far from the density α^k of k-term arithmetic progressions expected in a random subset of density α, then $\|f_A\|_{U^{k-1}}$ is large. To make use of this information, we need an *inverse theorem* for the U^{k-1}-norm, i.e., a structural result for bounded functions with large U^{k-1}-norm. When $k=3$ in this analysis, so that $\|f_A\|_{U^2}$ is large, then it is easy to deduce that 1_A must have some large nontrivial Fourier coefficient.

Lemma 2.3 (U^2-inverse theorem) *Let $f : \mathbf{F}_p^n \to \mathbf{C}$ be a 1-bounded function. Then*
$$\|f\|_{U^2}^2 \leq \|\widehat{f}\|_{\ell^\infty}.$$

Proof By the last statement of Lemma 2.1,
$$\|f\|_{U^2}^4 = \sum_{\xi \in \mathbf{F}_p^n} \left|\widehat{f}(\xi)\right|^4.$$

So,
$$\|f\|_{U^2}^4 \leq \max_{\xi \in \mathbf{F}_p^n} \left|\widehat{f}(\xi)\right|^2 \sum_{\xi \in \mathbf{F}_p^n} \left|\widehat{f}(\xi)\right|^2 \leq \max_{\xi \in \mathbf{F}_p^n} \left|\widehat{f}(\xi)\right|^2 \cdot \mathbf{E}_{x \in \mathbf{F}_p^n} |f(x)|^2 \leq \|\widehat{f}\|_{\ell^\infty}^2,$$

since f is 1-bounded. Taking the square root of both sides gives the conclusion of the lemma. □

To prove Szemerédi's theorem for sets lacking progressions of length greater than three, Gowers proved a "local" inverse theorem for the U^s-norm on cyclic groups. This says, roughly speaking, that if f is 1-bounded and $\|f\|_{U^s}$ is large, then there exists a partition of $\mathbf{Z}/N\mathbf{Z}$ into long arithmetic progressions $I_1, \ldots, I_k$ such that, on average, f has large correlation on I_j with a polynomial phase $e(P(x))$ of degree $\deg P \leq s - 1$. A "global" inverse theorem on cyclic groups was not proved until several years later, first in the case $s = 3$ by Green and Tao [49] and then in general by Green, Tao, and Ziegler [55, 56], who showed that f must have large correlation over the entire group with an $(s-1)$-step nilsequence of bounded complexity. This work of Green, Tao, and Ziegler was purely qualitative, and, more recently, Manners [82] proved the first fully general quantitative version of the global inverse theorems for the Gowers norms. Defining nilsequences and discussing the integer setting further would take us too far from the scope of this article, so for the remainder of this section we will discuss the U^s-inverse theorems in the setting of high-dimensional vector spaces over finite fields.

2.3 The U^3-Inverse Theorem and the Polynomial Freiman–Ruzsa Conjecture

Based on the early arguments of Gowers in cyclic groups, Samorodnitsky [105] (for $p = 2$) and Green and Tao [49] (for $p > 2$) proved the following U^3-inverse theorem in the finite field model setting.

Theorem 2.4 *Fix a prime p. There exists a constant $c > 0$ such that if $f : \mathbf{F}_p^n \to \mathbf{C}$ is 1-bounded and $\|f\|_{U^3} \geq \delta$, then there exists a polynomial $Q : \mathbf{F}_p^n \to \mathbf{F}_p$ of degree at most 2 for which*
$$\left|\mathbf{E}_{x \in \mathbf{F}_p^n} f(x) e_p(Q(x))\right| \geq \exp(-c\delta^{-c}).$$

Sanders [109] improved the lower bound for the correlation of f with a quadratic phase in Theorem 2.4 to
$$\left|\mathbf{E}_{x \in \mathbf{F}_p^n} f(x) e_p(Q(x))\right| \geq \exp\left(-c(\log(2/\delta))^c\right)$$

for some constant $c > 0$ by proving quasipolynomial bounds in Bogolyubov's theorem, a key ingredient in the proof of Theorem 2.4. Green and Tao [53] and, independently, Lovett [77] showed that Theorem 2.4 holding with a lower bound depending only polynomially on δ is equivalent to the polynomial Freiman–Ruzsa conjecture, one of the most important conjectures in additive combinatorics.

Conjecture 2.5 *There exists an absolute constant $c > 0$ such that the following holds. If $A \subset \mathbf{F}_p^n$ has small doubling $|A + A| \leq K|A|$, then there exists an affine subspace $V \leq \mathbf{F}_p^n$ size $|V| \ll K^c|A|$ such that $|A \cap V| \gg K^{-c}|A|$.*

Several equivalent formulations of Conjecture 2.5 can be found at the end of Green's survey [42], with proofs of the equivalences located in Green's accompanying notes [43]. In addition to implying polynomial bounds in Theorem 2.4, the $p = 2$ case of the polynomial Freiman–Ruzsa conjecture has numerous applications in theoretical computer science, a list of which can be found in Lovett's exposition [78] of Sanders's quasipolynomial Bogolyubov theorem.

In a spectacular breakthrough posted to the arXiv right before the due date for this survey article, Gowers, Green, Manners, and Tao [35] have proved Conjecture 2.5 in the case $p = 2$ and announced a forthcoming proof of the conjecture for all odd primes as well. Further developing the theory of sumsets and entropy studied in [48], [104], and [118], their argument proceeds by proving an entropic version of the polynomial Freiman–Ruzsa conjecture, which the latter three authors showed is equivalent to Conjecture 2.5 in (also very recent) earlier work [48]. Time and space constraints unfortunately prevent us from giving an exposition of the elegant argument of Gowers, Green, Manners, and Tao in this survey, so the reader is encouraged to read their very well-written and almost self-contained preprint [35].

Though the polynomial Freiman–Ruzsa conjecture is now settled, the argument in [35] does not seem to adapt to prove a polynomial Bogolyubov theorem, so the question of whether such a result holds is still open.

Question 2.6 *Fix a prime p. Is it the case that there exists a constant $C > 0$ such that, whenever $A \subset \mathbf{F}_p^n$ has density α, $2A - 2A$ contains a subspace of codimension at most $C \log \alpha^{-1}$?*

2.4 Inverse Theorems for Higher Degree Uniformity Norms

Bergelson, Tao, and Ziegler [9, 121] proved inverse theorems for the U^s-norm in the finite field model setting for all $s \geq 4$, provided that p is sufficiently large in terms of s.

Theorem 2.7 *Let s be a natural number and $p \geq s$. If $f : \mathbf{F}_p^n \to \mathbf{C}$ is 1-bounded and $\|f\|_{U^s} \geq \delta$, then there exists a polynomial $P : \mathbf{F}_p^n \to \mathbf{F}_p$ of degree at most $s - 1$ such that*
$$\left|\mathbf{E}_{x \in \mathbf{F}_p^n} f(x) e_p(P(x))\right| \gg_{\delta,s,p} 1.$$

The proof of Theorem 2.7 proceeds via ergodic theory, and thus produces no quantitative bounds for the correlation of f with a polynomial phase.

While Theorem 2.4 holds for all primes p, it turns out that Theorem 2.7 is false for $p < s$ as soon as $s \geq 4$, as was observed by Green and Tao [50] and Lovett,

Meshulam, and Samorodnitsky [80]. Despite this, Tao and Ziegler [122] showed that Theorem 2.7 can be modified to a statement that holds for all primes by enlarging the class of polynomial phases to include those coming from *non-classical polynomials* of degree at most $s-1$, i.e., functions $P : \mathbf{F}_p^n \to \mathbf{T}$ satisfying

$$\partial_{h_1,\ldots,h_s} P(x) = 0$$

for all $x, h_1, \ldots, h_s \in \mathbf{F}_p^n$.

Theorem 2.8 *Let s be a natural number. If $f : \mathbf{F}_p^n \to \mathbf{C}$ is 1-bounded and $\|f\|_{U^s} \geq \delta$, then there exists a non-classical polynomial $P : \mathbf{F}_p^n \to \mathbf{T}$ of degree at most $s-1$ such that*

$$\left|\mathbf{E}_{x \in \mathbf{F}_p^n} f(x) e(P(x))\right| \gg_{\delta,s,p} 1.$$

Theorem 2.7 says that when $p \geq s$, the P in Theorem 2.8 can be assumed to take values in $\frac{1}{p}\mathbf{Z}$ (mod 1).

At the time of Wolf's survey, it was a major open problem to prove quantitative versions of the inverse theorem for the U^s-norms when $s \geq 4$. Since then, Gowers and Milićević [36, 37] have proved a quantitative version of Theorem 2.7 with reasonable bounds.

Theorem 2.9 *Let s be a natural number and $p \geq s$. There exists a natural number $m = m(s)$ and a constant $c = c(s,p) > 0$ such that the following holds. If $f : \mathbf{F}_p^n \to \mathbf{C}$ is 1-bounded and $\|f\|_{U^s} \geq \delta$, then there exists a polynomial $P : \mathbf{F}_p^n \to \mathbf{F}_p$ of degree at most $s-1$ such that*

$$\left|\mathbf{E}_{x \in \mathbf{F}_p^n} f(x) e_p(P(x))\right| \gg_{s,p} \frac{1}{\exp^{(m)}(c\delta^{-1})}.$$

Quantitative bounds in the low characteristic case are only known for $s \leq 6$, due to work of Tidor [123] and Milićević [86], so the following problem is still open.

Problem 2.10 *Prove a version of Theorem 2.8 with reasonable quantitative bounds when $s > 6$.*

The value of m obtained in Theorem 2.9 by Gowers and Milićević's [37] argument grows like $3^s s!$, in contrast to the inverse theorems of Manners [82], which are of the same quality for all $s \geq 4$. This prompts the following natural problem.

Problem 2.11 *Improve the bounds in Theorem 2.9.*

Any progress on this problem would automatically improve any result, such as the main theorem in [97], that depends on Theorem 2.9. It would be very interesting to see a proof of Theorem 2.9 yielding a constant tower height $m \ll 1$, independent of s. Even obtaining $m \ll s$ will require several new ideas. Kim, Li, and Tidor [65] have obtained improved bounds in the inverse theorem for the U^4-norm in all characteristics, showing that

$$\left|\mathbf{E}_{x \in \mathbf{F}_p^n} f(x) e_p(P(x))\right| \gg \frac{1}{\exp^{(2)}(c_p \log(2/\delta)^{c_p})}$$

for some constants $c_p > 0$ depending only on p. Considering that quasipolynomial bounds in the bilinear Bogolyubov theorem, a key ingredient in the proof of the U^4-inverse theorem in the finite field model setting, are known thanks to work of Hosseini and Lovett [59], quasipolynomial bounds in the U^4-inverse theorem should be within reach, though more will have to be done to improve the quantitative aspects of other parts of the argument. Given the resolution of the polynomial Freiman–Ruzsa conjecture, it could be that polynomial bounds are even within reach.

3 The Polynomial Szemerédi Theorem

In 1977, Furstenberg [29] gave an alternative proof of Szemerédi's theorem via ergodic theory, in which he introduced his now-famous correspondence principle and created the field of ergodic Ramsey theory. Furstenberg's argument has now been extended to prove very broad generalizations of Szemerédi's theorem, most notably a multidimensional generalization due to Furstenberg and Katznelson [30] and a polynomial generalization due to Bergelson and Leibman [7].

Theorem 3.1 (Multidimensional Szemerédi Theorem) *Let $S \subset \mathbf{Z}^d$ be finite and nonempty. If $A \subset [N]^d$ contains no nontrivial homothetic copies*

$$a + b \cdot S \qquad (b \neq 0)$$

of S, then $|A| = o_S(N^d)$.

There are now multiple proofs of Theorem 3.1 using hypergraph regularity methods [34, 88, 101, 117], which all give a savings over the trivial bound $|A| \leq N^d$ of inverse Ackermann-type.

Theorem 3.2 (Polynomial Szemerédi Theorem) *Let $P_1, \ldots, P_m \in \mathbf{Z}[y]$, all satisfying $P_i(0) = 0$. If $A \subset [N]$ contains no nontrivial polynomial progressions*

$$x, x + P_1(y), \ldots, x + P_m(y) \qquad (y \neq 0), \tag{3.1}$$

then $|A| = o_{P_1, \ldots, P_m}(N)$.

The assumption that $P_i(0) = 0$ for $i = 1, \ldots, m$ in Theorem 3.2 is there to prevent local obstructions to the result being true. For example, note that the even integers contain no configurations of the form $x, x + 2y + 1$, since $2y + 1$ is always odd when y is an integer. The only known proofs of Theorem 3.2 in full generality are via ergodic theory, and give no quantitative bounds at all.

Following his proof of reasonable bounds in Szemerédi's theorem, Gowers [32] posed the problem of proving reasonable bounds in Theorems 3.1 and 3.2.

Problem 3.3 *Prove quantitative versions of the multidimensional and polynomial generalizations of Szemerédi's theorem, with reasonable bounds.*

This problem has turned out to be very difficult, though a good amount of progress has finally been made in the past few years on proving a quantitative version of the polynomial Szemerédi theorem. We will devote the remainder of this section to discussing this progress, and say a bit about the multidimensional Szemerédi theorem in Section 4.

Prior to a couple of years ago, quantitative bounds were known in Theorem 3.2 in only three special cases:

1. when $m = 1$, as discussed in Subsection 1.2,

2. when all of the P_i are linear, which follows from Gowers's quantitative proof of Szemerédi's theorem,

3. and when the P_i are all monomials of the same fixed degree, due to work of Prendiville [100].

There are insurmountable obstructions to applying the methods used to handle the above three special cases to any additional polynomial progressions, such as the *nonlinear Roth configuration*,

$$x, x+y, x+y^2,$$

which is the simplest nonlinear polynomial progression of length greater than 2. Indeed, Sárközy's argument only works for two-term polynomial progressions, as its starting point is the fact that the count of such progressions in a set can be written as the inner product of two functions, one of which is a convolution. Gowers's argument crucially relies on the fact that arithmetic progressions are translation- and dilation-invariant, and Prendiville was able to generalize Gowers's proof by using the fact that arithmetic progressions with common difference equal to a perfect d^{th} power are invariant under dilations by a perfect d^{th} power. No other polynomial progressions (including the nonlinear Roth configuration) are anywhere close to being dilation-invariant, so the configurations considered by Prendiville in [100] are exactly those that can be handled using Gowers's methods.

For every $P_1, \ldots, P_m \in \mathbf{Z}[y]$ and S equal to either $[N]$ or $\mathbf{F}_p$, let $r_{P_1,\ldots,P_m}(S)$ denote the size of the largest subset of S lacking the nontrivial progression (3.1). Observe that $r_{P_1,\ldots,P_m}(\mathbf{F}_p) \leq r_{P_1,\ldots,P_m}([p])$, since reducing a nontrivial polynomial progression in $[p]$ modulo p produces a nontrivial polynomial progression in $\mathbf{F}_p$. Thus, any bounds obtained in the integer setting automatically give bounds in the finite field setting. As was discussed in Subsection 1.2, the finite field setting is strictly easier than the integer setting when $m = 1$. This is true also for longer polynomial progressions, as evidenced by the following result of Bourgain and Chang [18] giving power-saving bounds for subsets of finite fields lacking the nonlinear Roth configuration.

Theorem 3.4 *We have*

$$r_{y,y^2}(\mathbf{F}_p) \ll p^{14/15}.$$

The argument of Bourgain and Chang was very specific to the progression $x, x+y, x+y^2$, as it used the fact that one can explicitly evaluate quadratic Gauss sums. Their method could only possibly generalize to progressions involving one linear polynomial and one quadratic polynomial, so they asked whether a similar power-saving bound holds in general for three-term polynomial progression involving linearly independent polynomials. Using a different approach, I answered their question in the affirmative, with a slightly worse exponent [93].

Finite Field Models in Arithmetic Combinatorics   179

Theorem 3.5 *Let $P_1, P_2 \in \mathbf{Z}[y]$ be linearly independent and satisfy $P_1(0) = P_2(0) = 0$. Then $r_{P_1,P_2}(\mathbf{F}_p) \ll_{P_1,P_2} p^{23/24}$.*

Dong, Li, and Sawin [22] very shortly after improved the exponent in Theorem 3.5, obtaining the bound $r_{P_1,P_2}(\mathbf{F}_p) \ll_{P_1,P_2} p^{11/12}$.

None of the arguments in [18], [22], or [93] seem to adapt to prove quantitative bounds for subsets of finite fields lacking longer polynomial progressions or to the integer setting. Because of this, I introduced a new technique, now known as "degree-lowering", and used it to prove power-saving bounds for subsets of finite fields lacking arbitrarily long polynomial progressions, provided the polynomials are linearly independent [94].

Theorem 3.6 *Let $P_1, \ldots, P_m \in \mathbf{Z}[y]$ be linearly independent polynomials, all satisfying $P_i(0) = 0$. There exists $\gamma_{P_1,\ldots,P_m} > 0$ such that $r_{P_1,\ldots,P_m}(\mathbf{F}_p) \ll p^{1-\gamma_{P_1,\ldots,P_m}}$.*

The degree-lowering technique is robust enough that it works in a wide variety of settings, including the integer, continuous, and ergodic settings. In the integer setting, Prendiville and I used it to prove quantitative bounds for subsets of $[N]$ lacking the nonlinear Roth configuration [98, 99], and I then generalized our argument to handle arbitrarily long polynomial progressions with distinct degrees [95]. Kuca [68, 70] and Leng [75] have also used the degree-lowering method to prove good quantitative bounds in more special cases of the polynomial Szemerédi theorem in finite fields. More applications relevant to the topic of this survey will be discussed in the last subsection of this section.

3.1 Degree-Lowering and the Non-Linear Roth Configuration

We will now give an illustration of the degree-lowering method by using it to prove a power-saving bound for sets lacking the nonlinear Roth configuration. The following argument is an adaptation of the proof in [98] to the finite field setting, which greatly simplifies it.

Fix a prime $p > 2$ and define, for all $f_0, f_1, f_2 : \mathbf{F}_p \to \mathbf{C}$, the trilinear average

$$\Lambda(f_0, f_1, f_2) := \mathbf{E}_{x,y \in \mathbf{F}_p} f_0(x) f_1(x+y) f_2(x+y^2).$$

Lemma 3.7 *Let $f_0, f_1, f_2 : \mathbf{F}_p \to \mathbf{C}$ be 1-bounded. Then*

$$|\Lambda(f_0, f_1, f_2)|^4 \ll \|f_2\|_{U^3} + \frac{1}{p}.$$

Proof First, write

$$\Lambda(f_0, f_1, f_2) = \mathbf{E}_{x \in \mathbf{F}_p} f_0(x) \left(\mathbf{E}_{y \in \mathbf{F}_p} f_1(x+y) f_2(x+y^2) \right),$$

so that, by the Cauchy–Schwarz inequality,

$$|\Lambda(f_0, f_1, f_2)|^2 \le \mathbf{E}_{x,y,a \in \mathbf{F}_p} f_1(x+y) \overline{f_1(x+y+a)} f_2(x+y^2) \overline{f_2(x+(y+a)^2)},$$

since f_0 is 1-bounded. Making the change of variables $x \mapsto x - y$, we can write the right-hand side of the above as

$$\mathbf{E}_{x,y,a \in \mathbf{F}_p} f_1(x) \overline{f_1(x+a)} f_2(x+y^2-y) \overline{f_2(x+(y+a)^2-y)},$$

and then apply the Cauchy–Schwarz inequality again and make the change of variables $x \mapsto x + y$ to get that $|\Lambda(f_0, f_1, f_2)|^4$ is bounded above by

$$\mathbb{E}_{x,y,a,b \in \mathbf{F}_p} f_2(x+y^2)\overline{f_2(x+(y+a)^2)}f_2(x+(y+b)^2-b)\overline{f_2(x+(y+a+b)^2-b)}.$$

Making the change of variables $x \mapsto x - y^2$ rewrites the above as

$$\mathbb{E}_{a,b \in \mathbf{F}_p}\mathbb{E}_{x,y \in \mathbf{F}_p} f_2(x)\overline{g_{1,a,b}(x+2ay)}g_{2,a,b}(x+2by)\overline{g_{3,a,b}(x+2(a+b)y)}, \qquad (3.2)$$

where, for each $i = 1, 2, 3$, $g_{i,a,b}(x) = f_2(x + \phi_i(a,b))$ for some function $\phi_i : \mathbf{F}_p^2 \to \mathbf{F}_p$.

It is a standard fact, which we will soon prove, that the inner average in (3.2) is controlled by the U^3-norm of f_2 whenever a, b, and $a + b$ are nonzero. But, first observe that

$$\#\left\{(a,b) \in \mathbf{F}_p^2 : a = 0,\ b = 0,\ \text{or}\ a + b = 0\right\} = 3p - 2,$$

so that the total contribution to (3.2) coming from pairs (a, b) such that a, b, and $a + b$ are nonzero is at most $\frac{3p-2}{p^2}$. Thus, (3.2) is

$$\ll \mathbb{E}'_{a,b \in \mathbf{F}_p} \left| \mathbb{E}_{x,y \in \mathbf{F}_p} f_2(x)\overline{g_{1,a,b}(x+2ay)}g_{2,a,b}(x+2by)\overline{g_{3,a,b}(x+2(a+b)y)} \right| + \frac{1}{p}, \qquad (3.3)$$

where $\mathbb{E}'_{a,b \in \mathbf{F}_p}$ denotes the average over pairs (a, b) for which a, b, and $a + b$ are nonzero.

Now, we will bound

$$\left| \mathbb{E}_{x,y \in \mathbf{F}_p} f_2(x)\overline{g_{1,a,b}(x+2ay)}g_{2,a,b}(x+2by)\overline{g_{3,a,b}(x+2(a+b)y)} \right| \qquad (3.4)$$

whenever a, b, and $a + b$ are nonzero. By another application of the Cauchy–Schwarz inequality, the square of (3.4) is at most

$$\mathbb{E}_{x,y,h_1 \in \mathbf{F}_p} \Delta_{2ah_1}\overline{g_{1,a,b}(x+2ay)}\Delta_{2bh_1}\overline{g_{2,a,b}(x+2by)}\Delta_{2(a+b)h_1}\overline{g_{3,a,b}(x+2(a+b)y)},$$

which equals

$$\mathbb{E}_{x,y,h_1 \in \mathbf{F}_p} \Delta_{2ah_1}\overline{g_{1,a,b}(x)}\Delta_{2bh_1}\overline{g_{2,a,b}(x+2(b-a)y)}\Delta_{2(a+b)h_1}\overline{g_{3,a,b}(x+2by)},$$

by making the change of variables $x \mapsto x - 2ay$. We apply the Cauchy–Schwarz inequality again to bound the square of the above by

$$\mathbb{E}_{x,y,h_1,h_2 \in \mathbf{F}_p} \Delta_{2bh_1, 2(b-a)h_2}\overline{g_{2,a,b}(x+2(b-a)y)}\Delta_{2(a+b)h_1, 2bh_2}\overline{g_{3,a,b}(x+2by)},$$

which, by the change of variables $x \mapsto x - 2(b-a)y$, equals

$$\mathbb{E}_{x,y,h_1,h_2 \in \mathbf{F}_p} \Delta_{2bh_1, 2(b-a)h_2}\overline{g_{2,a,b}(x)}\Delta_{2(a+b)h_1, 2bh_2}\overline{g_{3,a,b}(x+2ay)}.$$

A final application of the Cauchy–Schwarz inequality and a change of variables bounds the square of the above by

$$\mathbb{E}_{x,h_1,h_2,h_3 \in \mathbf{F}_p} \Delta_{2(a+b)h_1, 2bh_2, 2ah_3} \overline{g_{3,a,b}(x)}.$$

Since $p > 2$ and a, b, and $a + b$ are all nonzero, making the change of variables $h_1 \mapsto \frac{h_1}{2(a+b)}$, $h_2 \mapsto \frac{h_2}{2b}$, $h_3 \mapsto \frac{h_3}{2a}$, and $x \mapsto x - \phi_3(a,b)$ reveals that

$$\mathbf{E}_{x, h_1, h_2, h_3 \in \mathbf{F}_p} \Delta_{2(a+b)h_1, 2bh_2, 2ah_3} g_{3,a,b}(x) = \|f_2\|_{U^3}^8.$$

Combining this with (3.3) yields

$$|\Lambda(f_0, f_1, f_2)|^4 \ll \|f_2\|_{U^3} + \frac{1}{p}.$$

□

A simple application of the Cauchy–Schwarz inequality shows that not only is $|\Lambda(f_0, f_1, f_2)|$ bounded by the U^3-norm of f_2, but it is also bounded by the U^3-norm of a *dual function*.

Lemma 3.8 *Let $f_0, f_1, f_2 : \mathbf{F}_p \to \mathbf{C}$ be 1-bounded, and define the dual function $F_2 : \mathbf{F}_p \to \mathbf{C}$ by*

$$F_2(x) := \mathbf{E}_{z \in \mathbf{F}_p} f_0(x - z^2) f_1(x - z^2 + z).$$

Then

$$|\Lambda(f_0, f_1, f_2)|^8 \ll \|F_2\|_{U^3} + \frac{1}{p}.$$

Proof By making the change of variables $x \mapsto x - y^2$, we can write

$$\Lambda(f_0, f_1, f_2) = \mathbf{E}_{x,y \in \mathbf{F}_p} f_0(x - y^2) f_1(x + y - y^2) f_2(x),$$

so that, by the Cauchy–Schwarz inequality and the change of variables $x \mapsto x + y^2$,

$$|\Lambda(f_0, f_1, f_2)|^2 \leq \mathbf{E}_{x,y,z \in \mathbf{F}_p} f_0(x - z^2) f_1(x + z - z^2) \overline{f_0(x - y^2) f_1(x + y - y^2)}$$
$$= \mathbf{E}_{x,y,z \in \mathbf{F}_p} \overline{f_0(x)} f_1(x+y) F_2(x + y^2)$$
$$= \Lambda\left(\overline{f_0}, \overline{f_1}, F_2\right).$$

Noting that F_2 is 1-bounded (being the average of 1-bounded functions), the desired bound now follows from applying Lemma 3.7. □

The key insight of [94] is that the U^s-norm of a polynomial dual function involving linearly independent polynomials can be bounded in terms of the U^{s-1}-norm of the dual function. The following lemma presents the simplest nontrivial instance of this phenomenon.

Theorem 3.9 (One step degree-lowering) *Let $f, g : \mathbf{F}_p \to \mathbf{C}$ be 1-bounded, and set*

$$F(x) := \mathbf{E}_{z \in \mathbf{F}_p} f(x - z^2) g(x + z - z^2).$$

If $\|F\|_{U^3}^8 \geq 4/p$, then

$$\|F\|_{U^3}^8 \ll \|F\|_{U^2}.$$

Theorem 3.9 is proved by combining two lemmas.

Lemma 3.10 (Difference-dual interchange) *Let $K \subset \mathbf{F}_p$, $f, g : \mathbf{F}_p \to \mathbf{C}$ be 1-bounded, and set*

$$F(x) := \mathbf{E}_{z \in \mathbf{F}_p} f(x - z^2) g(x + z - z^2).$$

Then, for every function $\phi : \mathbf{F}_p \to \mathbf{F}_p$, we have that

$$\mathbf{E}_{k \in K} \left| \mathbf{E}_x \Delta_k F(x) e_p(\phi(k) x) \right|^2$$

is bounded above by

$$\mathbf{E}_{k,k' \in K} \left| \mathbf{E}_{x,y \in \mathbf{F}_p} \Delta_{k'-k} f(x) \Delta_{k'-k} g(x+y) e_p([\phi(k) - \phi(k')][x + y^2]) \right|^2.$$

Proof Expanding the definition of the dual function F, we have that

$$\mathbf{E}_{k \in K} \left| \mathbf{E}_{x \in \mathbf{F}_p} \Delta_k F(x) e_p(\phi(k) x) \right|^2$$

equals

$$\mathbf{E}_{k \in K} \mathbf{E}_{x, x', z_1, z_2, z_3, z_4 \in \mathbf{F}_p} \Big[f(x - z_1^2) g(x + z_1 - z_1^2) \\ \overline{f(x - z_2^2 + k) g(x + z_2 - z_2^2 + k)} e_p(\phi(k) x) \\ \overline{f(x' - z_3^2) g(x' + z_3 - z_3^2)} \\ f(x' - z_4^2 + k) g(x' + z_4 - z_4^2 + k) e_p(-\phi(k) x') \Big],$$

which can be written as

$$\mathbf{E}_{x, x', z_1, z_2, z_3, z_4 \in \mathbf{F}_p} \Big[f(x - z_1^2) g(x + z_1 - z_1^2) \overline{f(x' - z_3^2) g(x' + z_3 - z_3^2)} \\ \Big(\mathbf{E}_{k \in K} \overline{f(x - z_2^2 + k) g(x + z_2 - z_2^2 + k)} \\ f(x' - z_4^2 + k) g(x' + z_4 - z_4^2 + k) e_p(\phi(k)[x - x']) \Big) \Big],$$

Thus, by the Cauchy–Schwarz inequality, $\mathbf{E}_{k \in K} \left| \mathbf{E}_{x \in \mathbf{F}_p} \Delta_k F(x) e_p(\phi(k) x) \right|^2$ is bounded above by

$$\mathbf{E}_{k,k' \in K} \left| \mathbf{E}_{x,z \in \mathbf{F}_p} \Delta_{k'-k} f(x - z^2 + k) \Delta_{k'-k} g(x + z - z^2 + k) e_p([\phi(k) - \phi(k')] x) \right|^2,$$

which, by the change of variables $x \mapsto x - k + z^2$, is bounded above by

$$\mathbf{E}_{k,k' \in K} \left| \mathbf{E}_{x,z \in \mathbf{F}_p} \Delta_{k'-k} f(x) \Delta_{k'-k} g(x + z) e_p([\phi(k) - \phi(k')][x + z^2]) \right|^2.$$

$\square$

Lemma 3.11 (Major arc lemma) *Let $f, g : \mathbf{F}_p \to \mathbf{C}$ be 1-bounded. If $\xi \in \mathbf{F}_p$ is nonzero, then*

$$\left| \mathbf{E}_{x,y \in \mathbf{F}_p} f(x) g(x + y) e_p(\xi[x + y^2]) \right| \leq \frac{1}{\sqrt{p}}.$$

Proof Plugging in the Fourier inversion formula for f and g and using orthogonality of characters yields

$$\mathbf{E}_{x,y\in\mathbf{F}_p}f(x)g(x+y)e_p(\xi[x+y^2]) = \sum_{\zeta,\eta\in\mathbf{F}_p}\widehat{f}(\zeta)\widehat{g}(\eta)\mathbf{E}_{x,y\in\mathbf{F}_p}e_p([\zeta+\eta+\xi]x+\eta y+\xi y^2)$$
$$= \sum_{\eta\in\mathbf{F}_p}\widehat{f}(-\eta-\xi)\widehat{g}(\eta)\mathbf{E}_{y\in\mathbf{F}_p}e_p(\eta y+\xi y^2).$$

Thus, since $\left|\mathbf{E}_{y\in\mathbf{F}_p}e_p(\eta y+\xi y^2)\right| \leq p^{-1/2}$ whenever $\xi \neq 0$, we have

$$\left|\mathbf{E}_{x,y\in\mathbf{F}_p}f(x)g(x+y)e_p(\xi[x+y^2])\right| \leq \frac{1}{\sqrt{p}}\sum_{\eta\in\mathbf{F}_p}|\widehat{f}(-\eta-\xi)||\widehat{g}(\eta)| \leq \frac{1}{\sqrt{p}}\|f\|_{L^2}\|g\|_{L^2},$$

and the conclusion of the lemma follows from the 1-boundedness of f and g. □

Now we can prove Theorem 3.9.

Proof By the U^2-inverse theorem, we have

$$\|F\|_{U^3}^8 = \mathbf{E}_{k\in\mathbf{F}_p}\|\Delta_k F\|_{U^2}^4 \leq \mathbf{E}_{k\in\mathbf{F}_p}\|\widehat{\Delta_k F}\|_{\ell^\infty}^2.$$

Set

$$K := \left\{k\in\mathbf{F}_p : \|\widehat{\Delta_k F}\|_{\ell^\infty}^2 \geq \|F\|_{U^3}^8/2\right\},$$

so that $|K|/p \geq \|F\|_{U^3}^8/2$ and

$$\frac{\|F\|_{U^3}^8}{2} \leq \mathbf{E}_{k\in K}\|\widehat{\Delta_k F}\|_{\ell^\infty}^2$$

Let $\phi : \mathbf{F}_p \to \mathbf{F}_p$ be such that $\left|\widehat{\Delta_k F}(\phi(k))\right| = \|\widehat{\Delta_k F}\|_{\ell^\infty}$ for all $k \in \mathbf{F}_p$. Then Lemma 3.10 says that $\mathbf{E}_{k\in K}\|\widehat{\Delta_k F}\|_{\ell^\infty}^2$ is bounded above by

$$\mathbf{E}_{k,k'\in K}\left|\mathbf{E}_{x,y\in\mathbf{F}_p}\Delta_{k'-k}f(x)\Delta_{k'-k}g(x+y)e_p([\phi(k)-\phi(k')][x+y^2])\right|^2. \quad (3.5)$$

Applying Lemma 3.11 to (3.5) therefore yields

$$\frac{\|F\|_{U^3}^8}{2} \leq \frac{\#\left\{(k,k')\in K^2 : \phi(k)=\phi(k')\right\}}{|K|^2} + \frac{1}{p},$$

which implies that

$$\frac{\|F\|_{U^3}^8}{4} \leq \frac{\#\left\{(k,k')\in K^2 : \phi(k)=\phi(k')\right\}}{|K|^2},$$

by our lower bound assumption on $\|F\|_{U^3}$. By the pigeonhole principle, there must exist a $k' \in K$ for which

$$\frac{\#\{k\in K : \phi(k)=\phi(k')\}}{|K|} \geq \frac{\|F\|_{U^3}^8}{4}.$$

Fix this k', and set $\beta := \phi(k')$. Then, since $|K|/p \geq \|F\|_{U^3}^8/2$, the above implies that

$$\frac{\#\{k \in \mathbf{F}_p : \phi(k) = \beta\}}{p} \geq \frac{\|F\|_{U^3}^8}{8}.$$

Thus,

$$\mathbf{E}_{k \in \mathbf{F}_p} \left|\mathbf{E}_{x \in \mathbf{F}_p} \Delta_k F(x) e_p(\beta x)\right|^2 \geq \frac{\|F\|_{U^3}^8}{8} \mathbf{E}_{k \in \mathbf{F}_p} \left\|\widehat{\Delta_k F}\right\|_{\ell^\infty}^2 \geq \frac{\|F\|_{U^3}^{16}}{8}.$$

Now, expanding the square on the left-hand side of the above,

$$\mathbf{E}_{k \in \mathbf{F}_p} \left|\mathbf{E}_{x \in \mathbf{F}_p} \Delta_k F(x) e_p(\beta x)\right|^2 = \mathbf{E}_{x,h,k \in \mathbf{F}_p} \Delta_h F(x) \overline{\Delta_h F(x+k)} e_p(-\beta h),$$

so that, by the Cauchy–Schwarz inequality,

$$\mathbf{E}_{k \in \mathbf{F}_p} \left|\mathbf{E}_{x \in \mathbf{F}_p} \Delta_k F(x) e_p(\beta x)\right|^2 \leq \|F\|_{U^2}^2.$$

Putting everything together then gives

$$\frac{\|F\|_{U^3}^{16}}{8} \leq \|F\|_{U^2}^2.$$

$\square$

Combining control of $\Lambda(f_0, f_1, f_2)$ by $\|F_2\|_{U^3}$ with the degree-lowering lemma quickly yields not only a power-saving bound on sets lacking the nonlinear Roth configuration, but that any subset of $\mathbf{F}_p$ of sufficiently large density contains very close to the number of non-linear Roth configurations one would expect in a random set of the same density.

Theorem 3.12 *If $A \subset \mathbf{F}_p$ has density α, then*

$$\#\left\{(x,y) \in \mathbf{F}_p^2 : x, x+y, x+y^2 \in A\right\} = \alpha^3 p^2 + O\left(p^{2-1/144}\right).$$

Thus, if $\alpha > p^{-1/500}$, then A contains a nontrivial nonlinear Roth configuration.

This recovers the main result of Bourgain and Chang from [18], but with a weaker exponent.

Proof First observe that

$$\Lambda(1, 1_A, 1_A) = \mathbf{E}_{x,y \in \mathbf{F}_p} 1_A(x) 1_A(x + y^2 - y)$$
$$= \sum_{\xi \in \mathbf{F}_p} \left|\widehat{1_A}(\xi)\right|^2 \left(\mathbf{E}_{y \in \mathbf{F}_p} e_p(\xi[y^2 - y])\right)$$
$$= \alpha^2 + O\left(\frac{1}{\sqrt{p}}\right)$$

by Fourier inversion, orthogonality of characters, and the fact that $|\mathbf{E}_{y \in \mathbf{F}_p} e_p(\xi[y^2 - y])| \leq p^{-1/2}$ whenever $\xi \neq 0$. Setting $f_A := 1_A - \alpha$, we therefore have

$$\left|\Lambda(1_A, 1_A, 1_A) - \alpha^3\right| \leq |\Lambda(f_A, 1_A, 1_A)| + O\left(\frac{1}{\sqrt{p}}\right).$$

By Lemmas 3.8 and 3.9,

$$|\Lambda(f_0, f_1, f_2)|^8 \ll \|F_2\|_{U^2}^{1/8} + \frac{1}{p}.$$

where $F_2(x) := \mathbf{E}_{z \in \mathbf{F}_p} f_A(x - z^2) 1_A(x + z - z^2)$. But, by the U^2-inverse theorem, there exists a $\xi \in \mathbf{F}_p$ such that

$$\|F_2\|_{U^2}^2 \leq \left|\mathbf{E}_{x \in \mathbf{F}_p} F_2(x) e_p(\xi x)\right| = \left|\mathbf{E}_{x, z \in \mathbf{F}_p} f_A(x) 1_A(x + z) e_p(\xi[x + z^2])\right|.$$

If $\xi \neq 0$, then $\|F_2\|_{U^2}^2 \leq p^{-1/2}$ by Lemma 3.11, and if $\xi = 0$, then $\|F_2\|_{U^2}^2 = 0$ because f_A has mean zero. So, in either case, we have

$$|\Lambda(f_0, f_1, f_2)| \ll \frac{1}{p^{1/144}},$$

from which we conclude that

$$\left|\Lambda(1_A, 1_A, 1_A) - \alpha^3\right| \ll \frac{1}{p^{1/144}}.$$

$\square$

3.2 Further Results and Questions

Bourgain and Chang [18] also asked whether power-saving bounds hold for subsets of finite fields lacking three-term progressions involving linearly independent rational functions.

Question 3.13 *Let $P_1, P_2, Q_1, Q_1 \in \mathbf{Z}[y]$ with $P_1(0) = P_2(0) = 0$ be such that P_1/Q_1 and P_2/Q_2 are linearly independent rational functions. Does there exist a $\gamma = \gamma_{P_1, P_2, Q_1, Q_2} > 0$ such that if $A \subset \mathbf{F}_p$ contains no nontrivial progressions*

$$x, x + \frac{P_1(y)}{Q_1(y)}, x + \frac{P_2(y)}{Q_2(y)}$$

then

$$|A| \ll p^{1-\gamma}?$$

This question is still open, and it is not clear how to adapt the arguments of [18], [22], [93], or [94] to attack it.

Based on the success story of the degree-lowering method, a reasonable roadmap to a fully general quantitative polynomial Szemerédi theorem may be to first try to prove such a result in finite fields, and then try to adapt the proof to the integer setting. But, I think that even the finite field setting is likely to be very difficult, and one should instead start with the function field setting, where the inverse theorems for the U^s-norms produce genuine polynomial phases instead of nilsequences. Thus, I propose the following problem.

Problem 3.14 *Prove reasonable bounds in the polynomial Szemerédi theorem in function fields.*

All that is known towards this problem is the work of Le–Liu [74], Green [46], and Li–Sauermann [76] mentioned in Subsection 1.2.

The main result of [7] is actually a joint multidimensional-polynomial generalization of Szemerédi's theorem.

Theorem 3.15 *Let $v_1, \ldots, v_m \in \mathbf{Z}^d$ be nonzero vectors and $P_1, \ldots, P_m \in \mathbf{Z}[y]$ be polynomials with zero constant term. If $A \subset [N]^d$ contains no nontrivial multidimensional polynomial progressions*

$$x, x + P_1(y)v_1, \ldots, x + P_m(y)v_m,$$

then $|A| = o_{P_1,\ldots,P_m,v_1,\ldots,v_m}(N^d)$.

Thus, Problem 3.3 is the combination of two special cases of the following, even more difficult, problem.

Problem 3.16 *Prove a quantitative version of Theorem 3.15 with reasonable bounds.*

Not a single nonlinear, genuinely multidimensional (i.e., with $v_1, \ldots, v_m$ not all contained in a line) case of this problem is known in the integer setting. Until very recently, none were known in the finite field setting either. Adapting the methods of Dong, Li, and Sawin [22], Han, Lacey, and Yang [57] proved a power-saving bound for subsets of $\mathbf{F}_p^2$ lacking nontrivial instances of the two-dimensional "polynomial corner"

$$(x,y), (x, y + P_1(z)), (x + P_2(z), y),$$

whenever P_1 and P_2 have distinct degrees. Using the degree-lowering method, Kuca [69] proved power-saving bounds for subsets of $\mathbf{F}_p^d$ lacking polynomial corners

$$(x_1, \ldots, x_d), (x_1 + P_1(y), \ldots, x_d), \ldots, (x_1, \ldots, x_d + P_d(y)),$$

of arbitrarily high dimension, again provided that $P_1, \ldots, P_d$ have distinct degrees. Replacing the distinct degree condition on the polynomials with a linear independence condition in these results turned out to be a significant challenge, and was only done very recently by Kuca [71]. Perhaps the next most attackable problem is to find a joint generalization of Prendiville's work on arithmetic progressions with perfect power common difference and Shkredov's result on corners, in both the integer and finite field settings.

Problem 3.17 *Prove, in both $\mathbf{F}_p$ and $[N]$, reasonable quantitative bounds for sets lacking the multidimensional-polynomial configurations*

$$(x,y), (x, y + z^d), (x + z^d, y).$$

4 Additional Topics

4.1 Comparing Different Notions of Rank

Following the emergence of the slice rank method, there has been a great deal of interest in trying to find quantitative relations between various definitions of rank. One reason for this is that, so far, the notion of slice rank does not seem sufficient to prove power-saving bounds for sets lacking four-term arithmetic progressions in the finite field model setting. It could be easier to show that the diagonal tensor of a set free of four-term arithmetic progressions has small rank for some other definition of rank different from slice rank. If we knew that this other notion of rank was roughly quantitatively equivalent to slice rank, then we could obtain bounds for the size of sets lacking four-term arithmetic progressions. Another reason for understanding the relations between different definitions of rank is that many arguments in additive combinatorics, particularly in the finite field model setting, split into a high rank case (the "pseudorandom" case) and a low rank case (the "structured" case), and proving stronger quantitative relationships between analytic rank and other notions of rank can improve these arguments.

We will now introduce the notion of analytic rank. In connection with the inverse theorem for the Gowers uniformity norms for polynomial phases in the finite field model setting, Green and Tao [50] studied the distribution of values of polynomials over finite fields. They showed that if the values of a polynomial $P : \mathbf{F}_p^n \to \mathbf{F}_p$ are far from uniformly distributed in $\mathbf{F}_p$ as the input ranges over $\mathbf{F}_p^n$, then P can be written as a low-complexity combination of polynomials of strictly smaller degree. More precisely, the failure of equidistribution of P can be measured by the correlation of P with any nontrivial additive character of $\mathbf{F}_p$: this is called the *bias* of P,

$$\mathrm{bias}(P) := \left| \mathbf{E}_{\mathbf{x} \in \mathbf{F}_p^n} e_p(P(\mathbf{x})) \right|.$$

Gowers and Wolf [39] defined the *analytic rank* of a polynomial with nonzero bias to be -1 times the logarithm of its bias:

$$\mathrm{arank}(P) := -\log_p(\mathrm{bias}(P)).$$

Note that, since $|\mathrm{bias}(P)| \leq 1$, the analytic rank of a nonconstant polynomial is always positive.

The extent to which P can be written as a low-complexity combination of polynomials of strictly smaller degree is measured by its *rank*, which is the smallest integer m for which there exist polynomials $P_1, \ldots, P_m : \mathbf{F}_p^n \to \mathbf{F}_p$ of degree strictly less than $\deg P$ and a function $Q : \mathbf{F}_p^m \to \mathbf{F}_p$ such that $P = Q(P_1, \ldots, P_m)$. Green and Tao [50] proved the following inverse theorem for biased polynomials.

Theorem 4.1 *Let $P : \mathbf{F}_p^n \to \mathbf{F}_p$ be a polynomial of degree strictly less than p. For all $\delta > 0$, there exists an $R > 0$ depending only on δ and p such that if $\mathrm{bias}(P) \geq \delta$, then $\mathrm{rank}(P) \leq R$.*

Kaufman and Lovett [62] removed the high characteristic assumption from Theorem 4.1. Both of these arguments produce an upper bound for R that has Ackermann-type dependence on $1/\delta$ and p. This was improved to tower-type dependence for fixed p by Janzer [60], and then to polynomial dependence by Janzer [61] and Milićević [85]. Theorem 4.1 and its quantitative improvements says that the rank of P can be bounded in terms of the analytic rank of P.

Another useful notion of rank is partition rank, which was first defined by Naslund [90] in work extending the slice rank method.

Definition 4.2 *Let $M : (\mathbf{F}_p^n)^m \to \mathbf{F}_p$ be a multilinear form.*

1. We say that M has *partition rank* 1 if there is a nonempty subset of indices $I \subset [m]$ and multilinear forms $M' : (\mathbf{F}_p^n)^{|I|} \to \mathbf{F}_p$ and $M'' : (\mathbf{F}_p^n)^{m-|I|} \to \mathbf{F}_p$ such that

$$M(x_1, \ldots, x_m) := M'(x_i : i \in I) M''(x_j : j \in [m] \setminus I)$$

for all $x = (x_1, \ldots, x_m) \in (\mathbf{F}_p^n)^m$.

2. The *partition rank* of M, denoted by $\operatorname{prank} M$, is the smallest nonnegative integer k such that M can be written as the sum of k multilinear forms of partition rank 1.

In [89], Naslund used a variant of the slice rank method adapted to partition rank to obtain the first power-saving bounds for the Erdős–Ginzburg–Ziv constant for high-dimensional vector spaces over finite fields.

Lovett [79] and, independently, Kazhdan and Ziegler [63] showed that the analytic rank of a multilinear form is bounded above by its partition rank and slice rank. Janzer [61] and Milićević [85] showed that, in the reverse direction, the partition rank of a multilinear form can be bounded in terms of its analytic rank, with polynomial dependence. Recently, Moshkovitz and Zhu [87] proved almost-linear dependence of partition rank on analytic rank: $\operatorname{prank}(M) \ll_m \operatorname{arank}(M)(\log(1 + \operatorname{arank}(M)))^{m-1}$.

4.2 The Multidimensional Szemerédi Theorem

Reasonable bounds in Theorem 3.1 are only known for only one genuinely multidimensional configuration: two-dimensional corners, due to work of Shkredov [113, 114], who showed that if $A \subset [N]^2$ contains no nontrivial corners, then

$$|A| = O\left(\frac{N^2}{(\log \log N)^c}\right)$$

for some $c > 0$. Shkredov proved this result in 2006, and more than fifteen years later, no reasonable bounds are known for sets lacking any multidimensional four-point configuration. I have, however, managed to make a small amount of progress in the finite field model setting [97].

Theorem 4.3 *There exists an absolute constant $m \in \mathbf{N}$ such that the following holds. Fix $p \geq 11$. If $A \subset (\mathbf{F}_p^n)^2$ contains no nontrivial L-shapes,*

$$(x, y), (x, y + z), (x, y + 2z), (x + z, y) \qquad (z \neq 0),$$

then

$$|A| \ll \frac{p^{2n}}{\log_{(m)} p^n}.$$

The size of m in this result depends on the number of iterated exponentials appearing in the Gowers–Milićević inverse theorem for the U^{10}-norm, and can be taken to be 24 trillion. Any improvement in the quantitative aspect of Gowers and Milićević's theorem would automatically improve the size of m. The argument used to prove this result is robust enough that it adapts to the integer setting, though with significant technical difficulties arising from the need to work relative to Bohr sets.

Proving reasonable bounds for any other genuinely multidimensional four-point configuration that is not a linear image of an L-shape seems to be an extremely difficult problem, even in the finite field model setting. Such configurations include axis-aligned squares and three-dimensional corners.

Problem 4.4 *Fix a sufficiently large prime p. Prove reasonable bounds for subsets of $(\mathbf{F}_p^n)^2$ containing no nontrivial axis-aligned squares,*

$$(x,y), (x, y+z), (x+z, y), (x+z, y+z) \qquad (z \neq 0).$$

The count of axis-aligned squares in a set is controlled by the following two-dimensional analogue of a Gowers uniformity norm:

$$\|f\|_\bullet := \left(\mathbf{E}_{x,y,h_1,h_2,h_3 \in \mathbf{F}_p^n} \Delta_{(h_1,0),(0,h_2),(h_3,h_3)} f(x,y)\right)^{1/8}.$$

So far, no one has been able to prove an inverse theorem for $\|\cdot\|_\bullet$.

Problem 4.5 *Prove an inverse theorem, preferably with reasonable bounds, for the norm $\|\cdot\|_\bullet$.*

This norm also shows up in the analysis of three-dimensional corners.

Problem 4.6 *Fix a sufficiently large prime p. Prove reasonable bounds for subsets of $(\mathbf{F}_p^n)^3$ containing no nontrivial three-dimensional corners,*

$$(x,y,z), (x,y,z+w), (x, y+w, z), (x+w, y, z).$$

By a projection argument, any solution to this problem would yield a solution to Problem 4.4, but this problem is strictly harder since one must also find a way to get around using Szemerédi's regularity lemma (or related results with very weak bounds) when "pseudorandomizing" the objects produced by various inverse theorems.

4.3 True Complexity

The *true complexity* of a system of linear forms $\phi_1, \ldots, \phi_m \in \mathbf{Z}[x_1, \ldots, x_d]$ in d variables is the smallest natural number s such that the following statement holds: for every $\varepsilon > 0$, there exists a $\delta > 0$ such that if $p \gg_{\phi_1,\ldots,\phi_m} 1$ and $f : \mathbf{F}_p^n \to \mathbf{C}$ is a 1-bounded function on $G := \mathbf{F}_p^n$ satisfying $\|f\|_{U^{s+1}} < \delta$, then

$$\left| \mathbf{E}_{\mathbf{x} \in G^d} \prod_{i=1}^m f(\phi_i(\mathbf{x})) \right| < \varepsilon.$$

The notion of true complexity, defined by Gowers and Wolf in [38], is important both in additive combinatorics and in ergodic theory. Gowers and Wolf conjectured that the true complexity of a system of linear equations is equal to the smallest integer s such that the polynomials

$$\phi_1(\mathbf{x})^{s+1}, \ldots, \phi_m(\mathbf{x})^{s+1}$$

are linearly independent over $\mathbf{Q}$, and proved their conjecture in the finite field model setting [40, 39] (i.e., in the regime where p is fixed and $n \to \infty$) in full generality and in the setting of cyclic groups [41] (i.e., in the regime where $n = 1$ and $p \to \infty$) when $s = 2$. Green and Tao [52] proved the conjecture of Gowers and Wolf in cyclic groups for systems of linear forms satisfying a technical condition known as the "flag condition", and Altman [2] recently proved the conjecture in full generality in the cyclic group setting.

All of the aforementioned proofs use the inverse theorems for the Gowers uniformity norms and some variant of a regularity lemma, and thus mostly produce very poor dependence of δ on ε. In a pair of amazing papers, Manners [81, 83] has given proofs of the conjecture of Gowers and Wolf in both the finite field model and integer settings using only repeated applications of the Cauchy–Schwarz inequality (along with essentially trivial moves like changes of variables and applying the pigeonhole principle), which produces polynomial dependence of δ on ε (though the polynomial dependence, necessarily, depends on the system of linear forms).

One can also define the true complexity of a polynomial progressions. Let $P_1, \ldots, P_m \in \mathbf{Z}[y]$ be a collection of polynomials and $0 \le i \le m$ be a natural number. We say that the polynomial progression $x, x+P_1(y), \ldots, x+P_m(y)$ has *true complexity* s *at* i if s is the smallest natural number for which the following statement holds: for every $\varepsilon > 0$, there exists a $\delta > 0$ such that if $p \gg_{P_1,\ldots,P_m} 1$ and $f_0, \ldots, f_m : \mathbf{F}_p \to \mathbf{C}$ are 1-bounded functions such that $\|f_i\|_{U^{s+1}} < \delta$, then

$$\left| \mathbf{E}_{x,y \in \mathbf{F}_p} f_0(x) \prod_{j=1}^m f_j(x + P_j(y)) \right| < \varepsilon.$$

A polynomial progression of length $m+1$ has true complexity s if its maximum true complexity at i over $0 \le i \le m$ is s. Very little is known about the true complexity of a general polynomial progression. Bergelson, Leibman, and Lesigne [8] asked whether any polynomial progression of length at most $m+1$ must have true complexity at most $m-1$, and this question is still entirely open for $m > 3$ (Frantzikinakis [28] verified it when $m \le 3$). In contrast, a simple modification of the proof of Lemma 2.2 shows that the true complexity of an arbitrary system of $m+1$ linear forms of finite complexity has true complexity at most $m-1$.

The main difficulty in extending Frantzikinakis's argument to longer progressions lies in analyzing the distribution of certain polynomial sequences on nilmanifolds. Thus, since the inverse theorems for the U^s-norms in the finite field model setting give correlation with a genuine polynomial phase, this problem may be more approachable in the function field setting.

Problem 4.7 *Resolve the question of Bergelson, Leibman, and Lesigne in the function field setting.*

In [72], Kuca formulated a polynomial progression analogue of the conjecture of Gowers and Wolf, and verified some special cases of his conjecture.

Conjecture 4.8 *Let $P_1, \ldots, P_m \in \mathbf{Z}[y]$ be a collection of polynomials and $0 \le i \le m$ be a natural number. The polynomial progression $x, x+P_1(y), \ldots, x+P_m(y)$ has true complexity s at i if s is the smallest natural number such that for any algebraic relation*

$$Q_0(x) + Q_1(x + P_1(y)) + \cdots + Q_m(x + P_m(y)) = 0$$

with $Q_0, \ldots, Q_m \in \mathbf{Z}[z]$ satisfied by $x, x + P_1(y), \ldots, x + P_m(y)$, the degree of Q_i is at most s.

Similarly to the question of Bergelson, Leibman, and Lesigne, Kuca's conjecture may be more attackable in the function field setting.

Problem 4.9 *Prove Kuca's conjecture in the function field setting.*

Finally, it may be possible to prove true complexity statements for polynomial progressions just using repeated applications of the Cauchy–Schwarz inequality (along with trivial moves). Manners has posed the following concrete problem.

Problem 4.10 *Prove that the nonlinear Roth configuration has true complexity zero using only the Cauchy–Schwarz inequality.*

In Subsection 3.1, we proved that the nonlinear Roth configuration has true complexity zero by combining many applications of the Cauchy–Schwarz inequality and the pigeonhole principle with some basic Fourier analysis.

Acknowledgements

I would like to thank Thomas Bloom and Julia Wolf for helpful comments on earlier drafts and Dan Altman for helpful discussions. I was supported by the NSF Mathematical Sciences Postdoctoral Research Fellowship Program under Grant No. DMS-1903038 while writing the bulk of this survey.

References

[1] N. Alon and M. Dubiner, *Zero-sum sets of prescribed size*, Combinatorics, Paul Erdős is eighty, Vol. 1, Bolyai Soc. Math. Stud., János Bolyai Math. Soc., Budapest, 1993, pp. 33–50. MR 1249703

[2] D. Altman, *On a conjecture of Gowers and Wolf*, Discrete Anal. (2022), Paper No. 10, 13. MR 4481407

[3] N. Arala, *A maximal extension of the Bloom-Maynard bound for sets with no square differences*, preprint (2023), arXiv:2303.03345.

[4] A. Balog, J. Pelikán, J. Pintz, and E. Szemerédi, *Difference sets without κth powers*, Acta Math. Hungar. **65** (1994), no. 2, 165–187. MR 1278767

[5] M. Bateman and N. H. Katz, *New bounds on cap sets*, J. Amer. Math. Soc. **25** (2012), no. 2, 585–613. MR 2869028

[6] F. A. Behrend, *On sets of integers which contain no three terms in arithmetical progression*, Proc. Nat. Acad. Sci. U. S. A. **32** (1946), 331–332. MR 0018694

[7] V. Bergelson and A. Leibman, *Polynomial extensions of van der Waerden's and Szemerédi's theorems*, J. Amer. Math. Soc. **9** (1996), no. 3, 725–753. MR 1325795

[8] V. Bergelson, A. Leibman, and E. Lesigne, *Complexities of finite families of polynomials, Weyl systems, and constructions in combinatorial number theory*, J. Anal. Math. **103** (2007), 47–92. MR 2373264

[9] V. Bergelson, T. Tao, and T. Ziegler, *An inverse theorem for the uniformity seminorms associated with the action of* $\mathbb{F}_p^\infty$, Geom. Funct. Anal. **19** (2010), no. 6, 1539–1596. MR 2594614

[10] J. Blasiak, T. Church, H. Cohn, J. A. Grochow, E. Naslund, W. F. Sawin, and C. Umans, *On cap sets and the group-theoretic approach to matrix multiplication*, Discrete Anal. (2017), Paper No. 3, 27. MR 3631613

[11] T. F. Bloom, *A quantitative improvement for Roth's theorem on arithmetic progressions*, J. Lond. Math. Soc. (2) **93** (2016), no. 3, 643–663. MR 3509957

[12] T. F. Bloom and J. Maynard, *A new upper bound for sets with no square differences*, Compos. Math. **158** (2022), no. 8, 1777–1798. MR 4490931

[13] T. F. Bloom and O. Sisask, *Breaking the logarithmic barrier in Roth's theorem on arithmetic progressions*, preprint (2020), arXiv:2007.03528.

[14] _____, *An improvement to the Kelley-Meka bounds on three-term arithmetic progressions*, preprint (2023), arXiv:2309.02353.

[15] _____, *The Kelley–Meka bounds for sets free of three-term arithmetic progressions*, preprint (2023), arXiv:2302.07211.

[16] J. Bourgain, *On triples in arithmetic progression*, Geom. Funct. Anal. **9** (1999), no. 5, 968–984. MR 1726234

[17] _____, *Roth's theorem on progressions revisited*, J. Anal. Math. **104** (2008), 155–192. MR 2403433

[18] J. Bourgain and M.-C. Chang, *Nonlinear Roth type theorems in finite fields*, Israel J. Math. (2017).

[19] T. C. Brown and J. P. Buhler, *A density version of a geometric Ramsey theorem*, J. Combin. Theory Ser. A **32** (1982), no. 1, 20–34. MR 640624

[20] E. Croot, V. F. Lev, and P. P. Pach, *Progression-free sets in $\mathbb{Z}_4^n$ are exponentially small*, Ann. of Math. (2) **185** (2017), no. 1, 331–337. MR 3583357

[21] E. Croot and O. Sisask, *A probabilistic technique for finding almost-periods of convolutions*, Geom. Funct. Anal. **20** (2010), no. 6, 1367–1396. MR 2738997

[22] D. Dong, X. Li, and W. Sawin, *Improved estimates for polynomial Roth type theorems in finite fields*, J. Anal. Math. **141** (2020), no. 2, 689–705. MR 4179774

[23] J. S. Ellenberg and D. Gijswijt, *On large subsets of $\mathbb{F}_q^n$ with no three-term arithmetic progression*, Ann. of Math. (2) **185** (2017), no. 1, 339–343. MR 3583358

[24] J. Fox, *A new proof of the graph removal lemma*, Ann. of Math. (2) **174** (2011), no. 1, 561–579. MR 2811609

[25] J. Fox and L. M. Lovász, *A tight bound for Green's arithmetic triangle removal lemma in vector spaces*, Adv. Math. **321** (2017), 287–297. MR 3715712

[26] J. Fox, L. M. Lovász, and L. Sauermann, *A polynomial bound for the arithmetic k-cycle removal lemma in vector spaces*, J. Combin. Theory Ser. A **160** (2018), 186–201. MR 3846201

[27] P. Frankl, R. L. Graham, and V. Rödl, *On subsets of abelian groups with no 3-term arithmetic progression*, J. Combin. Theory Ser. A **45** (1987), no. 1, 157–161. MR 883900

[28] N. Frantzikinakis, *Multiple ergodic averages for three polynomials and applications*, Trans. Amer. Math. Soc. **360** (2008), no. 10, 5435–5475. MR 2415080

[29] H. Furstenberg, *Ergodic behavior of diagonal measures and a theorem of Szemerédi on arithmetic progressions*, J. Analyse Math. **31** (1977), 204–256. MR 0498471

[30] H. Furstenberg and Y. Katznelson, *An ergodic Szemerédi theorem for commuting transformations*, J. Analyse Math. **34** (1978), 275–291 (1979). MR 531279

[31] W. T. Gowers, *A new proof of Szemerédi's theorem for arithmetic progressions of length four*, Geom. Funct. Anal. **8** (1998), no. 3, 529–551. MR 1631259

[32] _____, *Arithmetic progressions in sparse sets*, Current developments in mathematics, 2000, Int. Press, Somerville, MA, 2001, pp. 149–196. MR 1882535

[33] _____, *A new proof of Szemerédi's theorem*, Geom. Funct. Anal. **11** (2001), no. 3, 465–588. MR 1844079

[34] _____, *Hypergraph regularity and the multidimensional Szemerédi theorem*, Ann. of Math. (2) **166** (2007), no. 3, 897–946. MR 2373376

[35] W. T. Gowers, B. Green, F. Manners, and T. Tao, *On a conjecture of marton*, preprint (2023), arXiv:2311.05762.

[36] W. T. Gowers and L. Milićević, *A quantitative inverse theorem for the U^4 norm over finite fields*, preprint (2017), arXiv:1712.00241.

[37] _____, *An inverse theorem for Freiman multi-homomorphisms*, preprint (2020), arXiv:2002.11667.

[38] W. T. Gowers and J. Wolf, *The true complexity of a system of linear equations*, Proc. Lond. Math. Soc. (3) **100** (2010), no. 1, 155–176. MR 2578471

[39] _____, *Linear forms and higher-degree uniformity for functions on $\mathbb{F}_p^n$*, Geom. Funct. Anal. **21** (2011), no. 1, 36–69. MR 2773103

[40] _____, *Linear forms and quadratic uniformity for functions on $\mathbb{F}_p^n$*, Mathematika **57** (2011), no. 2, 215–237. MR 2825234

[41] _____, *Linear forms and quadratic uniformity for functions on $\mathbb{Z}_N$*, J. Anal. Math. **115** (2011), 121–186. MR 2855036

[42] B. Green, *Finite field models in additive combinatorics*, Surveys in combinatorics 2005, London Math. Soc. Lecture Note Ser., vol. 327, Cambridge Univ. Press, Cambridge, 2005, pp. 1–27. MR 2187732

[43] _____, *Notes on the polynomial freiman–ruzsa conjecture*, unpublished notes (2005), http://people.maths.ox.ac.uk/greenbj/papers/PFR.pdf.

[44] _____, *A Szemerédi-type regularity lemma in abelian groups, with applications*, Geom. Funct. Anal. **15** (2005), no. 2, 340–376. MR 2153903

[45] _____, *Montréal notes on quadratic Fourier analysis*, Additive combinatorics, CRM Proc. Lecture Notes, vol. 43, Amer. Math. Soc., Providence, RI, 2007, pp. 69–102. MR 2359469

[46] _____, *Sárközy's theorem in function fields*, Q. J. Math. **68** (2017), no. 1, 237–242. MR 3658291

[47] _____, *On Sárközy's theorem for shifted primes*, preprint (2022), arXiv:2206.08001.

[48] B. Green, F. Manners, and T. Tao, *Sumsets and entropy revisited*, preprint (2023), arXiv:2306.13403.

[49] B. Green and T. Tao, *An inverse theorem for the Gowers $U^3(G)$ norm*, Proc. Edinb. Math. Soc. (2) **51** (2008), no. 1, 73–153. MR 2391635

[50] _____, *The distribution of polynomials over finite fields, with applications to the Gowers norms*, Contrib. Discrete Math. **4** (2009), no. 2, 1–36. MR 2592422

[51] _____, *New bounds for Szemerédi's theorem. I. Progressions of length 4 in finite field geometries*, Proc. Lond. Math. Soc. (3) **98** (2009), no. 2, 365–392. MR 2481952

[52] _____, *An arithmetic regularity lemma, an associated counting lemma, and applications*, An irregular mind, Bolyai Soc. Math. Stud., vol. 21, János Bolyai Math. Soc., Budapest, 2010, pp. 261–334. MR 2815606

[53] _____, *An equivalence between inverse sumset theorems and inverse conjectures for the U^3 norm*, Math. Proc. Cambridge Philos. Soc. **149** (2010), no. 1, 1–19. MR 2651575

[54] _____, *New bounds for Szemerédi's theorem, Ia: Progressions of length 4 in finite field geometries revisited*, preprint (2012), arXiv:1205.1330.

[55] B. Green, T. Tao, and T. Ziegler, *An inverse theorem for the Gowers U^4-norm*, Glasg. Math. J. **53** (2011), no. 1, 1–50. MR 2747135

[56] _____, *An inverse theorem for the Gowers $U^{s+1}[N]$-norm*, Ann. of Math. (2) **176** (2012), no. 2, 1231–1372. MR 2950773

[57] R. Han, M. T. Lacey, and F. Yang, *A polynomial Roth theorem for corners in finite fields*, Mathematika **67** (2021), no. 4, 885–896. MR 4304416

[58] D. R. Heath-Brown, *Integer sets containing no arithmetic progressions*, J. London Math. Soc. (2) **35** (1987), no. 3, 385–394. MR 889362

[59] K. Hosseini and S. Lovett, *A bilinear Bogolyubov-Ruzsa lemma with polylogarithmic bounds*, Discrete Anal. (2019), Paper No. 10, 14. MR 3975362

[60] O. Janzer, *Low analytic rank implies low partition rank for tensors*, preprint (2018), arXiv:1809.10931.

[61] _____, *Polynomial bound for the partition rank vs the analytic rank of tensors*, Discrete Anal. (2020), Paper No. 7, 18. MR 4107323

[62] T. Kaufman and S. Lovett, *Worst case to average case reductions for polynomials*, 49th Annual IEEE symposium on Foundations of Computer Science, 2008, pp. 166–175.

[63] D. Kazhdan and T. Ziegler, *Approximate cohomology*, Selecta Math. (N.S.) **24** (2018), no. 1, 499–509. MR 3769736

[64] Z. Kelley and R. Meka, *Strong bounds for 3-progressions*, preprint (2023), arXiv:2302.05537.

[65] D. Kim, A. Li, and J. Tidor, *Cubic Goldreich-Levin*, Proceedings of the 2023 Annual ACM-SIAM Symposium on Discrete Algorithms (SODA), SIAM, Philadelphia, PA, 2023, pp. 4846–4892. MR 4538136

[66] R. Kleinberg, W. Sawin, and D. E. Speyer, *The growth of tri-colored sum-free sets*, Discrete Anal. (2018), Paper No. 12, 10. MR 3827120

[67] D. Král, O. Serra, and L. Vena, *A combinatorial proof of the removal lemma for groups*, J. Combin. Theory Ser. A **116** (2009), no. 4, 971–978. MR 2513645

[68] B. Kuca, *Further bounds in the polynomial Szemerédi theorem over finite fields*, Acta Arith. **198** (2021), no. 1, 77–108. MR 4214350

[69] _____, *Multidimensional polynomial Szemerédi theorem in finite fields for polynomials of distinct degrees*, preprint (2021), arXiv:2103.12606.

[70] _____, *True complexity of polynomial progressions in finite fields*, Proc. Edinb. Math. Soc. (2) **64** (2021), no. 3, 448–500. MR 4330272

[71] _____, *Multidimensional polynomial patterns over finite fields: bounds, counting estimates and Gowers norm control*, preprint (2023), arXiv:2304.10793.

[72] _____, *On several notions of complexity of polynomial progressions*, Ergodic Theory Dynam. Systems **43** (2023), no. 4, 1269–1323. MR 4555828

[73] M. T. Lacey and W. McClain, *On an argument of Shkredov on two-dimensional corners*, Online J. Anal. Comb. (2007), no. 2, Art. 2, 21. MR 2289954

[74] T. H. Lê and Y.-R. Liu, *On sets of polynomials whose difference set contains no squares*, Acta Arith. **161** (2013), no. 2, 127–143. MR 3141915

[75] J. Leng, *A quantitative bound for Szemerédi's theorem for a complexity one polynomial progression over $\mathbb{Z}/N\mathbb{Z}$*, preprint (2019), arXiv:1903.02592.

[76] A. Li and L. Sauermann, *Sárközy's theorem in various finite field settings*, preprint (2022), arXiv:2212.12754.

[77] S. Lovett, *Equivalence of polynomial conjectures in additive combinatorics*, Combinatorica **32** (2012), no. 5, 607–618. MR 3004811

[78] ———, *An exposition of Sanders' quasi-polynomial Freiman-Ruzsa theorem*, Graduate Surveys, no. 6, Theory of Computing Library, 2015.

[79] ———, *The analytic rank of tensors and its applications*, Discrete Anal. (2019), Paper No. 7, 10. MR 3964143

[80] S. Lovett, R. Meshulam, and A. Samorodnitsky, *Inverse conjecture for the Gowers norm is false*, Theory Comput. **7** (2011), 131–145. MR 2862496

[81] F. Manners, *Good bounds in certain systems of true complexity one*, Discrete Anal. (2018), Paper No. 21, 40. MR 3900336

[82] ———, *Quantitative bounds in the inverse theorem for the Gowers U^{s+1}-norms over cyclic groups*, preprint (2018), arXiv:1811.00718.

[83] ———, *True complexity and iterated Cauchy–Schwarz*, preprint (2021), arXiv:2109.05731.

[84] R. Meshulam, *On subsets of finite abelian groups with no 3-term arithmetic progressions*, J. Combin. Theory Ser. A **71** (1995), no. 1, 168–172. MR 1335785

[85] L. Milićević, *Polynomial bound for partition rank in terms of analytic rank*, Geom. Funct. Anal. **29** (2019), no. 5, 1503–1530. MR 4025518

[86] L. Milićević, *Quantitative inverse theorem for Gowers uniformity norms U^5 and U^6 in $\mathbb{F}_2^n$*, preprint (2022), arXiv:2207.01591.

[87] G. Moshkovitz and D. G. Zhu, *Quasi-linear relation between partition and analytic rank*, preprint (2022), arXiv:2211.05780.

[88] B. Nagle, V. Rödl, and M. Schacht, *The counting lemma for regular k-uniform hypergraphs*, Random Structures Algorithms **28** (2006), no. 2, 113–179. MR 2198495

[89] E. Naslund, *Exponential bounds for the Erdős-Ginzburg-Ziv constant*, J. Combin. Theory Ser. A **174** (2020), 105185, 19. MR 4078996

[90] _____, *The partition rank of a tensor and k-right corners in $\mathbb{F}_q^n$*, J. Combin. Theory Ser. A **174** (2020), 105190, 25. MR 4078997

[91] S. Norin, *A distribution on triples with maximum entropy marginal*, Forum Math. Sigma **7** (2019), e46, 12. MR 4061969

[92] L. Pebody, *Proof of a conjecture of Kleinberg-Sawin-Speyer*, Discrete Anal. (2018), Paper No. 13, 7. MR 3827119

[93] S. Peluse, *Three-term polynomial progressions in subsets of finite fields*, Israel J. Math. **228** (2018), no. 1, 379–405. MR 3874848

[94] _____, *On the polynomial Szemerédi theorem in finite fields*, Duke Math. J. **168** (2019), no. 5, 749–774. MR 3934588

[95] _____, *Bounds for sets with no polynomial progressions*, Forum Math. Pi **8** (2020), e16, 55. MR 4199235

[96] _____, *Recent progress on bounds for sets with no three terms in arithmetic progression*, Astérisque (2022), no. 438, No. 1196, 581. MR 4576028

[97] _____, *Subsets of $\mathbb{F}_n^p \times \mathbb{F}_n^p$ without L-shaped configurations*, Compos. Math. **160** (2024), no. 1, 176–236.

[98] S. Peluse and S. Prendiville, *Quantitative bounds in the non-linear Roth theorem*, preprint (2019), arXiv:1903.02592.

[99] _____, *A polylogarithmic bound in the nonlinear Roth theorem*, Int. Math. Res. Not. IMRN (2020), rnaa261.

[100] S. Prendiville, *Quantitative bounds in the polynomial Szemerédi theorem: the homogeneous case*, Discrete Anal. (2017), no. 5.

[101] V. Rödl and J. Skokan, *Regularity lemma for k-uniform hypergraphs*, Random Structures Algorithms **25** (2004), no. 1, 1–42. MR 2069663

[102] K. F. Roth, *On certain sets of integers*, J. London Math. Soc. **28** (1953), 104–109. MR 0051853

[103] I. Z. Ruzsa, *Difference sets without squares*, Period. Math. Hungar. **15** (1984), no. 3, 205–209. MR 756185

[104] _____, *Sumsets and entropy*, Random Structures & Algorithms **34** (2009), no. 1, 1–10.

[105] A. Samorodnitsky, *Low-degree tests at large distances*, STOC'07—Proceedings of the 39th Annual ACM Symposium on Theory of Computing, ACM, New York, 2007, pp. 506–515. MR 2402476

[106] T. Sanders, *Additive structures in sumsets*, Math. Proc. Cambridge Philos. Soc. **144** (2008), no. 2, 289–316. MR 2405891

[107] _____, *On Roth's theorem on progressions*, Ann. of Math. (2) **174** (2011), no. 1, 619–636. MR 2811612

[108] _____, *On certain other sets of integers*, J. Anal. Math. **116** (2012), 53–82. MR 2892617

[109] _____, *On the Bogolyubov-Ruzsa lemma*, Anal. PDE **5** (2012), no. 3, 627–655. MR 2994508

[110] A. Sárközy, *On difference sets of sequences of integers. III*, Acta Math. Acad. Sci. Hungar. **31** (1978), 355–386.

[111] T. Schoen, *Improved bound in Roth's theorem on arithmetic progressions*, Adv. Math. **386** (2021), Paper No. 107801, 20. MR 4266746

[112] T. Schoen and O. Sisask, *Roth's theorem for four variables and additive structures in sums of sparse sets*, Forum Math. Sigma **4** (2016), Paper No. e5, 28. MR 3482282

[113] I. D. Shkredov, *On a generalization of Szemerédi's theorem*, Proc. London Math. Soc. (3) **93** (2006), no. 3, 723–760. MR 2266965

[114] _____, *On a problem of Gowers*, Izv. Ross. Akad. Nauk Ser. Mat. **70** (2006), no. 2, 179–221. MR 2223244

[115] E. Szemerédi, *On sets of integers containing no k elements in arithmetic progression*, Acta Arith. **27** (1975), 199–245, Collection of articles in memory of JuriĭVladimirovič Linnik. MR 0369312

[116] E. Szemerédi, *Integer sets containing no arithmetic progressions*, Acta Math. Hungar. **56** (1990), no. 1-2, 155–158. MR 1100788

[117] T. Tao, *A variant of the hypergraph removal lemma*, J. Combin. Theory Ser. A **113** (2006), no. 7, 1257–1280. MR 2259060

[118] _____, *Sumset and inverse sumset theory for Shannon entropy*, Comb. Probab. Comput. **19** (2010), no. 4, 603–639.

[119] _____, *Higher order Fourier analysis*, Graduate Studies in Mathematics, vol. 142, American Mathematical Society, Providence, RI, 2012. MR 2931680

[120] _____, *A symmetric formulation of the Croot–Lev–Pach–Ellenberg–Gijswijt capset bound*, https://terrytao.wordpress.com/2016/05/18/a-symmetric-formulation-of-the-croot-lev-pach-ellenberg-gijswijt-capset-bound/, 2016, Accessed: 2021-05-17.

[121] T. Tao and T. Ziegler, *The inverse conjecture for the Gowers norm over finite fields via the correspondence principle*, Anal. PDE **3** (2010), no. 1, 1–20. MR 2663409

[122] _____, *The inverse conjecture for the Gowers norm over finite fields in low characteristic*, Ann. Comb. **16** (2012), no. 1, 121–188. MR 2948765

[123] J. Tidor, *Quantitative bounds for the U^4-inverse theorem over low characteristic finite fields*, Discrete Anal. (2022), Paper No. 14, 17. MR 4503221

[124] F. Tyrrell, *New lower bounds for cap sets*, preprint (2022), arXiv:2209.10045.

[125] J. Wolf, *Finite field models in arithmetic combinatorics—ten years on*, Finite Fields Appl. **32** (2015), 233–274. MR 3293412

Department of Mathematics
University of Michigan
East Hall, 530 Church Street
Ann Arbor, Michigan 48109, USA
speluse@umich.edu

The Slice Rank Polynomial Method – A Survey a Few Years Later

Lisa Sauermann

Abstract

The slice rank polynomial method was introduced by Tao in 2016 following the breakthrough of Ellenberg and Gijswijt on the famous Cap-Set Problem, which in turn was building on work of Croot, Lev and Pach. This survey gives an introduction to the slice rank polynomial method, shows some of its early applications, and discusses the developments since then.

1 Introduction

Bounding the size of progression-free subsets of $\{1, \dots, N\}$ or of $\mathbb{F}_p^n$ is one of the most fundamental problems in additive combinatorics. A three-term arithmetic progression in $\{1, \dots, N\}$ or in $\mathbb{F}_p^n$ for a prime $p \geq 3$ consists of x, y, z such that $y - x = z - y$, or equivalently $x - 2y + z = 0$. It is called non-trivial if x, y, z are distinct. More generally, a non-trivial k-term arithmetic progression consists of distinct $x_1, \dots, x_k$ such that $x_2 - x_1 = x_3 - x_2 = \dots = x_k - x_{k-1}$.

The problem of studying the maximum possible size of a subset of $\{1, \dots, N\}$ not containing a non-trivial three-term arithmetic progression (or, more generally, not containing a non-trivial k-term arithmetic progression for some fixed $k \geq 3$) has a very long history. It was posed by Erdős and Turán [27] in 1936 and has been one of the most important questions in additive combinatorics since then. In 1946, Behrend [6] gave a construction for a subset of $\{1, \dots, N\}$ without a non-trivial three-term arithmetic progression of size $\exp(-C(\log N)^{1/2}) \cdot N$ (for some absolute constant $C > 0$, and all N), improving earlier lower bounds of Erdős–Turán [27] and of Salem–Spencer [67]. Concerning upper bounds, Roth [65] showed in 1953 that every subset of $\{1, \dots, N\}$ without a non-trivial three-term arithmetic progression must have size at most $N/\log \log N$. There has been a lot of work on improving the upper bound [10, 11, 13, 14, 41, 68, 69, 74, 77], and until very recently the best upper bound was $N/(\log N)^{1+c}$ for some absolute constant $c > 0$. In a spectacular breakthrough in early 2023, Kelley and Meka finally showed an upper bound of the form $\exp(-c(\log N)^{1/11}) \cdot N$ for the size of any subset of $\{1, \dots, N\}$ without a non-trivial three-term arithmetic progression (for an absolute constant $c > 0$ and all $N \geq 3$). This upper bound matches the shape of Behrend's lower bound, and so it resolves the long-standing open problem of understanding the growth behavior of the maximum size of such a subset.

For k-term progressions with fixed $k \geq 4$, the analogous problem is still widely open. By a very famous result of Szemerédi [76] from 1975, for every fixed $k \geq 3$, any subset of $\{1, \dots, N\}$ without a non-trivial k-term arithmetic progression has size $o(N)$. In the context of this result, Szemerédi developed his celebrated *regularity lemma* (now a very important tool in graph theory), and this was also the main result described the citation for Szemerédi's Abel Prize in 2012.

This survey paper is not concerned with these questions in $\{1, \dots, N\}$, but in the setting of $\mathbb{F}_p^n$ for a fixed prime $p \geq 3$ and large n. Specifically for $p = 3$, the question of estimating the maximum possible size of a subset of $\mathbb{F}_3^n$ without a non-trivial three-term arithmetic progression, is also very famous, and is called the *Cap-Set Problem*. The best known lower bound for this problem is roughly 2.218^n and is due to Tyrrell [79], improving on a much older construction of Edel [20] giving a bound of roughly 2.217^n. As an upper bound, Meshulam [49] proved a bound of the form $O(3^n/n)$ for this problem, which was improved in 2012 by Bateman and Katz [5] to $3^n/n^{1+c}$ for some absolute constant $c > 0$. Both of these papers used Fourier-Analytic techniques, which are also the predominant techniques for the analogous problem in $\{1, \ldots, N\}$ discussed above.

A few years ago, Ellenberg and Gijswijt [22] obtained a drastically better bound (which is, surprisingly, of a very different shape than the answer to corresponding problem in $\{1, \ldots, N\}$). They proved that any subset of $\mathbb{F}_3^n$ without a non-trivial three-term arithmetic progression must have size *exponentially smaller* than 3^n, as stated in the following theorem.

Theorem 1.1 *For any positive integer n, any subset $X \subseteq \mathbb{F}_3^n$ not containing a non-trivial three-term arithmetic progression has size $|X| \leq 2.756^n$.*

This was a radical improvement of the previously known upper bounds for this problem for large n. Let us remark that for small n, the upper bound in Theorem 1.1 is not particularly good. For example for $n = 4$, the theorem gives an upper bound of 57, while it is known since 1970 that the maximum size of such a subset for $n = 4$ is actually equal to 20 [64]. In the case of $n = 4$, this problem is actually equivalent to asking about the maximum number of cards in the popular card game "SET" without a so-called "set" (indeed, the cards in the deck of this game can be identified with the vectors in $\mathbb{F}_3^4$ and a "set" then corresponds to a non-trivial three-term arithmetic progression in $\mathbb{F}_3^4$).

The point of Theorem 1.1 is, however, to understand the growth behavior for large n of the maximum possible size of a subset of $\mathbb{F}_3^n$ without a non-trivial three-term arithmetic progression. The shape of the bound in Theorem 1.1 matches the shape of the lower bounds of Tyrrell [79] and Edel [20] up to the value of the absolute constant in the base.

Ellenberg and Gijswijt actually proved a more general result in $\mathbb{F}_p^n$, for any fixed prime $p \geq 3$ (not just $p = 3$ as in Theorem 1.1). Again, they proved that any subset of $\mathbb{F}_p^n$ without a non-trivial three-term arithmetic progression must have size *exponentially smaller* than p^n. To state their result more precisely, let us define

$$\Gamma_p = \inf_{0 < \gamma < 1} \frac{1 + \gamma + \cdots + \gamma^{p-1}}{\gamma^{(p-1)/3}} \tag{1.1}$$

for any prime $p \geq 2$. It is not hard to see that $\Gamma_p < p$ (indeed, for $\gamma = 1$ the term on the right-hand side has value p and positive derivative, so the term must have value less than p for some $0 < \gamma < 1$). It turns out that $0.841p \leq \Gamma_p \leq 0.945p$ (see [9, p. 20]).

The result of Ellenberg and Gijswijt [22] for $\mathbb{F}_p^n$ can now be stated as follows.

Theorem 1.2 *Fix a prime $p \geq 3$. For any positive integer n, any subset $X \subseteq \mathbb{F}_p^n$ not containing a non-trivial three-term arithmetic progression (i.e. not containing three distinct vectors $x, y, z \in \mathbb{F}_p^n$ with $x - 2y + z = 0$) has size $|X| \leq (\Gamma_p)^n$.*

For $p = 3$, one has $\Gamma_3 \approx 2.756$ (and, strictly speaking, $\Gamma_3 < 2.756$), so Theorem 1.1 corresponds to the case $p = 3$ in Theorem 1.2.

In contrast to the Fourier-Analysis techniques used in the previous upper bounds for the Cap-Set Problem, the proof of Ellenberg and Gijswijt relied on completely different arguments. They used a new polynomial method that had been introduced by Croot, Lev, and Pach [18] for the analogous problem in $\mathbb{Z}_4^n$ shortly before (the preprint of Ellenberg–Gijswijt [22] appeared just two weeks after the preprint of Croot–Lev–Pach [18]). A few days later, the Ellenberg–Gijswijt proof was reformulated and generalized in a blog post by Tao [78] in a way that is now referred to as the *slice rank polynomial method*.

Various other polynomial methods were used for combinatorial problems already several decades ago (see the books [4] and [38]), but this polynomial method introduced by Croot–Lev–Pach [18] and further developed by Tao [78] is substantially different.

The aim of this survey paper is to give an introduction to the slice rank polynomial method, show some of its early applications (in particular, a proof of Theorems 1.1 and 1.2 above is given in Section 3.1, and a proof of the Erdős–Szemerédi Sunflower Conjecture in Section 4.1), and discuss some more recent developments (in particular related to the Erdős–Ginzburg–Ziv Problem, see Section 6).

We will also discuss some of the limitations of the method (see Section 5), which may be the reason that progress around the slice rank polynomial method has slowed down after an extremely active period right in the beginning (indeed, in the first few months right after the preprint versions of the papers of Croot–Lev–Pach [18] and Ellenberg–Gijswijt [22] and the blog post by Tao [78] appeared, there were many other applications, but after that period new applications have appeared much more scarcely).

There was already a survey paper on the proofs of Theorems 1.1 and 1.2 written by Grochow [37] shortly after the paper of Ellenberg and Gijswijt appeared. Here, we take a slightly more distanced view, looking back several years later and commenting on the further development of the method. As in any survey paper, the particular angle and selection of precise topics is heavily biased by the author's own interests and expertise. Hopefully the survey is still of interest to readers not yet familiar with the slice rank polynomial method, and as a reference collecting various material on this topic scattered around the literature.

2 Tao's Slice Rank

2.1 Definition

The slice rank polynomial method relies on a new notion of rank for tensors. In this section, will define this notion and prove some of its important properties.

A k-dimensional *tensor* is a function $F : X^k \to \mathbb{F}$, where X is some finite set and $\mathbb{F}$ is a field. Thinking of the set $X = \{1, \ldots, m\}$ for some positive integer m, we can view a tensor $F : \{1, \ldots, m\}^k \to \mathbb{F}$ as a k-dimensional hypermatrix (specifically, for $k = 2$, a function $F : \{1, \ldots, m\}^2 \to \mathbb{F}$ corresponds to an $m \times m$ matrix with entries in the field $\mathbb{F}$, and for larger k one can think of the entries as being arranged in some higher-dimensional box rather than a 2-dimension matrix).

Alternatively, a tensor $F : X^k \to \mathbb{F}$ can be viewed as an element of the k-fold tensor product $V \otimes V \otimes \cdots \otimes V$, where V is a vector space over $\mathbb{F}$ of dimension $|X|$ (with a distinguished basis whose basis vectors are in correspondence with the elements of X). This perspective lead to the name "tensor", but is less relevant for our purposes here.

There are various notions of *rank* for tensors. The slice rank polynomial method relies on a fairly new such notion, introduced by Tao [78] in 2016 specifically for the purposes of this method. To motivate Tao's notion of *slice rank*, let us first define the rank of a 2-dimensional tensor.

Definition 2.1 A 2-dimensional tensor $F : X \times X \to \mathbb{F}$ has rank 1, if and only if it can be factorized as

$$F(x,y) = G(x) \cdot H(y) \qquad \text{for all } x, y \in X$$

for some non-zero functions $G : X \to \mathbb{F}$ and $H : X \to \mathbb{F}$.

The rank of any 2-dimensional tensor $F : X \times X \to \mathbb{F}$ is the minimum number r such that F can be written as the sum of r tensors of rank 1 (from $X \times X$ to $\mathbb{F}$).

Intuitively speaking, a tensor $F : X \times X \to \mathbb{F}$ has rank 1 if it can be factored into a product of two functions, each depending only on one of the two variables. In general, the rank of a tensor $F : X \times X \to \mathbb{F}$ is the smallest number of rank 1 tensors needed to "build" the tensor F (by summing up these rank 1 tensors).

A 2-dimensional tensor $F : X \times X \to \mathbb{F}$ corresponds to a matrix with entries in $\mathbb{F}$, where the rows and columns are indexed by X (so, for $X = \{1, \ldots, m\}$ we obtain just an ordinary $m \times m$ matrix). Definition 2.1 therefore gives a (non-standard) definition of the rank of a (square) matrix. It turns out that this definition is in fact equivalent to the standard definition of matrix rank.

Tao's notion of slice rank (as well as many of the other existing rank notions for tensors) now generalizes this definition of rank from 2-dimensional tensors to higher-dimensional tensors:

Definition 2.2 For $k \geq 2$, a k-dimensional tensor $F : X^k \to \mathbb{F}$ has *slice rank* 1, if and only if for some index $i \in \{1, \ldots, k\}$ it can be factorized as

$$F(x_1, \ldots, x_k) = G(x_i) \cdot H(x_1, \ldots, x_{i-1}, x_{i+1}, \ldots, x_k) \qquad \text{for all } x_1, \ldots, x_k \in X \tag{2.1}$$

for some non-zero functions $G : X \to \mathbb{F}$ and $H : X^{k-1} \to \mathbb{F}$.

The *slice rank* of any k-dimensional tensor $F : X^k \to \mathbb{F}$ is the minimum number r such that F can be written as the sum of r tensors of slice rank 1 (from X^k to $\mathbb{F}$).

Intuitively speaking, a tensor $F : X^k \to \mathbb{F}$ has slice rank 1 if it can be factored into a product of a function depending only on one of the variables and another function depending on the remaining variables. And again, the slice rank of a tensor $F : X \times X \to \mathbb{F}$ is the smallest number of slice rank 1 tensors needed to "build" the tensor F (by summing up these slice rank 1 tensors).

Note that in the case of dimension $k = 2$, the slice rank of a tensor $F : X \times X \to \mathbb{F}$ agrees with the rank defined in Definition 2.1. In this sense, slice rank is a higher-dimensional generalization of the ordinary notion of matrix rank.

The Slice Rank Polynomial Method

Also note that any tensor $F : X^k \to \mathbb{F}$ has slice rank at most $|X|$. Indeed, we can write

$$F(x_1, \ldots, x_k) = \sum_{x \in X} G_x(x_1) \cdot H_x(x_2, x_3, \ldots, x_k) \quad \text{for all } x_1, \ldots, x_k \in X, \quad (2.2)$$

where for each $x \in X$ the function $G_x : X \to \mathbb{F}$ is the indicator function of x (i.e. the function given by $G_x(x_1) = 1$ if $x_1 = x$ and $G_x(x_1) = 0$ for all $x_1 \in X \setminus \{x\}$), and $H_x : X^{k-1} \to \mathbb{F}$ is the function given by $H_x(x_2, x_3, \ldots, x_k) = F(x, x_2, x_3 \ldots, x_k)$ for all $x_2, x_3, \ldots, x_k \in X$. Each of the summands $G_x(x_i) \cdot H_x(x_1, \ldots, x_{i-1}, x_{i+1}, \ldots, x_k)$ above is by definition a slice rank 1 tensor, so we obtain a representation of F as the sum of $|X|$ tensors of slice rank 1. Thus, $F : X^k \to \mathbb{F}$ indeed has slice rank at most $|X|$. Visually, this represents F as a sum of "slices" (where each slice corresponds to a value for x_1, i.e. to a layer in the visualization of F as a hypermatrix), which motivates the name "slice rank".

The upper bound that every tensor $F : X^k \to \mathbb{F}$ has slice rank at most $|X|$ turns out to be tight for various classes of tensors, in particular for diagonal tensors (with non-zero entries on the diagonal). This is discussed in the next subsection.

2.2 The Slice Rank of Diagonal Tensors

It is well-known (and covered in most courses on linear algebra) that diagonal matrices (with non-zero entries on the diagonal) have full rank. The same is true for diagonal tensors (with non-zero entries on the diagonal) with respect to slice rank, as the following lemma due to Tao [78] shows.

Lemma 2.3 *Let X be a non-empty finite set, $k \geq 2$ be an integer, and $\mathbb{F}$ be a field. Suppose that $f : X^k \to \mathbb{F}$ is a k-dimensional tensor such that*

$$F(x_1, \ldots, x_k) \neq 0 \quad \text{for all } x_1, \ldots, x_k \in X \text{ with } x_1 = \cdots = x_k,$$

and

$$F(x_1, \ldots, x_k) = 0 \quad \text{for all } x_1, \ldots, x_k \in X \text{ which are not all equal.}$$

Then the slice rank of the tensor F is equal to $|X|$.

Proof. First, we may assume that $X = \{1, \ldots, m\}$ for some positive integer m (indeed, the statement in the lemma is unchanged when renaming the elements of the set X).

We have already seen in the previous subsection via the decomposition in (2.2) that the slice rank of F is at most $|X| = m$ (this is true for any tensor $X^k \to \mathbb{F}$). So we only need to show that under the assumptions in the lemma the slice rank of F is at least m, i.e. that F cannot be written as a sum of fewer than m tensors of slice rank 1.

So suppose for contradiction that there is a representation of the form

$$F(x_1, \ldots, x_k) = \sum_{i=1}^{k} \sum_{\lambda \in \Lambda_i} G_\lambda(x_i) H_\lambda(x_1, \ldots, x_{i-1}, x_{i+1}, \ldots, x_k) \text{ for all } x_1, \ldots, x_k \in X$$

(2.3)

for some disjoint index sets $\Lambda_1,\ldots,\Lambda_k$ with $|\Lambda_1|+\cdots+|\Lambda_k|<m$ and (non-zero) functions $G_\lambda : X \to \mathbb{F}$ and $H_\lambda : X^{k-1} \to \mathbb{F}$ for all $\lambda \in \Lambda_1 \cup \cdots \cup \Lambda_k$ (in simple words, this is a representation of F as the sum of slice rank 1 tensors of the form $G_\lambda(x_i) \cdot H_\lambda(x_1,\ldots,x_{i-1},x_{i+1},\ldots,x_k)$ for some $i \in \{1,\ldots,k\}$, where λ is used to index these different slice rank 1 tensors and Λ_i is the set of those indices λ, where G_λ is applied to the variable x_i).

We will now choose functions $Q_i : X \to \mathbb{F}$ for $i = 1,\ldots,k$ satisfying

$$\sum_{x \in X} Q_i(x) G_\lambda(x) = 0 \tag{2.4}$$

for all $i = 1,\ldots,k$ and all $\lambda \in \Lambda_i$, as well as some other conditions. First, note that for every index $i = 1,\ldots,k$ the space of functions $Q_i : X \to \mathbb{F}$ satisfying (2.4) for all indices $\lambda \in \Lambda_i$ has dimension at least $|X| - |\Lambda_i| = m - |\Lambda_i|$. We can think of the conditions imposed by (2.4) for all $\lambda \in \Lambda_i$ as a homogeneous system of linear equations over the field $\mathbb{F}$ in the variables $Q_i(1),\ldots,Q_i(m)$ (recalling that $X = \{1,\ldots,m\}$).

Let us first consider the index $i = 1$. Then this system of $|\Lambda_1|$ linear equations in the variables $Q_1(1),\ldots,Q_1(m)$ can be written in row echelon form, meaning that we obtain a subset $T_1 \subseteq \{1,\ldots,m\}$ of size $|T_1| \geq m - |\Lambda_1|$ such that the variables $Q_i(t)$ for $t \in T_i$ are "free" variables. More precisely, we obtain a subset $T_1 \subseteq \{1,\ldots,m\}$ of size $|T_1| \geq m - |\Lambda_1|$ such that for each $t \in T_1$ there exists a function $Q_1 : \{1,\ldots,m\} \to \mathbb{F}$ satisfying (2.4) for $i = 1$ and all $\lambda \in \Lambda_1$, with the property that $Q_1(t) \neq 0$ but $Q_1(t+1) = Q_1(t+2) = \cdots = Q_1(k) = 0$.

Similarly, for $i = 2,\ldots,k-1$ we obtain a subset $T_i \subseteq \{1,\ldots,m\}$ of size $|T_i| \geq m - |\Lambda_i|$ such that for each $t \in T_i$ there exists a function $Q_i : \{1,\ldots,m\} \to \mathbb{F}$ satisfying (2.4) for all $\lambda \in \Lambda_i$, with the property that $Q_i(t) \neq 0$ but $Q_i(t+1) = Q_i(t+2) = \cdots = Q_i(k) = 0$.

Finally, for $i = k$, we again consider the system of $|\Lambda_k|$ linear equations given by the conditions imposed by (2.4) for all $\lambda \in \Lambda_i$, but now we view the variables $Q_1(1),\ldots,Q_1(m)$ in the opposite order when writing the system in row echelon form. Then we obtain a subset $T_k \subseteq \{1,\ldots,m\}$ of size $|T_k| \geq m - |\Lambda_k|$ such that for each $t \in T_k$ there exists a function $Q_k : \{1,\ldots,m\} \to \mathbb{F}$ satisfying (2.4) for $i = k$ and all $\lambda \in \Lambda_k$, with the property that $Q_1(t) \neq 0$ but $Q_k(t-1) = Q_k(t-2) = \cdots = Q_k(1) = 0$.

Now, $T_1,\ldots,T_k \subseteq \{1,\ldots,m\}$ are subsets with $|T_i| \geq m - |\Lambda_i|$ for $i = 1,\ldots,k$. Hence

$$|T_1 \cap \cdots \cap T_k| \geq m - (m - |T_1|) - \cdots - (m - |T_k|) \geq m - |\Lambda_1| - \cdots - |\Lambda_k| > 0,$$

recalling that $|\Lambda_1| + \cdots + |\Lambda_k| < m$. Thus, there exists an index $t \in T_1 \cap \cdots \cap T_k$ and for this index we can find functions $Q_1,\ldots,Q_k : \{1,\ldots,m\} \to \mathbb{F}$ as above. In particular, we then have (2.4) for all $i = 1,\ldots,k$ and all $\lambda \in \Lambda_i$. Furthermore, we have $Q_i(t) \neq 0$ for $i = 1,\ldots,k$, but for every $s \in \{1,\ldots,m\} \setminus \{t\}$ we have $Q_1(s) \cdots Q_k(s) = 0$ (since $Q_1(s) = 0$ if $s > t$ and $Q_k(s) = 0$ if $s < t$).

To now obtain the desired contradiction, let us multiply both sides of the equation (2.3) by $Q_1(x_1) Q_2(x_2) \cdots Q_k(x_k)$ and add this up for all choices of $x_1,\ldots,x_k \in X$.

We then obtain

$$\sum_{x_1 \in X} \cdots \sum_{x_k \in X} Q_1(x_1) Q_2(x_2) \cdots Q_k(x_k) \cdot F(x_1, \ldots, x_k)$$

$$= \sum_{i=1}^{k} \sum_{\lambda \in \Lambda_i} \sum_{x_1 \in X} \cdots \sum_{x_k \in X} Q_1(x_1) Q_2(x_2) \cdots Q_k(x_k) G_\lambda(x_i) H_\lambda(x_1, \ldots, x_{i-1}, x_{i+1}, \ldots, x_k)$$

$$= \sum_{i=1}^{k} \sum_{\lambda \in \Lambda_i} \Big(\sum_{x_i \in X} Q_i(x_i) G_\lambda(x_i) \Big) \Big(\sum_{x_1} \cdots \sum_{x_{i-1}} \sum_{x_{i+1}} \cdots \sum_{x_k} H_\lambda^*(x_1, \ldots, x_{i-1}, x_{i+1}, \ldots, x_k) \Big),$$

where the term abbreviated by $H_\lambda^*(x_1, \ldots, x_{i-1}, x_{i+1}, \ldots, x_k)$ is given by the expression $Q_1(x_1) \cdots Q_{i-1}(x_{i-1}) Q_{i+1}(x_{i+1}) \cdots Q_k(x_k) H_\lambda(x_1, \ldots, x_{i-1}, x_{i+1}, \ldots, x_k)$. Note that by (2.4) the first factor above is zero for all $i = 1, \ldots, k$ and all $\lambda \in \Lambda_i$, and so we can conclude that

$$\sum_{x_1 \in X} \cdots \sum_{x_k \in X} Q_1(x_1) Q_2(x_2) \cdots Q_k(x_k) \cdot F(x_1, \ldots, x_k) = 0.$$

On the other hand, by the second assumption in the lemma statement we have $F(x_1, \ldots, x_k) = 0$ whenever $x_1, \ldots, x_k \in X$ are not all equal. Recalling that $X = \{1, \ldots, m\}$, we can therefore rewrite this equation as

$$\sum_{s=1}^{m} Q_1(s) \cdots Q_k(s) \cdot F(s, \ldots, s) = 0.$$

Also recalling that $Q_1(s) \cdots Q_k(s) = 0$ for every $s \in \{1, \ldots, m\} \setminus \{t\}$, we can conclude that

$$Q_1(t) \cdots Q_k(t) \cdot F(t, \ldots, t) = 0.$$

However, we have $Q_i(t) \neq 0$ for $i = 1, \ldots, k$ and also $F(t, t, \ldots, t) \neq 0$ by the first assumption in the lemma statement. This gives the desired contradiction. □

Besides diagonal tensors, there are also other tensors $F : X^k \to \mathbb{F}$ with full slice rank $|X|$. The following exercises (adapted from a blog-post of Sawin and Tao [73]) establishes this property for a more general class of tensors.

Exercise 2.4 Let X be a finite set and $k \geq 2$ be an integer. Consider k complete linear orders $\succeq^{(1)}, \ldots, \succeq^{(k)}$ on the set X (i.e. consider k rankings of the set X ordering the elements of X). By considering the product of these k linear orders, we obtain a partial order $\succeq$ on X^k where for any $(x_1, \ldots, x_k), (y_1, \ldots, y_k) \in X^k$ by definition we have $(x_1, \ldots, x_k) \succeq (y_1, \ldots, y_k)$ if and only if $x_i \succeq^{(i)} y_i$ for $i = 1, \ldots, k$.

Now, for some field $\mathbb{F}$ consider a k-dimensional tensor $F : X^k \to \mathbb{F}$. Let

$$\text{supp}\, F = \{(x_1, \ldots, x_k) \in X^k \mid F(x_1, \ldots, x_k) \neq 0\} \subseteq X^k$$

be the support of F (i.e. the set of k-tuples in X^k where the tensor F takes a non-zero value). Suppose that $\text{supp}\, F$ forms an anti-chain with respect to the partial order $\succeq$ considered above (i.e. there do not exist distinct k-tuples $(x_1, \ldots, x_k), (y_1, \ldots, y_k) \in \text{supp}\, F$ with $(x_1, \ldots, x_k) \succeq (y_1, \ldots, y_k)$). Then

$$\text{slice rank of } F = \min_{\text{supp}\, F = S_1 \cup \cdots \cup S_k} |\pi_1(S_1)| + \cdots + |\pi_k(S_k)|, \quad (2.5)$$

where the minimum is taken over all partitions $\operatorname{supp} F = S_1 \cup \cdots \cup S_k$. Here, for $i = 1, \ldots, k$, the map $\pi_i : X^k \to X$ is the projection to the i-th coordinate (i.e. $\pi_i(x_1, \ldots, x_k) = x_i$ for all $(x_1, \ldots, x_k) \in X^k$).

Remark Using a representation as a sum of "slices" similar to (2.2), one can show that one always has an inequality replacing "=" in (2.5) by "$\leq$" (and this inequality does not require the assumption that $\operatorname{supp} F$ is an anti-chain). To show the opposite inequality (using the assumption that $\operatorname{supp} F$ is an anti-chain), one can use similar arguments as in the proof of Lemma 2.3 (choosing the ordering of the variables for the row echelon form according to the complete linear orders $\succeq^{(1)}, \ldots, \succeq^{(k)}$ on the set X). In fact, even without the assumption that $\operatorname{supp} F$ is an anti-chain with respect to the partial order $\succeq$, one can show a lower bound for the slice rank of F of a similar form as the right-hand side of (2.5), but where the minimum is instead taken over all partitions $\max \operatorname{supp} F = S_1 \cup \ldots, \cup S_k$, where $\max \operatorname{supp} F$ denotes the set of maximal elements in $\operatorname{supp} F$ with respect to the partial order $\succeq$.

It is not hard to see that for a diagonal tensor $F : X^k \to \mathbb{F}$ (with $k \geq 2$) one can always choose partial orders $\succeq^{(1)}, \ldots, \succeq^{(k)}$ on the set X such that the support $\operatorname{supp} F$ is anti-chain with respect to the product partial order $\succeq$ considered in Exercise 2.4. It is also not hard to show in the case of a diagonal tensor $F : X^k \to \mathbb{F}$ (with non-zero entries on the diagonal) the right-hand side of (2.5) is equal to $|X|$. Hence Exercise 2.4 implies Lemma 2.3 above.

In general, it is quite difficult to establish lower bounds for the slice rank of tensors with some given structure. We will see later that such lower bounds play a critical role in the slice rank polynomial method. Most applications of the slice rank polynomial method so far are based on the slice rank lower bound for diagonal tensors in Lemma 2.3 (maybe partly because the somewhat complicated condition in Exercise 2.4 is difficult to use in practice, and no other useful general lower bounds are known besides the ones in Lemma 2.3 and in Exercise 2.4 and the remark following it).

2.3 Related Notions of Rank

For $k \geq 3$, the standard notion of rank for a k-dimensional tensor is different from the notion of slice rank in Definition 2.2. In the standard definition of rank, a k-dimensional tensor $F : X^k \to \mathbb{F}$ has rank 1 if it can be factorized as

$$F(x_1, \ldots, x_k) = G_1(x_1) \cdot G_2(x_2) \cdots G_k(x_k) \qquad \text{for all } x_1, \ldots, x_k \in X$$

for some non-zero functions $G_1, \ldots, G_k : X \to \mathbb{F}$. Then, the rank of any k-dimensional tensor $F : X^k \to \mathbb{F}$ is again defined to be the minimum number r such that F can be written as the sum of r tensors of rank 1 (from X^k to $\mathbb{F}$). Since every tensor with rank 1 (according to this definition) also has slice rank 1, the slice rank of any k-dimensional tensor is upper-bounded by its rank according to this standard definition (and for $k = 2$, the two notions agree).

Another notion of rank, similar to slice rank, is the so-called *partition rank* introduced by Naslund [56]. For $k \geq 2$, a k-dimensional tensor $F : X^k \to \mathbb{F}$ has partition rank 1 if it can be factorized as

$$F(x_1, \ldots, x_k) = G((x_i)_{i \in I}) \ldots H((x_j)_{j \in J}) \qquad \text{for all } x_1, \ldots, x_k \in X$$

for some partition $\{1,\ldots,k\} = I \cup J$ into non-empty subsets I and J and some non-zero functions $G : X^I \to \mathbb{F}$ and $H : X^J \to \mathbb{F}$. In other words, F has partition rank 1 if it can be written as a product of a function depending only on the variables with indices in I multiplied by a function depending only on the variables with indices in J, for some partition $\{1,\ldots,k\} = I \cup J$ into non-empty subsets. Again, the partition rank of any k-dimensional tensor $F : X^k \to \mathbb{F}$ is defined to be the minimum number r such that F can be written as the sum of r tensors of partition rank 1 (from X^k to $\mathbb{F}$).

For every $k \geq 2$, every k-dimensional tensor of slice rank 1 also has partition rank 1 (indeed given a factorization as in (2.1), one can consider the partition $\{1,\ldots,k\} = I \cup J$ given by $I = \{i\}$ and $J = \{1,\ldots,k\} \setminus \{i\}$). Hence the partition rank is always at most the slice rank for every k-dimensional tensor (with $k \geq 2$). For $k \leq 3$ the notions of partition rank and slice rank actually coincide, but for $k \geq 4$ they are different. Indeed, for any $k \geq 4$, any infinite set X, and any field $\mathbb{F}$, there are examples of tensors $F : X^k \to \mathbb{F}$ with partition rank 1, but slice rank $|X|$ (an example of such a tensor is given by defining $F(x_1,\ldots,x_k) = 1$ if and only if both $x_1 = x_2$ and $x_3 = \cdots = x_k$ hold, and defining $F(x_1,\ldots,x_k) = 0$ otherwise; by definition, this tensor has partition rank 1, but by Exercise 2.4 it has slice rank $|X|$ if $k \geq 4$).

Interestingly, Lemma 2.3 still holds with slice rank replaced by partition rank, see [56, Lemma 11]. This stronger version of the lemma has been used in some applications [55, 56].

The partition rank of a k-dimensional tensor can also be related to the so-called *analytic rank*, a notion introduced by Gowers and Wolf [33] in 2011 in the context of higher-order Fourier-Analysis. The analytic rank of any k-dimensional tensor is at most its partition rank, and in the opposite direction the partition rank can also be bounded by an almost-linear function of the analytic rank (for fixed k). The latter was shown by Moshkovitz and Zhu [53] improving upon a long sequence of earlier work (see e.g. [15, 43, 50]).

Furthermore, there is the so-called notion of *max-flattening rank*, which on the first glance seems almost identical to the notion of slice rank, but differs in the variable quantification. For a k-dimensional tensor $F : X^k \to \mathbb{F}$ and an index $i \in \{1,\ldots,k\}$, the tensor F has i-flattening rank 1 if there exists a factorization as in (2.1) for this particular index i. The i-flattening rank of any tensor $F : X^k \to \mathbb{F}$ is the minimum number r such that F can be written as the sum of r tensors $X^k \to \mathbb{F}$ of i-flattening rank 1. Finally, the max-flattening rank of F is the maximum of the i-flattening rank of F for $i = 1,\ldots,k$. Several combinatorial applications of max-flattening rank were given by Munhá Correia, Sudakov, and Tomon [54].

3 The Slice Rank Polynomial Method in Action: Proving the Ellenberg–Gijswijt Bound for the Cap-Set Problem

3.1 Tao's Proof

In this section, we prove the result of Ellenberg and Gijswijt [22] bounding the size of three-term-progression free subsets of $\mathbb{F}_p^n$ stated in Theorem 1.2 (this in particular proves their bound for the Cap-Set Problem in Theorem 1.1). The proof presented here is due to Tao [78] and relies on the notion of slice rank for tensors, discussed in the previous section. The proof technique that Tao introduced in this proof is now called the *slice rank polynomial method*. The original proof of Ellenberg and Gijswijt used a similar method, but not utilizing the notion of slice rank (which had not been introduced yet at that time). Both the Ellenberg–Gijswijt proof and Tao's proof with the slice rank polynomial method were building on arguments due to Croot, Lev and Pach [18], see also Remark 3.1 below.

The strategy to prove Theorem 1.2 using the slice rank polynomial method is as follows. We consider a subset $X \subseteq \mathbb{F}_p^n$ not containing a non-trivial three-term arithmetic progression. We then define a 3-dimensional tensor $F : X \times X \times X \to \mathbb{F}_p$ via a smartly chosen polynomial expression. Using the assumption that X does not contain a non-trivial three-term arithmetic progression, we will show that this tensor F is a diagonal tensor (with non-zero entries on the diagonal), i.e. that it satisfies the assumption of Lemma 2.3. Then by Lemma 2.3 we know that the slice rank of the tensor F is equal to $|X|$. On the other hand, from the polynomial expression defining the tensor F we will be able to derive a representation of F as a sum of tensors of slice rank 1. This gives us an upper bound for the slice rank of F, and hence we obtain an upper bound for $|X|$.

We will now present the proof in detail. Recall that for a prime p, the constant $\Gamma_p < p$ has been defined in (1.1). Rather than proving Theorem 1.2 directly, we will first prove the following slightly weaker statement (with a slightly worse bound on $|X|$):

Proposition 3.1 *Fix a prime $p \geq 3$. For any positive integer n, any subset $X \subseteq \mathbb{F}_p^n$ not containing a non-trivial three-term arithmetic progression (i.e. not containing three distinct vectors $x, y, z \in \mathbb{F}_p^n$ with $x - 2y + z = 0$) has size $|X| \leq 3 \cdot (\Gamma_p)^n$.*

Note that the upper bound for $|X|$ in this proposition differs from the upper bound in Theorem 1.2 only by the constant factor 3. In particular, the proposition is already exponentially better than the trivial bound $|X| \leq p^n$ (as $\Gamma_p < p$), and so the proposition is basically (almost) as good as the theorem. However, it is actually not hard to deduce Theorem 1.2 from Proposition 3.1 by removing the factor 3 with a product trick (see below).

Proof of Proposition 3.1. Assume that $X \subseteq \mathbb{F}_p^n$ does not contain a non-trivial three-term arithmetic progression. This means that for any three distinct vectors $x, y, z \in X$ we have $x - 2y + z \neq 0$. Note that we also have $x - 2y + z \neq 0$ if $x, y, z \in X$ are three vectors such that two of them are equal to each other, but the third is different. On the other hand, if $x, y, z \in X$ are three vectors that are all equal, then we clearly have $x - 2y + z = 0$. In summary, this means that for any vectors $x, y, z \in X$ we have $x - 2y + z = 0$ if and only if $x = y = z$.

Let us now define a 3-dimensional tensor $F : X \times X \times X \to \mathbb{F}_p$ by setting

$$F(x, y, z) = \prod_{j=1}^{n} \left((x_j - 2y_j + z_j)^{p-1} - 1 \right) \qquad \text{for all } x, y, z \in X, \qquad (3.1)$$

where $x_1, \ldots, x_n \in \mathbb{F}_p$ denote the coordinates of the vector $x \in \mathbb{F}_p^n$ (and similarly for the vectors $y \in \mathbb{F}_p^n$ and $z \in \mathbb{F}_p^n$).

First, we claim that F is a diagonal tensor (with non-zero entries on the diagonal), i.e. that the assumptions in Lemma 2.3 are satisfied. Indeed, if $x, y, z \in X$ are such that $x = y = z$, then for $j = 1, \ldots, n$ we have $x_j - 2y_j + z_j = 0$ and hence $F(x, y, z) = (-1)^n \neq 0$. On the other hand, if $x, y, z \in X$ are not all equal, then by the discussion above we have $x - 2y + z \neq 0$, so there exists an index $j \in \{1, \ldots, n\}$ with $x_j - 2y_j + z_j \neq 0$. This means that $(x_j - 2y_j + z_j)^{p-1} = 1$ in $\mathbb{F}_p$, and hence the product on the right-hand side of (3.1) has a zero factor, implying that $F(x, y, z) = 0$.

Thus, the assumptions in Lemma 2.3 indeed hold for the tensor F, and by the lemma we can conclude that the slice rank of F is equal to $|X|$.

To now obtain an upper bound for the slice rank of F, let us examine the polynomial expression in the definition of F in (3.1). The right-hand side of (3.1) is a polynomial in the variables $x_1, \ldots, x_n, y_1, \ldots, y_n, z_1, \ldots, z_n$. In each individual variable, the degree of the polynomial is at most $p - 1$, and furthermore the total degree of the polynomial is at most $(p - 1)n$. Let us imagine multiplying out this polynomial, this gives a representation of $F(x, y, z)$ as a sum of monomials. Each of these monomials is of the form

$$x_1^{a_1} x_2^{a_2} \cdots x_n^{a_n} y_1^{b_1} y_2^{b_2} \cdots y_n^{b_n} z_1^{c_1} z_2^{c_2} \cdots z_n^{c_n}$$

with $a_1, \ldots, a_n, b_1, \ldots, b_n, c_1, \ldots, c_n \in \{0, 1, \ldots, p - 1\}$ and $a_1 + \cdots + a_n + b_1 + \cdots + b_n + c_1 + \cdots + c_n \leq (p - 1)n$. In particular, for each such monomial we have $a_1 + \cdots + a_n \leq (p - 1)n/3$ or $b_1 + \cdots + b_n \leq (p - 1)n/3$ or $c_1 + \cdots + c_n \leq (p - 1)n/3$.

Thus, $F(x, y, z)$ can be written as a sum of monomials of the form

$$x_1^{d_1} x_2^{d_2} \cdots x_n^{d_n} \cdot \mathrm{monomial}(y_1, \ldots, y_n, z_1, \ldots, z_n)$$
$$\text{or} \quad y_1^{d_1} y_2^{d_2} \cdots y_n^{d_n} \cdot \mathrm{monomial}(x_1, \ldots, x_n, z_1, \ldots, z_n)$$
$$\text{or} \quad z_1^{d_1} z_2^{d_2} \cdots z_n^{d_n} \cdot \mathrm{monomial}(x_1, \ldots, x_n, y_1, \ldots, y_n)$$

for $d_1, \ldots, d_n \in \{0, 1, \ldots, p - 1\}$ with $d_1 + \cdots + d_n \leq (p - 1)n/3$ (here "monomial" stands for some monomial function). Using the distributive law, we can now combine terms with the same first factor, i.e. terms belonging to the same of the three cases with the same n-tuple $(d_1, \ldots, d_n)$. We then obtain a representation

$$F(x, y, z) = \sum_{d_1, \ldots, d_n} x_1^{d_1} x_2^{d_2} \cdots x_n^{d_n} \cdot \mathrm{polynomial}(y_1, \ldots, y_n, z_1, \ldots, z_n)$$
$$+ \sum_{d_1, \ldots, d_n} y_1^{d_1} y_2^{d_2} \cdots y_n^{d_n} \cdot \mathrm{polynomial}(x_1, \ldots, x_n, z_1, \ldots, z_n)$$
$$+ \sum_{d_1, \ldots, d_n} z_1^{d_1} z_2^{d_2} \cdots z_n^{d_n} \cdot \mathrm{polynomial}(x_1, \ldots, x_n, y_1, \ldots, y_n),$$

where each of the three sums is over all n-tuples $(d_1, \ldots, d_n) \in \{0, 1, \ldots, p-1\}^n$ with $d_1 + \cdots + d_n \leq (p-1)n/3$ (and "polynomial" stands for some polynomial function, note that this may be a different polynomial in each of the summands).

Note that each of the summands above is a 3-dimensional tensor from $X \times X \times X$ to $\mathbb{F}_p$ of slice rank 1 (for example, $x_1^{d_1} x_2^{d_2} \cdots x_n^{d_n} \cdot \mathrm{polynomial}(y_1, \ldots, y_n, z_1, \ldots, z_n)$ is a product of a function only depending on x and a function depending jointly on y and z). Thus, we obtained a representation of F as a sum of tensors of slice rank

1. The slice rank of F is therefore by definition at most the number of summands in this representation, which is

$$3 \cdot |\{(d_1, \ldots, d_n) \in \{0, 1, \ldots, p-1\}^n \mid d_1 + \cdots + d_n \leq (p-1)n/3\}|.$$

Hence, since the slice rank of F is equal to $|X|$, we obtain

$$|X| \leq 3 \cdot |\{(d_1, \ldots, d_n) \in \{0, 1, \ldots, p-1\}^n \mid d_1 + \cdots + d_n \leq (p-1)n/3\}|.$$

This gives the desired bound $|X| \leq 3 \cdot (\Gamma_p)^n$ via the following simple lemma. □

Lemma 3.2 *For any prime $p \geq 2$, and any positive integer n, we have*

$$|\{(d_1, \ldots, d_n) \in \{0, 1, \ldots, p-1\}^n \mid d_1 + \cdots + d_n \leq (p-1)n/3\}| \leq (\Gamma_p)^n$$

for Γ_p as defined in (1.1).

Proof. Note that the total number of n-tuples $(d_1, \ldots, d_n) \in \{0, 1, \ldots, p-1\}^n$ is p^n. Thus, we need to show that for a uniformly random n-tuple $(d_1, \ldots, d_n) \in \{0, 1, \ldots, p-1\}^n$ (i.e. for independent uniformly random $d_1, \ldots, d_n \in \{0, \ldots, p-1\}$) the probability of having $d_1 + \cdots + d_n \leq (p-1)n/3$ is at most $(\Gamma_p/p)^n$. Indeed, for every $0 < \gamma < 1$, we have

$$\Pr[d_1 + \cdots + d_n \leq (p-1)n/3] = \Pr[\gamma^{d_1 + \cdots + d_n} \geq \gamma^{(p-1)n/3}] \leq \frac{\mathbb{E}[\gamma^{d_1 + \cdots + d_n}]}{\gamma^{(p-1)n/3}}$$

by Markov's inequality. Noting that

$$\mathbb{E}[\gamma^{d_1 + \cdots + d_n}] = \prod_{j=1}^{n} \mathbb{E}[\gamma^{d_j}] = \left(\frac{1 + \gamma + \cdots + \gamma^{p-1}}{p}\right)^n,$$

this implies

$$\Pr[d_1 + \cdots + d_n \leq (p-1)n/3] \leq \left(\frac{1}{p} \cdot \frac{1 + \gamma + \cdots + \gamma^{p-1}}{\gamma^{(p-1)/3}}\right)^n$$

for every $0 < \gamma < 1$. Thus, we obtain

$$\Pr[d_1 + \cdots + d_n \leq (p-1)n/3] \leq \left(\frac{1}{p} \cdot \inf_{0 < \gamma < 1} \frac{1 + \gamma + \cdots + \gamma^{p-1}}{\gamma^{(p-1)/3}}\right)^n = \left(\frac{1}{p} \cdot \Gamma_p\right)^n$$

by the definition of Γ_p in (1.1). This means that the number of n-tuples $(d_1, \ldots, d_n) \in \{0, 1, \ldots, p-1\}^n$ with $d_1 + \cdots + d_n \leq (p-1)n/3$ is

$$\Pr[d_1 + \cdots + d_n \leq (p-1)n/3] \cdot p^n \leq \left(\frac{1}{p} \cdot \Gamma_p\right)^n \cdot p^n = (\Gamma_p)^n,$$

as desired. □

We remark that in Lemma 3.2 it is not surprising that the number of n-tuples $(d_1, \ldots, d_n) \in \{0, 1, \ldots, p-1\}^n$ with $d_1 + \cdots + d_n \leq (p-1)n/3$ is exponentially smaller than p^n. Indeed, for a random n-tuple $(d_1, \ldots, d_n) \in \{0, 1, \ldots, p-1\}^n$ the expected value for the sum $d_1 + \cdots + d_n$ is $(p-1)n/2$. It now follows from standard probabilistic arguments (e.g. Hoeffding's inequality) that the probability of having $d_1 + \cdots + d_n \leq (p-1)n/3$ is exponentially small in n, meaning that the number of n-tuples $(d_1, \ldots, d_n) \in \{0, 1, \ldots, p-1\}^n$ with $d_1 + \cdots + d_n \leq (p-1)n/3$ is indeed exponentially smaller than p^n. The bound in Lemma 3.2 is also not the new part of the proof of Theorem 1.2 (such bounds were known long before).

Let us now deduce Theorem 1.2 fro Proposition 3.1.

Proof of Theorem 1.2. Again, let $X \subseteq \mathbb{F}_p^n$ be a subset not containing a non-trivial three-term arithmetic progression, and recall that we need to show that $|X| \leq \Gamma_p^n$.

For every positive integer m, we can consider the m-fold product space $(\mathbb{F}_p^n)^m \cong \mathbb{F}_p^n \times \cdots \times \mathbb{F}_p^n \cong \mathbb{F}_p^{nm}$. The m-fold product $X^m = X \times \cdots \times X$ is a subset of this space and does not contain a non-trivial three-term arithmetic progression. Indeed, suppose $(x^{(1)}, \ldots, x^{(m)}), (y^{(1)}, \ldots, y^{(m)}), (z^{(1)}, \ldots, z^{(m)}) \in X^m$ form an arithmetic progression, then for each $i = 1, \ldots, m$ the vectors $x^{(i)}, y^{(i)}, z^{(i)} \in X$ would form an arithmetic progression and we would therefore need to have $x^{(i)} = y^{(i)} = z^{(i)}$ for $i = 1, \ldots, m$. Hence $(x^{(1)}, \ldots, x^{(m)}) = (y^{(1)}, \ldots, y^{(m)}) = (z^{(1)}, \ldots, z^{(m)}) \in X^m$, so there cannot be any non-trivial three-term arithmetic progression in $X^m \subseteq \mathbb{F}_p^{nm}$. By Proposition 3.1, this implies that $|X|^m = |X^m| \leq 3 \cdot (\Gamma_p)^{nm}$ for all positive integers m.

Thus, we obtain
$$|X| \leq 3^{1/m} \cdot (\Gamma_p)^n$$
for all positive integers m and therefore
$$|X| \leq \lim_{m \to \infty} 3^{1/m} \cdot (\Gamma_p)^n = (\Gamma_p)^n.$$

This finishes the proof of Theorem 1.2. □

Remark The original proof of Ellenberg and Gijswijt [22] of Theorem 1.2 was not phrased in terms of slice rank (this notion was only introduced by Tao [78] later). Instead, Ellenberg and Gijswijt argued with the rank of a certain diagonal $|X| \times |X|$ matrix (namely, a matrix whose (x, y)-entry is $P(2y - x)$, where $P: \mathbb{F}_p^n \to \mathbb{F}_p$ is a polynomial vanishing on all points $\mathbb{F}_p^n \setminus X$). They obtained an upper bound for the rank of this matrix with exactly the same idea also appearing in Tao's proof of Proposition 3.1 above, splitting the polynomial defining the matrix entries up in exactly the same way and then obtaining a decomposition of the matrix into matrices of rank 1 (which gives an upper bound on the rank of the original matrix). This idea was actually not introduced by Ellenberg and Gijswijt, but by Croot, Lev and Pach [18] in their paper bounding the size of three-term progression-free subsets of $\mathbb{Z}_4^n$ that appeared a few weeks earlier. So the idea for upper-bounding the slice rank of the tensor F in the proof of Proposition 3.1 (as well as in many other applications of the slice rank polynomial method) is due to Croot, Lev and Pach [18] (in the context of matrix rank).

3.2 Generalizations and Extensions

Recall that a three-term arithmetic progression is characterized by the equation $x - 2y + z = 0$. Theorem 1.2 states that for any subset $X \subseteq \mathbb{F}_p^n$ without a non-trivial three-term arithmetic progression, its size $|X|$ is exponentially smaller than p^n. As stated in the following exercise, this result can be generalized to other suitable linear equations instead of $x - 2y + z = 0$, and even to certain systems of linear equations. The proof for this more general statement is basically the same as Tao's proof for three-term progressions presented above (one just needs to change the definition of the tensor F according to the given equations).

Exercise 3.3 *Given a prime p, consider a system of m linear equations in k variables $x_1, \ldots, x_k$ of the form*

$$a_{1,1}x_1 + \cdots + a_{1,k}x_k = 0$$
$$\vdots \qquad\qquad\qquad (3.2)$$
$$a_{m,1}x_1 + \cdots + a_{m,k}x_k = 0$$

with coefficients $a_{j,i} \in \mathbb{F}_p$ for $j = 1, \ldots, m$ and $i = 1, \ldots, k$. Suppose that for every $j = 1, \ldots, m$ we have $a_{j,1} + \cdots + a_{j,k} = 0$ (i.e. the sum of the coefficients appearing in each of the equations in (3.2) is zero). Also suppose that $m < k/2$.

Furthermore, let n be a positive integer, and consider a subset $X \subseteq \mathbb{F}_p^n$. Suppose that for any vectors $x_1, \ldots, x_k \in X \subseteq \mathbb{F}_p^n$ such that $(x_1, \ldots, x_k)$ forms a solution to the system (3.2), we have $x_1 = \cdots = x_k$. Then the size of the set X is bounded by

$$|X| \leq \left(\inf_{0 < \gamma < 1} \frac{1 + \gamma + \cdots + \gamma^{p-1}}{\gamma^{(p-1)m/k}} \right)^n. \qquad (3.3)$$

Let us make a few remarks concerning the statement in Exercise 3.3. Firstly, and very importantly, by the assumption $m < k/2$ the infimum on the right-hand side of (3.3) is strictly less than p (this follows from a very similar argument to the discussion below (1.1)). Thus, the bound on $|X|$ in Exercise 3.3 is again exponentially smaller than p^n.

The assumption to have $a_{j,1} + \cdots + a_{j,k} = 0$ for $j = 1, \ldots, m$ in Exercise 3.3 is quite natural, and actually necessary in order to obtain a bound on $|X|$ of this form. Indeed, if for some $j \in \{1, \ldots, m\}$ we have $a_{j,1} + \cdots + a_{j,k} \neq 0$ (i.e. if for some equation in the system (1.1) the coefficients do not sum to zero), then we can find a subset $X \subseteq \mathbb{F}_p^n$ of size $(1/p) \cdot p^n$ (i.e. of size linear in p^n for fixed p) without *any* solutions $(x_1, \ldots, x_k)$ to (1.1) with $x_1, \ldots, x_k \in X$. Such a set $X \subseteq \mathbb{F}_p^n$ can for example be obtained by taking all vectors in $\mathbb{F}_p^n$ whose first coordinate is 1 (then for any $x_1, \ldots, x_k \in X$ the first coordinate of $a_{j,1}x_1 + \cdots + a_{j,k}x_k \in \mathbb{F}_p^n$ is always $a_{j,1} + \cdots + a_{j,k} \neq 0$ and so $x_1, \ldots, x_k \in X$ cannot form a solution to the system (1.1)).

The assumption $m < k/2$ is needed in order for the infimum on the right-hand side of (3.3) to be strictly less than p (strictly speaking, the statement in the exercise as written would also be true without the assumption $m < k/2$, but for $m \geq k/2$ the conclusion would just be $|X| \leq p^n$, which is trivially true). If $k/2 \leq m \leq k - 2$, it would be plausible to still hope for a bound on $|X|$ which is exponentially smaller than p^n, but this seems to be out of reach of current methods (see the discussion in Section 5.3).

Finally, recall that the assumption in Exercise 3.3 for the set X is that for any solution $(x_1, \dots, x_k)$ to the system (3.2) with $x_1, \dots, x_k \in X$ we have $x_1 = \dots = x_k$. In other words, the assumption states that there is no solution $(x_1, \dots, x_k)$ to (3.2) in X where the vectors $x_1, \dots, x_k \in X$ not all equal. In contrast, in the setting of Theorem 1.2 the assumption was that there is no solution to $x - 2y + z = 0$ in X where the vectors $x, y, z \in X$ are distinct. Note, however, that for a solution to the equation $x - 2y + z = 0$ (in $\mathbb{F}_p^n$ with $p \geq 3$) the vectors x, y, z are distinct if and only if they are not all equal (since as soon as two of the vectors are equal, the equation $x - 2y + z = 0$ forces the third vector to also be the same). Thus, in the setting of three-term arithmetic progressions, i.e. in the setting of the equation $x - 2y + z = 0$, it is equivalent to demand for the vectors x, y, z to be distinct or to be not all equal. One therefore recovers the statement in Theorem 1.2 from Exercise 3.3 by taking $m = 1$, $k = 3$ and taking the system (3.2) to be the single equation $x_1 - 2x_2 + x_3 = 0$.

Of course, in Exercise 3.3 one can also ask what happens if in the assumption of having no solution $(x_1, \dots, x_k)$ to (3.2) in X where the vectors $x_1, \dots, x_k \in X$ not all equal, we replace "not all equal" by "distinct". As discussed above, this makes no difference in the case of a single equation in three variables (i.e. in the case $m = 1$ and $k = 3$). In general, however, the assumption of having no solution $(x_1, \dots, x_k)$ to (3.2) with distinct vectors $x_1, \dots, x_k \in X$ is strictly weaker. With this weaker assumption the proof in Section 3.1 breaks down, because one cannot ensure anymore that the relevant tensor is a diagonal tensor (and so one does not have a suitable lower bound for its slice rank anymore), see also the discussion in Section 5.2.

Another direction to generalize Theorem 1.2 is to consider a so-called *multi-colored* version of the statement. Specifically, the case $k = 3$ of the statement in the following exercise corresponds to the multi-colored generalization of Theorem 1.2 (see the discussion below the exercise). Again, the statement in the exercise follows from essentially the same proof as in Section 3.1.

Exercise 3.4 *Let $k \geq 3$ be an integer, let p be a prime, and let n be a positive integer. Let $(x_1^{(\ell)}, \dots, x_k^{(\ell)})$ for $\ell = 1, \dots, L$ be a list of k-tuples in $\mathbb{F}_p^n \times \dots \times \mathbb{F}_p^n$ (i.e. for $\ell = 1, \dots, L$ and $i = 1, \dots, k$ we have a vector $x_i^{(\ell)} \in \mathbb{F}_p^n$). Suppose that for all $\ell_1, \dots, \ell_k \in \{1, \dots, L\}$ we have*

$$x_1^{(\ell_1)} + x_2^{(\ell_2)} + \dots + x_k^{(\ell_k)} = 0 \quad \Leftrightarrow \quad \ell_1 = \ell_2 = \dots = \ell_k.$$

Then the given list of vectors has size

$$L \leq \left(\inf_{0 < \gamma < 1} \frac{1 + \gamma + \dots + \gamma^{p-1}}{\gamma^{(p-1)/k}} \right)^n. \tag{3.4}$$

Again, the infimum appearing on the right-hand side of (3.4) is strictly less than p. A collection of n-tuples $(x_1^{(\ell)}, \dots, x_k^{(\ell)})$ for $\ell = 1, \dots, L$ with the conditions in Exercise 3.4 is often called a *k-colored sum-free set*, and the statement in the exercise is called the *k-colored sum-free theorem*. In this terminology, one may think of all the vectors $x_1^{(\ell)}$ as colored with some color (the "first" color), all the vectors $x_2^{(\ell)}$ as colored with another color (the "second" color), and so on. We are then looking for a

k-tuple of vectors summing to zero, where the first vector has the first color (i.e. is of the form $x_1^{(\ell_1)}$ for some $\ell_1 \in \{1, \ldots, L\}$), the second vector has the second color (i.e. is of the form $x_2^{(\ell_2)}$ for some $\ell_2 \in \{1, \ldots, L\}$), and so on. The assumption in Exercise 3.4 states that these multi-colored k-tuples summing to zero are precisely the k-tuples $(x_1^{(\ell)}, \ldots, x_k^{(\ell)})$ in the original list (in particular, no further such multi-colored k-tuples summing to zero can be formed by selecting vectors with different indices ℓ, i.e. there does not exist a choice of $\ell_1, \ldots, \ell_k \in \{1, \ldots, L\}$ with $x_1^{(\ell_1)} + x_2^{(\ell_2)} + \cdots + x_k^{(\ell_k)} = 0$, where $\ell_1, \ldots, \ell_k$ are not all equal).

Another way to think about the statement in Exercise 3.4 is as follows: Let us think of all the vectors $x_i^{(\ell)} \in \mathbb{F}_p^n$ as being arranged in an $L \times k$ table (where the indices $\ell = 1, \ldots, L$ correspond to the rows and the indices $i = 1, \ldots, k$ correspond to the columns). The assumption in Exercise 3.4 then states the following: The vectors in each row of this table sum up to zero. Furthermore, there is no other way of selecting one vector from each column and obtain sum zero except if we select all vectors from the same row. Under these assumptions, the conclusion is that the number L of rows can be bounded as in (3.4). Note that this bound is again exponentially smaller than p^n.

Theorem 1.2 now follows directly from the statement in Exercise 3.4 for $k = 3$. Indeed, given a set $X \subseteq \mathbb{F}_p^n$ without a non-trivial three-term arithmetic progression, we can consider the list of 3-tuples $(x, -2x, x) \in \mathbb{F}_p^n \times \mathbb{F}_p^n \times \mathbb{F}_p^n$ for all $x \in X$. It is not hard to check that this list of 3-tuples satisfies the assumption in Exercise 3.4, and so we can conclude that the size $|X|$ of the list is bounded by $|X| \leq (\Gamma_p)^n$ (note that for $k = 3$, the infimum on the right-hand side of (3.4) agrees precisely with the definition of Γ_p in (1.1)). Thus, the 3-colored sum-free theorem (the case $k = 3$ of Exercise 3.4) is a generalization of Theorem 1.2 (and in the special case of $p = 3$ it is a generalization of Theorem 1.1).

The k-colored sum-free theorem stated in Exercise 3.4 may look strange at first, but it has been used in various applications (see for example [29, 70, 72]). Interestingly, the bound for L in (3.4) is essentially tight, see the discussion in Section 5.1.

4 Other Applications of the Slice Rank Polynomial Method

4.1 Sunflowers

The Erdős–Rado Sunflower Problem is a famous problem in extremal set theory. Surprisingly, as established by Alon–Shpilka–Umans [2], it has connections to the problem of fast matrix multiplication, see the discussion at the end of this subsection.

Definition 4.1 Three distinct sets A, B and C form a *sunflower* if their pairwise intersections agree, i.e. if $A \cap B = A \cap C = B \cap C$.

For example, three (non-empty) pairwise disjoint sets A, B, C always form a sunflower. In general, a sunflower consists of a "joint" part that belongs to all of the sets A, B, C and individual "petals" that belong only to one of the three sets (see Figure 4.1)

The Erdős–Rado Sunflower Problem asks how large a family of finite sets (all of the same size s) needs to be in order to ensure that within this family one can find

Figure 1: An illustration of a sunflower

three sets forming a sunflower. Specifically, in 1960, Erdős and Rado [25] made the following famous conjecture, stating that having a family of size C^s for a suitable absolute constant C suffices to find a sunflower.

Conjecture 4.2 *There exists an absolute constant C such that the following holds. Whenever $\mathcal{F}$ is a family of sets of size s, such that no three distinct sets $A, B, C \in \mathcal{F}$ form a sunflower, we have $|\mathcal{F}| \leq C^s$.*

Erdős and Rado also made a more general version of this conjecture for k-sunflowers, i.e. sunflowers consisting of k sets $A_1, \ldots, A_k$ rather than three sets. We say that k distinct sets $A_1, \ldots, A_k$ form a k-sunflower if their pairwise intersections all agree, i.e. if $A_i \cap A_j = A_1 \cap \cdots \cap A_k$ for all $1 \leq i < j \leq k$. Erdős and Rado conjectured that whenever $\mathcal{F}$ is a family of sets of size s, such that no k distinct sets $A_1, \ldots, A_k \in \mathcal{F}$ form a sunflower, we have $|\mathcal{F}| \leq (C_k)^s$ for some constant C_k depending only on k.

Conjecture 4.2 is still widely open despite of a lot of attention for more than 60 years. In a recent breakthrough, Alweiss, Lovett, Wu, and Zhang [3] improved the best known bound to $|\mathcal{F}| \leq (C \log s \log \log s)^s$ for some absolute constant C. More generally, they showed that any family $\mathcal{F}$ of sets of size s, such that no k distinct sets $A_1, \ldots, A_k \in \mathcal{F}$ form a sunflower, must have size $|\mathcal{F}| \leq (Ck^3 \log s \log \log s)^s$ for some absolute constant C.

In 1978, Erdős and Szemerédi [26] introduced a somewhat different version of the sunflower problem, where the sets inside the family $\mathcal{F}$ are no longer assumed to all have the same size. Instead, in the Erdős–Szemerédi Sunflower Problem, the family is restricted to some ground set of size n.

Question 4.3 *Given some positive integer $n \geq 2$. let $\mathcal{F}$ be a family of subsets of $\{1, \ldots, n\}$. How large can the family $\mathcal{F}$ be, if we assume that no three distinct sets $A, B, C \in \mathcal{F}$ form a sunflower?*

Erdős and Szemerédi [26] conjectured that $|\mathcal{F}| \leq c^n$ for some absolute constant $c < 2$. Note that the total number of subsets of the ground set $\{1, \ldots, n\}$ is 2^n, so the conjecture states that the size of the family $\mathcal{F}$ is "exponentially small" (i.e. exponentially smaller than 2^n). This conjecture is weaker than Conjecture 4.2 (it was shown by Erdős and Szemerédi [26] themselves that their conjecture is implied by Conjecture 4.2). Still, the Erdős–Szemerédi Sunflower Conjecture was open for a long time. In 2013, Alon, Shpilka and Umans [2] showed that the conjecture would follow from the 3-colored sum-free theorem in $\mathbb{F}_3^n$ (the case $k = 3$ and $p = 3$ of Exercise 3.4), but this theorem was not known at that time (in fact, it was stated as a conjecture by Alon–Shpilka–Umans [2, Conjecture 6], but they were unsure about whether to actually believe it, see the discussion in [2, Section 4]). The observation that the proof of the bound for the Cap-Set Problem in Theorem 1.2 due to Ellenberg–Gijswijt [22] implies the case $k = 3$ in Exercise 3.4 was first made in writing by Blasiak–Church–Cohn–Growchow–Naslund–Sawin–Umans [9]. This then established the Erdős–Szemerédi Sunflower Conjecture.

A more direct proof of the conjecture with a somewhat better bound was shortly afterwards given by Naslund and Sawin [60]. Specifically, they proved the following.

Theorem 4.4 *Let $\mathcal{F}$ be a family of subsets of $\{1, \ldots, n\}$ such that no three distinct subsets $A, B, C \in \mathcal{F}$ form a sunflower. Then the family $\mathcal{F}$ has size*

$$|\mathcal{F}| \leq 3(n+1) \sum_{0 \leq k \leq n/3} \binom{n}{k} = \left(\frac{3}{2^{2/3}}\right)^{n+o(n)}.$$

We remark that $3/2^{2/3} \approx 1.89$, so Theorem 4.4 gives a bound of roughly 1.89^n for Question 4.3.

We will now prove Theorem 4.4 using the slice rank polynomial method, following the proof of Naslund and Sawin [60]. We will first prove a similar bound for $|\mathcal{F}|$ under the additional assumption that the sets in $\mathcal{F}$ are all of the same size.

Proposition 4.5 *Let $\mathcal{F}$ be a family of subsets of $\{1, \ldots, n\}$, all of the same size, such that no three distinct subsets $A, B, C \in \mathcal{F}$ form a sunflower. Then the family $\mathcal{F}$ has size*

$$|\mathcal{F}| \leq 3 \sum_{0 \leq k \leq n/3} \binom{n}{k}.$$

Proof. For every set $A \in \mathcal{F}$, we consider the indicator vector $(A[1], \ldots, A[n]) \in \{0,1\}^n$ defined by

$$A[j] = \begin{cases} 1 & \text{if } j \in A \\ 0 & \text{if } j \notin A \end{cases}$$

for $j \in \{1, \ldots, n\}$. Let us now define the 3-dimensional tensor $F : \mathcal{F} \times \mathcal{F} \times \mathcal{F} \to \mathbb{R}$ by

$$F(A, B, C) = \prod_{j=1}^{n} (A[j] + B[j] + C[j] - 2) \tag{4.1}$$

for all $A, B, C \in \mathcal{F}$.

We claim that F is a diagonal tensor with non-zero entries on the diagonal. If $A = B = C$, then it is easy to see that indeed

$$F(A, B, C) = F(A, A, A) = \prod_{j=1}^{n} (3A[j] - 2) = 1^{|A|} \cdot (-2)^{n-|A|} \neq 0.$$

Furthermore, if $A, B, C \in \mathcal{F}$ are such that $A = B$ but $C \neq A$, then due to $|A| = |C|$ there must be an element $j \in A \setminus C$. We then have $A[j] = B[j] = 1$ but $C[j] = 0$. Consequently $A[j] + B[j] + C[j] - 2 = 0$ and so one of the factors of the product on the right-hand side of (4.1) is zero. Thus, we have $f(A, B, C) = 0$ if the sets A and

B agree, but C is different. Analogously, whenever any two of the sets $A, B, C \in \mathcal{F}$ agree but the third is different, we must have $f(A, B, C) = 0$.

Finally, if $A, B, C \in \mathcal{F}$ are distinct, then by assumption on $\mathcal{F}$ they cannot form a sunflower. Hence there must be an element $j \in \{1, \ldots, n\}$ that is contained in exactly two of the sets A, B, C. Then we again have $A[j] + B[j] + C[j] - 2 = 0$ and therefore $f(A, B, C) = 0$.

So we have indeed shown that F is a diagonal tensor with non-zero entries on the diagonal. By Lemma 2.3 the slice rank of the tensor F is therefore equal to the size of the family $|\mathcal{F}|$.

Now we will prove an upper bound for the slice rank of F by examining the polynomial expression on the right-hand side of (4.1), similarly to the arguments in Section 3.1. The right-hand side of (4.1) is a multi-linear polynomial in the variables $A[1], \ldots, A[n], B[1], \ldots, B[n], C[1], \ldots, C[n]$ of degree n. So every monomial in this polynomial is of the form

$$A[1]^{a_1} A[2]^{a_2} \cdots A[n]^{a_n} B[1]^{b_1} B[2]^{b_2} \cdots B[n]^{b_n} C[1]^{c_1} C[2]^{c_2} \cdots C[n]^{c_n}$$

with $a_1, \ldots, a_n, b_1, \ldots, b_n, c_1, \ldots, c_n \in \{0, 1\}$ and $a_1 + \cdots + a_n + b_1 + \cdots + b_n + c_1 + \cdots + c_n \leq n$. In particular, for each such monomial we have $a_1 + \cdots + a_n \leq n/3$ or $b_1 + \cdots + b_n \leq n/3$ or $c_1 + \cdots + c_n \leq n/3$.

Thus, $F(x, y, z)$ can be written as a sum of monomials of the form

$$A[1]^{d_1} A[2]^{d_2} \cdots A[n]^{d_n} \cdot \text{monomial}(B[1], \ldots, B[n], C[1] \ldots, C[n])$$
$$\text{or} \quad B[1]^{d_1} B[2]^{d_2} \cdots B[n]^{d_n} \cdot \text{monomial}(A[1], \ldots, A[n], C[1], \ldots, C[n])$$
$$\text{or} \quad C[1]^{d_1} C[2]^{d_2} \cdots C[n]^{d_n} \cdot \text{monomial}(A[1], \ldots, A[n], B[1], \ldots, B[n])$$

for $d_1, \ldots, d_n \in \{0, 1\}$ with $d_1 + \cdots + d_n \leq n/3$. Using the distributive law, we combine terms with the same first factor (i.e. terms belonging to the same of the three cases with the same n-tuple $(d_1, \ldots, d_n)$) and obtain a representation of the form

$$F(x, y, z) = \sum_{d_1, \ldots, d_n} A[1]^{d_1} \cdots A[n]^{d_n} \cdot \text{polynomial}(B[1], \ldots, B[n], C[1], \ldots, C[n])$$
$$+ \sum_{d_1, \ldots, d_n} B[1]^{d_1} \cdots B[n]^{d_n} \cdot \text{polynomial}(A[1], \ldots, A[n], C[1], \ldots, C[n])$$
$$+ \sum_{d_1, \ldots, d_n} C[1]^{d_1} \cdots C[n]^{d_n} \cdot \text{polynomial}(A[1], \ldots, A[n], B[1], \ldots, B[n]),$$

where each of the three sums is over all n-tuples $(d_1, \ldots, d_n) \in \{0, 1\}^n$ with $d_1 + \cdots + d_n \leq n/3$.

Now each of the summands above is a 3-dimensional tensor from $\mathcal{F} \times \mathcal{F} \times \mathcal{F}$ to $\mathbb{F}_p$ of slice rank 1 (e.g. $A[1]^{d_1} \cdots A[n]^{d_n} \cdot \text{polynomial}(B[1], \ldots, B[n], C[1] \ldots, C[n])$ is a product of a function only depending on A and a function depending jointly on B and C). Thus, we have a representation of F as a sum of tensors of slice rank 1, and so the slice rank of F is at most the number of summands in this representation, which is

$$3 \cdot |\{(d_1, \ldots, d_n) \in \{0, 1\}^n \mid d_1 + \cdots + d_n \leq n/3\}| = 3 \sum_{0 \leq k \leq n/3} \binom{n}{k}.$$

Hence, since the slice rank of F is equal to $|\mathcal{F}|$, we obtain

$$|\mathcal{F}| \leq 3 \sum_{0 \leq k \leq n/3} \binom{n}{k},$$

as desired. □

It is now relatively easy to deduce Theorem 4.4 from Proposition 4.5.

Proof of Theorem 4.4. There are $n+1$ possible sizes for subsets of the ground set $\{1,\ldots,n\}$, namely $0,\ldots,n$. Given a family $\mathcal{F}$ as in Theorem 4.4, we can define $\mathcal{F}_s = \{A \in \mathcal{F} \mid |A| = s\}$ for $s = 0,\ldots,n$ (i.e. $\mathcal{F}_s$ consists of the sets $A \in \mathcal{F}$ of size s). Then each of the families $\mathcal{F}_0,\ldots,\mathcal{F}_n$ satisfies the assumptions of Proposition 4.5, and so by the proposition we have

$$|\mathcal{F}_s| \leq 3 \sum_{0 \leq k \leq n/3} \binom{n}{k}$$

for $s = 0,\ldots,n$. Thus, we obtain

$$|\mathcal{F}| = |\mathcal{F}_0| + \cdots + |\mathcal{F}_n| \leq 3(n+1) \sum_{0 \leq k \leq n/3} \binom{n}{k},$$

showing the desired inequality. The fact that the bound on the right-hand side is of the form $(3/2^{2/3})^{n+o(n)}$ follows from Stirling's formula. □

It is not known whether the constant $3/2^{2/3} \approx 1.89$ in the bound in Theorem 4.4 is tight. It is easy to obtain lower bounds for the maximum possible size of a family $\mathcal{F}$ as in Theorem 4.4 that are of the form c^n for some constant $c > 1$ (for example, for even n one obtains a lower bound of $\sqrt{2}^n = 2^{n/2}$ by taking the family $\mathcal{F}$ of all sets $A \subseteq \{1,\ldots,n\}$ with $|A \cap \{2j-1,2j\}| = 1$ for $j = 1,\ldots,n/2$). However, the problem to determine the best possible constant is still open.

One can of course also ask an analogous version of Question 4.3 for k-sunflowers rather than 3-sunflowers. In other words, for a fixed number $k \geq 3$ and large n, one can ask about the largest size of a family $\mathcal{F}$ of subsets of $\{1,\ldots,n\}$ such that no k distinct sets $A_1,\ldots,A_k \in \mathcal{F}$ form a sunflower. In this setting, for each $k \geq 4$ the proof above breaks down (see also the discussion at the end of Section 5.3).

As mentioned above, Alon, Shpilka and Umans proved that the sunflower problems have a connection to certain approaches to finding fast matrix multiplication algorithms (more specifically, algorithms with running time $O(n^{2+\varepsilon})$ for any $\varepsilon > 0$ for multiplying two $n \times n$ matrices). The first of these approaches was by Coppersmith and Winograd [17] from 1990 and relied on the existence of some subset of some abelian group with a certain property (called the *no three disjoint equivoluminous subsets property*). Coppersmith and Winograd [17, Section 11] showed that if relatively large subsets with this property exist in some abelian groups, then one can obtain fast matrix multiplication algorithms. However, the Erdős–Szemerédi Sunflower Conjecture actually implies that such large subsets cannot exist, as shown by Alon–Shpilka–Umans in 2013 (before the Erdős–Szemerédi Sunflower Conjecture was known). So, since the Erdős–Szemerédi Sunflower Conjecture is now known to be true, this approach of Coppersmith and Winograd for fast matrix multiplication fails.

The second approach for fast matrix multiplication that is discussed in the paper of Alon–Shpilka–Umans is an approach due to Cohn, Kleinberg, Szegedy, and Umans [16] relying on certain combinatorial structures in $\mathbb{F}_3^n$. Cohn et al. [16, Conjecture 3.4] conjectured that such structures with certain sizes exist and showed that if this is true, one can use these structures to obtain fast matrix multiplication algorithms. But again Alon–Shpilka–Umans [2] showed that such structures of the desired sizes cannot exist if the 3-colored sum-free theorem in $\mathbb{F}_3^n$ is true (which is the case $k = 3$ and $p = 3$ of Exercise 3.4, but was not known at the time of [2]). Therefore, the approach of Cohn, Kleinberg, Szegedy, and Umans [16] via these structures in $\mathbb{F}_3^n$ also fails.

4.2 Arithmetic Removal Lemmas

Theorem 1.2 and Exercise 3.3 (in the case $m = 1$) state roughly speaking, that for every large enough set $X \subseteq \mathbb{F}_p^n$ one is able to find a (non-trivial) solution $(x_1, \ldots, x_k) \in X^k$ to some given linear equation of the form $a_1 x_1 + \cdots + a_k x_k = 0$ with $a_1 + \cdots + a_k = 0$. But what if we do not only want to find one solution, but want to show that there must be many solutions if the set $X \subseteq \mathbb{F}_p^n$ is large?

Also, what happens if we consider other linear equations, of the form $a_1 x_1 + \cdots + a_k x_k = b$ with $a_1 + \cdots + a_k \neq 0$ or with a vector $b \in \mathbb{F}_p^n \setminus \{0\}$? Then it is possible even for a very large subset $X \subseteq \mathbb{F}_p^n$ (of size $(1/p) \cdot p^n$, which is a constant fraction of all of $\mathbb{F}_p^n$ for fixed p) to contain no solution to the equation $a_1 x_1 + \cdots + a_k x_k = b$ at all (see also the discussion below Exercise 3.3).

If we take a (large) subset $X \subseteq \mathbb{F}_p^n$ containing no solution to some given linear equation, and we modify the set X a little bit by adding a few vectors in $\mathbb{F}_p^n$, then the resulting set still has only few solutions to the given linear equation. However, it turns out that this is the only way a subset $X \subseteq \mathbb{F}_p^n$ with few solutions can arise: Every subset $X \subseteq \mathbb{F}_p^n$ with few solutions to some given linear equation can be turned into a set without any solutions to the equation by deleting a few vectors from the set. This is the content of the *arithmetic removal lemma* for linear equations in $\mathbb{F}_p^n$.

This arithmetic removal lemma was introduced by Green [35] in 2005, motivated by removal lemmas in graph theory (which are a consequence of Szemerédi's celebrated regularity lemma). In fact, Green's statement is more general and allows us to consider different subsets $X_1, \ldots, X_k \subseteq \mathbb{F}_p^n$ for the different variables $x_1, \ldots, x_k$ in the equation (so we are looking for solutions $(x_1, \ldots, x_k) \in X_1 \times \cdots \times X_k$ to the equation). Green's arithmetic removal lemma for linear equations in $\mathbb{F}_p^n$ can then be stated as follows.

Theorem 4.6 *For any integer $k \geq 3$ and any $0 < \varepsilon < 1$, there exists $\delta = \delta(k, \varepsilon) > 0$ such that for any prime p, any positive integer n, and any subsets $X_1, \ldots, X_k \subseteq \mathbb{F}_p^n$ at least one of the following statements holds:*

(a) *The number of k-tuples $(x_1, \ldots, x_k) \in X_1 \times \cdots \times X_k$ with $x_1 + \cdots + x_k = 0$ is at least $\delta p^{(k-1)n}$, or*

(b) *it is possible to delete at most εp^n elements from each of the sets $X_1, \ldots, X_k$ such* that afterwards no k-tuples $(x_1, \ldots, x_k) \in X_1 \times \cdots \times X_k$ with $x_1 + \cdots + x_k = 0$ remain.

Green [35] actually proved a more general version of this statement, where $\mathbb{F}_p^n$ can be replaced by any abelian group (and the result was later also extended to non-abelian groups by Kral', Serra, and Vena [45]).

While Theorem 4.6 is phrased just in terms of the equation $x_1 + \cdots + x_k = 0$, it automatically implies the same statement for any equation of the form $a_1 x_1 + \cdots + a_k x_k = b$ with given coefficients $a_1, \ldots, a_k \in \mathbb{F}_p \setminus \{0\}$ and $b \in \mathbb{F}_p^n$. Indeed, for any given subsets $X_1, \ldots, X_k \subseteq \mathbb{F}_p^n$ studying solutions to $a_1 x_1 + \cdots + a_k x_k = b$ with $(x_1, \ldots, x_k) \in X_1 \times \cdots \times X_k$ is equivalent to studying solutions to $x'_1 + \cdots + x'_k = 0$ with $(x'_1, \ldots, x'_k) \in X'_1 \times \cdots \times X'_k$, where for $i = 1, \ldots, k-1$ the set $X'_i \subseteq \mathbb{F}_p^n$ is defined by $X'_i = \{a_i x_i \mid x_i \in X_i\}$ and the set $X'_k \subseteq \mathbb{F}_p^n$ is defined by $X'_k = \{a_k x_k - b \mid x_k \in X_k\}$. For this reason, there is no loss in generality by stating Theorem 4.6 just for the equation $x_1 + \cdots + x_k = 0$.

Theorem 4.6 has close connections to property testing in theoretical computer science, especially in the case $p = 2$ corresponding to property testing of Boolean functions. The query complexity of the resulting property testing algorithms is linked to the dependence of δ on ε (for fixed k and p). For this reason, the problem of improving the dependence of δ on ε has received a lot of attention (see e.g. [7, 8, 28, 29, 30, 40]).

Question 4.7 *For fixed $k \geq 3$ and a fixed prime p, what is the best possible dependence of δ on ε in Theorem 4.6?*

In other words, the question asks how large one can make δ as a function of ε in Theorem 4.6 (for fixed $k \geq 3$ and a fixed prime p). Green's proof of Theorem 4.6 used Fourier-Analytic methods and an arithmetic regularity lemma (introduced in the same paper), and gave only extremely poor bounds for δ in terms of ε (of so-called *tower type*, where $1/\delta$ is a tower of exponents $2^{2^{2^{\cdots}}}$ of height polynomial in $1/\varepsilon$). In 2011, Fox [28] improved the bounds for the graph theoretic removal lemma, and via a deduction due to Kral', Serra and Vena [45] this also gave a better dependence for the arithmetic removal lemma in Theorem 4.6, with $1/\delta$ now of the form $2^{2^{2^{\cdots}}}$ with tower height logarithmic in $1/\varepsilon$. This bound, still of tower type, was not improved despite considerable attention until Fox and Lovász [29] proved a polynomial dependence of δ in terms of ε in the case of $k = 3$ (and any fixed prime p). This came as a shock (for example, Green himself did not expect a polynomial dependence to be possible, see the remark below Problem 7.5 in [34]), and was made possible by the developments around the Ellenberg–Gijswijt bound for the Cap-Set Problem. In fact, the proof of Fox and Lovász [29] relied on the 3-colored sum-free theorem (the case $k = 3$ of Exercise 3.4), and the preprint version of their paper appeared only a few weeks after the paper of Ellenberg and Gijswijt.

Around a year later, the result of Fox and Lovász [29] for $k = 3$ was extended in joint work of Fox, Lovász, and the author [30] to any $k \geq 3$, as stated in the following theorem.

Theorem 4.8 *For any integer $k \geq 3$, and any prime p there exists a constant $C_{k,p}$ such that for every $0 < \varepsilon < 1$ one can take $\delta = \varepsilon^{C_{p,k}}$ in Theorem 4.6.*

It is actually necessary to fix the prime p in order to have a polynomial dependence of δ on ε (indeed, it follows from Behrend's construction [6] that even for $k = 3$ one cannot take δ to be some polynomial function of ε without restricting p).

We remark that Theorem 4.8 was proved with a completely different strategy compared to the earlier proof of Fox and Lovász [29] for the case $k = 3$. This earlier proof breaks down for $k \geq 4$. Instead, the proof of Theorem 4.8 in [30] proceeds by induction on k, relying the result for $k = 3$ in [29] as the base case. Unfortunately, however, this inductive approach leads to a (likely) loss in the value of the exponent $C_{p,k}$.

The proof of Theorem 4.8 in [30] implies that the statement in the theorem holds with
$$C_{p,k} = \frac{k-2}{1 - \log_p \Gamma_p} + 1, \tag{4.2}$$
where $\Gamma_p < p$ is defined as in (1.1). The best known lower bound for the optimal value of $C_{p,k}$ in Theorem 4.8 follows from a result of Lovász and the author [48] concerning tightness of the bound for the k-colored sum-free theorem in Exercise 3.4. This lower bound for the optimal value of $C_{p,k}$ is of the same form as the right-hand side of (4.2), but with Γ_p replaced by the infimum appearing on the right-hand side of (3.4) (in other words, with the 3 appearing in the definition of Γ_p in (1.1) replaced by k). It would be interesting to close the gap between these lower and upper bounds.

For $k = 3$, the optimal value of $C_{p,3}$ in Theorem 4.8 is actually known exactly. This value was determined by Fox and Lovász [29] in their paper proving Theorem 4.8 for $k = 3$, and the value is given by (4.2) for $k = 3$.

The proof strategy of Fox and Lovász [29] for $k = 3$ case of Theorem 4.8, relying on the 3-colored sum-free theorem (the case $k = 3$ of Exercise 3.4), is roughly as follows. In order to prove Theorem 4.8 for $k = 3$, one needs to show that for sets $X_1, X_2, X_3 \subseteq \mathbb{F}_p^n$ containing εp^n disjoint triples $(x_1, x_2, x_3) \in X_1 \times X_2 \times X_3$ satisfying $x_1 + x_2 + x_3 = 0$, there are at least δp^{2n} total such triples in $X_1 \times X_2 \times X_3$ for some δ depending polynomially on ε. By considering only the vectors forming the εp^n disjoint triples in $X_1 \times X_2 \times X_3$ with the desired property, one can assume that $|X_1| = |X_2| = |X_3| = \varepsilon p^n$. One can also assume that each vector $x_1 \in X_1$ has roughly the same number of extension to a triple $(x_1, x_2, x_3) \in X_1 \times X_2 \times X_3$ with $x_1 + x_2 + x_3 = 0$, and similarly for each vector $x_2 \in X_2$ and each vector $x_3 \in X_3$. The key idea of the argument is now to take a random subspace $U \subseteq \mathbb{F}_p^n$ of a certain carefully chosen dimension d, and to analyze which triples $(x_1, x_2, x_3) \in X_1 \times X_2 \times X_3$ with $x_1 + x_2 + x_3 = 0$ satisfy $x_1, x_2, x_3 \in U$. Because of the choice of d, each triple $(x_1, x_2, x_3) \in X_1 \times X_2 \times X_3$ with $x_1 + x_2 + x_3 = 0$ is reasonably likely to satisfy $x_1, x_2, x_3 \in U$ but to be disjoint from all other triples $(x_1', x_2', x_3') \in X_1 \times X_2 \times X_3$ with $x_1' + x_2' + x_3' = 0$ satisfying $x_1', x_2', x_3' \in U$ (this is because each such triple (x_1, x_2, x_3) does not "intersect" too many other triples $(x_1', x_2', x_3') \in X_1 \times X_2 \times X_3$ with $x_1' + x_2' + x_3' = 0$). The collection of all such triples $(x_1, x_2, x_3) \in X_1 \times X_2 \times X_3$ with $x_1 + x_2 + x_3 = 0$ and $x_1, x_2, x_3 \in U$ satisfying this disjointness condition is now a 3-colored sum-free set in $U \cong \mathbb{F}_p^d$. So the 3-colored sum-free theorem gives a bound for the number of such triples, which in turn gives a (polynomial) lower bound for δ in terms of ε.

A priori, this might also seem like a feasible proof strategy for Theorem 4.8 for $k > 3$, given that the k-colored sum-free theorem holds for any $k \geq 3$. The problem is, however, that the probabilistic subspace sampling argument breaks down for $k > 3$, due to problematic dependencies between the events $x_1, \ldots, x_k \in U$ for different k-tuples $(x_1, \ldots, x_k) \in X_1 \times \cdots \times X_3$ with $x_1 + \cdots + x_k = 0$ sharing one entry (these types of dependencies cannot occur in the case of $k = 3$). For this reason, a different proof approach was used in [30] to prove Theorem 4.8 for $k > 3$ (using the result in [29] for $k = 3$ as the base case for an induction on k).

Kral', Serra, and Vena [46] and independently Shapira [75] generalized Green's arithmetic removal lemma stated in Theorem 4.6 to systems of linear equations (instead of a single linear equation). Their proofs both rely on the hypergraph removal lemma, leading to bounds for the dependence of δ on ε which are even worse than the above-mentioned tower-type bounds for Theorem 4.6 proved by Green [35]. It would be interesting to prove a polynomial dependence for δ in terms of ε (for fixed p) for certain classes of systems of linear equations (for all systems, such a result seems completely out of reach, for example because it would immediately imply a positive answer to Question 5.5 about four-term arithmetic progressions, which is widely open and believed to be very difficult).

4.3 Further Applications

Section 6 discusses applications of the slice rank polynomial method to the Erdős–Ginzburg–Ziv Problem in discrete geometry. Other application of the slice rank polynomial, which cannot be covered in this survey paper due to space reasons, include the following. Naslund [56] applied the slice rank polynomial method (or, more precisely his partition rank version of it) to bound the size of subsets of $\mathbb{F}_p^n$ without three points forming a right angle, or without $k+1$ points forming a "k-right corner". In a similar diection, Omar [62] used the partition rank method to obtain upper bounds for size of subsets of $\mathbb{F}_p^n$ without three points forming an "obtuse angle", or without k points forming only right angles. Naslund also used the slice rank polynomial method to give an exponential lower bound for the number of colors needed to color the points in $\mathbb{R}^n$ in such a way that there is no monochromatic equilateral triangle of side length 1 [57] or (more generally) in such a way that there is no set of k points all of whose distances are in some prescribed set [58]. Finally, he [59] used Theorem 1.1 to prove an upper bound for the maximum size of a family of subsets of $\{1, \ldots, n\}$ not containing three distinct sets A, B, C with $|A \cap B| = |A \cap C| = |B \cap C|$ (Theorem 4.4 gives an upper bound of roughly 1.89^n for the size of such a family, but in [59] Naslund improved the bound to roughly 1.83^n).

The original proof method of Croot–Lev–Pach [18] that lead to the develpment of the slice rank polynomial method by Tao [78] has in addition also been used by Green [36] to prove a version of Sárközy's theorem in the setting of polynomial rings (with exponentially strong bounds), as well as in subsequent work by Li and the author [47]

5 Achievements and Barriers of the Slice Rank Polynomial Method

5.1 Is It Tight?

It is very natural to ask whether the bounds in Theorems 1.1 and 1.2, which led to the development of the slice rank polynomial method, are tight. The answer to this question is not known. For Theorem 1.1 the best known lower bound for the maximum possible size of a subset $X \subseteq \mathbb{F}_3^n$ without a non-trivial three-term arithmetic progression is roughly 2.218^n and was proved recently by Tyrrell [79] (the best previous lower bound was roughly 2.217^n due to Edel [20] from 2004). This bound is of the same shape as the upper bound 2.756^n in Theorem 1.1, apart from the value of the constant in the base (importantly, this constant is a number strictly less than 3).

Concerning Theorem 1.2, the best known lower bound for the maximum possible size of a subset $X \subseteq \mathbb{F}_p^n$ for a prime $p \geq 5$ without a non-trivial three-term arithmetic progression is of the form $c_p/\sqrt{n} \cdot ((p+1)/2)^n$ for some constant $c_p > 0$ and is due to Elsholtz and Pach [23] (a similar bound with the same exponential term has already been obtained by Salem and Spencer [67] in 1942). Again, this lower bound has the same shape as the upper bound $(\Gamma_p)^n$ in Theorem 1.2, recalling that $0.8414p \leq \Gamma_p \leq 0.9184p$.

Still, there is a gap between the upper and lower bounds for Theorems 1.1 and 1.2, and it is not clear whether the bounds obtained by the slice rank polynomial method in Section 3.1 are tight. Interestingly, however, in the multi-colored generalization of Theorems 1.1 and 1.2, called the 3-colored sum-free theorem (see the case $k=3$ of Exercise 3.4), the bound is essentially tight. More generally, the exponential base in the k-colored sum-free theorem (see Exercise 3.4) is tight for every fixed $k \geq 3$ and every fixed prime p.

Theorem 5.1 *For every fixed $k \geq 3$ and every fixed prime p, for all large n there exists a list of k-tuples $(x_1^{(\ell)}, \ldots, x_k^{(\ell)}) \in \mathbb{F}_p^n \times \cdots \times \mathbb{F}_p^n$ for $\ell = 1, \ldots, L$ satisfying the conditions in Exercise 3.4 with*

$$L \geq \left(\inf_{0 < \gamma < 1} \frac{1 + \gamma + \cdots + \gamma^{p-1}}{\gamma^{(p-1)/k}} \right)^{n - O(\sqrt{n})}.$$

Here, the implicit constant in the O-notation depends on k and p. This shows that in the k-colored sum-free theorem (stated in Exercise 3.4) the exponential base given by the infimum above is indeed tight (also note that for $k=3$, this infimum agrees with Γ_p as defined in (1.1)). The k-colored sum-free theorem can be proved with the slice rank polynomial method with essentially the same arguments as in Section 3.1. Thus, the natural "multi-colored" generalization of these arguments leads to tight results. This may be surprising, since the arguments may seem rather crude and lossy (it is natural to wonder whether one might get better bounds by considering a more careful decomposition of the tensor F into tensors of slice rank 1, but the tightness in the multi-colored case shows that this alone cannot lead to a better bound).

Theorem 5.1 concerning the tightness of the k-colored sum-free theorem was first proved in the case $k=3$ by Kleinberg, Sawin and Speyer [44], relying on a result proved independently by Norin [61] and Pebody [63] (this result had first been stated as a conjecture by Kleinberg–Sawin–Speyer [44]). In the case of general $k \geq 3$, Theorem 5.1 was later proved by Lovász and the author [48].

The fact that the "multi-colored" version of the arguments in Section 3.1 proving Theorem 1.2 leads to essentially tight bounds, means that the slice rank polynomial method has the feature of being able to produce tight bounds in some settings. At the same time, this fact also exhibits barrier of the slice rank polynomial method. Indeed, if one wants to improve upon the exponential base Γ_p in Theorem 1.2, one would need a proof which does not generalize to the multicolored setting (as the base Γ_p is optimal in the multicolored setting). Most arguments involving the slice rank polynomial methods, like in Section 3.1 (and also like in Section 4.1) do, however, naturally generalize to the multi-colored setting. Such arguments cannot improve the constant Γ_p in Theorem 1.2 (or the constant $\Gamma_3 \approx 2.756$ in Theorem 1.1). This is an important barrier: In order to improve these bounds one would need to modify the slice rank polynomial method approach in such a way that it does not generalize to the multi-colored setting.

5.2 The Fluke of Diagonal Tensors

Another important barrier when applying the slice rank polynomial method is the need for a diagonal tensor in order to obtain a lower bound for the slice rank via Tao's Lemma 2.3. There are lower bounds for the slice rank for certain classes of non-diagonal tensors (see Exercise 2.4 and the remark following it). However, the criteria for these classes of tensors make these results difficult to use in practice (in fact, to the author's knowledge, there only exists a single paper [71] applying the slice rank polynomial method to a combinatorial problem and avoiding the need to use a diagonal tensor for lower-bounding the slice rank by using another lower bound criterion). The difficulty of obtaining a lower bound for the slice rank if one does not have a diagonal tensor is a very significant obstruction for applying the slice rank polynomial method.

Let us illustrate this obstruction with a concrete example. The following statement can be straightforwardly proved with the slice rank polynomial method (it is actually a special case of Exercise 3.3 for $m = 1$ and $k = p$), as first observed by Naslund [55].

Theorem 5.2 *Let $p \geq 3$ be a fixed prime. For some positive integer n, let $X \subseteq \mathbb{F}_p^n$ be a subset not containing a solution to the equation $x_1 + \cdots + x_p = 0$ with vectors $x_1, \ldots, x_p \in X$ that are not all equal. Then*

$$|X| \leq \left(\inf_{0 < \gamma < 1} \frac{1 + \gamma + \cdots + \gamma^{p-1}}{\gamma^{(p-1)/p}} \right)^n < 4^n.$$

To see that the infimum above has value less than 4, one can simply plug in $\gamma = 1/2$. The assumption of not having a solution to the equation $x_1 + \cdots + x_p = 0$ in X where the vectors $x_1, \ldots, x_p$ are not all equal, means that the tensor $F : X^p \to \mathbb{F}_p$ given by

$$F(x_1, \ldots, x_p) = \prod_{j=1}^{n} \left((x_1[j] + \cdots + x_p[j])^{p-1} - 1 \right) \qquad \text{for all } x_1, \ldots, x_p \in X$$

is a diagonal tensor with non-zero entries on the diagonal (here, $x_i[j]$ denotes the j-th entry of the vector $x_i \in \mathbb{F}_p^n$ for $i = 1, \ldots, p$ and $j = 1, \ldots, n$).

However, what happens if we weaken the assumption of not having a solution to $x_1 + \cdots + x_p = 0$ in X where $x_1, \ldots, x_p$ are not all equal, replacing "not all equal" with "distinct"?

Question 5.3 *Let $p \geq 3$ be a fixed prime. How large can a subset $X \subseteq \mathbb{F}_p^n$ be if it does not contain a solution to the equation $x_1 + \cdots + x_p = 0$ with distinct vectors $x_1, \ldots, x_p \in X$?*

With this modified assumption replacing "not all equal" by "distinct", the slice rank polynomial method argument for Theorem 5.2 breaks down. Defining the tensor $F : X^p \to \mathbb{F}_p$ as before, we cannot ensure anymore that the tensor is diagonal. While we still know that $F(x_1, \ldots, x_p) = 0$ if $x_1, \ldots, x_p \in X$ are distinct, we can now also have non-zero entries $F(x_1, \ldots, x_p)$ in the tensor when some of $x_1, \ldots, x_p$ agree. Losing the diagonal structure of the tensor means that we cannot lower-bound its slice rank anymore by Lemma 2.3 and the proof fails.

It is plausible that a similar bound as in Theorem 5.2 still holds in the setting of Question 5.3 (up to constant factors depending on p). The best known bounds for Question 5.3 are, however, significantly weaker (see the discussion in Section 6). It is a very interesting open problem whether in the setting of Question 5.3 one has an upper bound of the form $C_p \cdot C^n$, where the exponential base C is an absolute constant (and only the constant factor C_p in front of the exponential term is allowed to depend on p).

Question 5.3 is closely connected to the Erdős–Ginzburg–Ziv Problem in discrete geometry, see Section 6. For this reason there has been a lot of work on this question [31, 42, 55, 70, 72, 80], especially after the slice rank polynomial method emerged.

More generally, there has also been a considerable amount of work (see [19, 32, 51, 52, 71]) bounding the size of subsets of $\mathbb{F}_p^n$ without a solution to a given system of linear equations consisting of distinct vectors (Question 5.3 corresponds to the case where the system consists of the single equation $x_1 + \cdots + x_p = 0$). For various classes of systems of linear equations, it has been proved that such subsets must be exponentially smaller than p^n. The most general result in this direction is the following theorem of Gijswijt [32]

Theorem 5.4 *Given a prime p, consider a system of m linearly independent equations in k variables $x_1, \ldots, x_k$ of the form*

$$a_{1,1} x_1 + \cdots + a_{1,k} x_k = 0$$
$$\vdots \tag{5.1}$$
$$a_{m,1} x_1 + \cdots + a_{m,k} x_k = 0$$

with coefficients $a_{j,i} \in \mathbb{F}_p$ for $j = 1, \ldots, m$ and $i = 1, \ldots, k$. Suppose that for every $j = 1, \ldots, m$, we have $a_{j,1} + \cdots + a_{j,k} = 0$ (i.e. the sum of the coefficients appearing in each of the equations in (5.1) is zero). Also suppose that for every $\ell = 1, \ldots, m$ and every choice of ℓ linearly independent linear equations that can be obtained as linear combinations of the equations in the system (5.1), the following condition holds: At least $2\ell + 1$ of the variables $x_1, \ldots, x_k$ appear with a non-zero coefficient in at least on of the ℓ chosen equations.

Then there exist constants $C > 0$ and $\Gamma < p$ (depending on p, m and k) such that for every positive integer n, every subset $X \subseteq \mathbb{F}_p^n$ of size $|X| \geq C \cdot \Gamma^n$ contains a solution $(x_1, \ldots, x_k)$ to (5.1) with $x_1, \ldots, x_k \in X$ such the affine span of $x_1, \ldots, x_k \in \mathbb{F}_p^n$ has dimension $k - m - 1$.

Recall that the affine span of vectors $x_1, \ldots, x_k \in \mathbb{F}_p^n$ is defined to be the smallest affine-linear subspace of $\mathbb{F}_p^n$ containing $x_1, \ldots, x_k$. For any vectors $x_1, \ldots, x_k \in \mathbb{F}_p^n$, this affine span has dimension at most $k - 1$. If in addition there are m (linearly independent) affine relationships imposed on $x_1, \ldots, x_k$ by the system (5.1), then the affine span has dimension at most $k - m - 1$. The conclusion of Theorem 5.4 is therefore that X contains a solution $(x_1, \ldots, x_k)$ to (5.1) such that the vectors $x_1, \ldots, x_k \in X$ have affine span "as large as theoretically possible". In other words, there are no additional affine dependencies between $x_1, \ldots, x_k$ besides those already imposed by the system (5.1). In particular, this means that $x_1, \ldots, x_k \in X$ are distinct (indeed, if $x_j - x_{j'} = 0$ for some $j \neq j'$, this would be an additional affine dependence relationship between $x_1, \ldots, x_k$ that is not imposed by the system (5.1), since every equation implied by the system (5.1) needs to have at least three variables with non-zero coefficients by the assumption for $\ell = 1$). So the conclusion in Theorem 5.4 implies in particular that every subset $X \subseteq \mathbb{F}_p^n$ of size $|X| \geq C \cdot \Gamma^n$ contains a solution $(x_1, \ldots, x_k)$ to (5.1) with distinct $x_1, \ldots, x_k \in X$. In other words, every subset $X \subseteq \mathbb{F}_p^n$ not containing a solution $(x_1, \ldots, x_k)$ to (5.1) with distinct $x_1, \ldots, x_k \in X$, must have size $|X| < C \cdot \Gamma^n$, i.e. size exponentially smaller than p^n.

The first assumption in Theorem 5.4, assuming that $a_{j,1} + \cdots + a_{j,k} = 0$ for $j = 1, \ldots, m$, is necessary for such a conclusion to hold, since otherwise there would be subsets $X \subseteq \mathbb{F}_p^n$ of size $(1/p) \cdot p^n$ not containing any solution to (5.1), see the discussion below Exercise 3.3. The second assumption corresponds to the assumption $m < k/2$ in Exercise 3.3 (in particular, taking $\ell = m$ in the assumption here implies that $k \geq 2m + 1$, meaning that $m < k/2$). This assumption is caused by degree constraints for a certain polynomial appearing in the proof of Theorem 5.4, see also the discussion in the next subsection. In this context, this assumption is actually very natural and is the best one could hope for with the current techniques. In this sense, Theorem 5.4 is the strongest possible result that one could hope to reach with the current methods (and it is quite amazing that Gijswijt [32] achieved this result, substantially strengthening all previous results in this direction).

Again, Theorem 5.4 cannot simply be proved by applying the slice rank polynomial method like in Sections 3.1 and 4.1, precisely due to the issue that the relevant tensor is not diagonal. Gijswijt's proof of Theorem 5.4 is quite involved and does actually not use the notion of slice rank, only rank for matrices. However, it does use the same splitting technique to upper-bound the rank as explained in Section 3.1 in the proof of Theorem 1.2. As discussed in the remark after the proof of Theorem 1.2, this splitting technique is due to Croot, Lev and Pach [18], and was also used by Ellenberg and Gijswijt in their original proof of Theorem 1.2 in [22].

5.3 The Degree of the Polynomial

Yet another obstruction for applying the slice rank polynomial method is related to the degree of the polynomial used to define the relevant tensor. In Section 3.1, the polynomial on the right-hand side of (3.1) has degree $(p-1)n$, leading to the bound of $d_1 + \cdots + d_n \leq (p-1)n/3$ for the n-tuples $(d_1, \ldots, d_n)$ counted in Lemma 3.2. It is very important that this bound $(p-1)n/3$ is (significantly) smaller than the average value $(p-1)n/2$ of $d_1 + \cdots + d_n$ for a random n-tuple $(d_1, \ldots, d_n) \in \{0, 1 \ldots, p-1\}^n$. Indeed, otherwise (for example, when counting the number n-tuples $(d_1, \ldots, d_n)$ with

The Slice Rank Polynomial Method   229

$d_1 + \cdots + d_n \leq (p-1)n/2$) we would obtain a bound of roughly p^n (which would be useless).

This degree obstruction means that one cannot apply the same approach as in Section 3.1 to bound the side a subset $X \subseteq \mathbb{F}_p^n$ (with $p \geq 5$) without a non-trivial four-term arithmetic progression. Indeed, a natural diagonal tensor to consider in this case would be $F : X^4 \to \mathbb{F}_p$ given by

$$F(x,y,z,w) = \prod_{j=1}^{n} \left((x_j - 2y_j + z_j)^{p-1} - 1\right) \cdot \left((y_j - 2z_j + w_j)^{p-1} - 1\right).$$

for all $x, y, z, w \in X$. The polynomial on the right-hand side here has degree $2(p-1)n$ (there are also various other ways to write down a natural diagonal tensor for this problem with the same degree for the polynomial). When continuing the argument in the same way as in Section 3.1, one then arrives at the bound

$$|X| \leq 4 \cdot |\{(d_1, \ldots, d_n) \in \{0, 1, \ldots, p-1\}^n \mid d_1 + \cdots + d_n \leq (p-1)n/2\}|.$$

(note that the tensor here is 4-dimensional rather than 3-dimensional, so the bound $d_1 + \cdots + d_n \leq (p-1)n/2$ arises via $d_1 + \cdots + d_n \leq 2(p-1)n/4$). But this bounds is useless, since it just amounts to $|X| \leq 2p^n$, which is worse than the trivial bound $|X| \leq p^n$.

Similarly, one cannot apply the slice rank polynomial method like in Section 3.1 to bound the size of a subset $X \subseteq \mathbb{F}_p^n$ without a non-trivial k-term arithmetic progression for any $k \geq 4$. It is a widely open question of whether the size of such subsets is again exponentially smaller than p^n, like for sets without a non-trivial three-term arithmetic progression. For concreteness, we state the question just for $k = 4$.

Question 5.5 *Let $p \geq 5$ be a prime. Does every subset $X \subseteq \mathbb{F}_p^n$ without a non-trivial four-term arithmetic progression have size at most $(\Gamma'_p)^n$ for some constant $\Gamma'_p < p$ (only depending on p)?*

Even just for $p = 5$ (or any other specific prime) this question is widely open. For any fixed $k \geq 4$ and any fixed prime $p > k$, it is known that $|X| = o(p^n)$ for any subset $X \subseteq \mathbb{F}_p^n$ without a non-trivial k-term arithmetic progression, but the quantitative bounds for the $o(p^n)$ term are fairly weak (especially for $k > 4$).

The polynomial degree obstruction is also the reason for having the assumption $m < k/2$ in Exercise 3.3. As discussed below the statement of the exercise, the bound (3.3) is trivial if $m \geq k/2$. This corresponds precisely to the fact that if the degree of the polynomial is too large (if $m \geq k/2$, the degree of the polynomial defining the relevant tensor would be at least $(p-1)k/2$) the slice rank polynomial method only gives trivial bounds.

Furthermore, the degree obstruction is the reason why the proof in Section 4.1 cannot be generalized to k-sunflowers for $k \geq 4$. For example, if $\mathcal{F}$ is a family of subsets of $\{1, \ldots, n\}$ such that no four distinct sets $A_1, \ldots, A_4 \in \mathcal{F}$ form a sunflower, it would be natural to consider the tensor $F : \mathcal{F}^4 \to \mathbb{R}$ given by

$$F(A, B, C, D) = \prod_{j=1}^{n} (A[j] + B[j] + C[j] + D[j] - 2)(A[j] + B[j] + C[j] + D[j] - 3)$$

for all $A, B, C, D \in \mathcal{F}$ (as long as all sets in $\mathcal{F}$ have the same size, this would be a diagonal tensor with non-zero entries on the diagonal). However, the polynomial on the right-hand side has degree $2n$, and so when following the arguments in Section 4.1 one would arrive at the bound

$$|\mathcal{F}| \leq 4 \cdot |\{(d_1, \ldots, d_n) \in \{0,1\}^n \mid d_1 + \cdots + d_n \leq n/2\}|,$$

which is again worse than the trivial bound $|\mathcal{F}| \leq 2^n$.

6 The Erdős–Ginzburg–Ziv Problem

The Erdős–Ginzburg–Ziv Problem is a classical extremal problem in discrete geometry. It can be phrased as follows.

Question 6.1 *Given positive integers $m \geq 2$ and $n \geq 1$, what is the minimum integer s such that among any s points in the integer lattice $\mathbb{Z}^n$ there are m points whose centroid is also a lattice point in $\mathbb{Z}^n$?*

The centroid of a collection of m points $p_1, \ldots, p_m \in \mathbb{Z}^n$ is simply their average $(p_1 + \cdots + p_m)/m$, so we are looking for m points (among the given s points) such that $(p_1 + \cdots + p_m)/m \in \mathbb{Z}^n$. Clearly, this only depends on the residues of the coordinates of the points $p_1, \ldots, p_m$ modulo m. Using this observation, one can project the points to $\mathbb{Z}_m^n$, which leads to the following equivalent question (after the projection to $\mathbb{Z}_m^n$ there can be repetitions among the points, so the question is phrased in terms of sequences in order to account for repetitions).

Question 6.2 *Given positive integers $m \geq 2$ and $n \geq 1$, what is the minimum integer s such that any sequence of length s with entries in $\mathbb{Z}^n$ contains a subsequence of length m whose entries have sum zero in $\mathbb{Z}^n$?*

The answer to Questions 6.1 and 6.2 is denoted by $\mathfrak{s}(\mathbb{Z}_m^n)$ and is called the *Erdős–Ginzburg–Ziv constant* of $\mathbb{Z}_m^n$. These questions have been studied intensively, but there are still only very few cases in which the value $\mathfrak{s}(\mathbb{Z}_m^n)$ is known exactly. It was proved by Erdős, Ginzburg, and Ziv [24] in 1961 that for $n = 1$ one has $\mathfrak{s}(\mathbb{Z}_m^1) = 2m - 1$ (they did not use this notation, and the problem was only named after them later). In 2007, Reiher [66] proved that for $n = 2$ the answer is $\mathfrak{s}(\mathbb{Z}_m^2) = 4m - 3$. The only other infinite family of known values is when m is a power of 2, and in this case one has $\mathfrak{s}(\mathbb{Z}_m^n) = (m-1)2^n + 1$ (first proved by Harborth [39]).

Alon and Dubiner [1] proved in 1995 that for every fixed dimension n, the Erdős–Ginzburg–Ziv constant $\mathfrak{s}(\mathbb{Z}_m^n)$ grows linearly with m. More specifically, they proved an upper bound of the form $\mathfrak{s}(\mathbb{Z}_m^n) \leq (cn \log n)^n \cdot m$ for some absolute constant c. Zakharov [80] improved this bound to $\mathfrak{s}(\mathbb{Z}_m^n) \leq 4^n \cdot m$ in the case where m is a prime that is sufficiently large with respect to n.

However, in the opposite regime, where m is fixed and the dimension n is large, the growth behavior of $\mathfrak{s}(\mathbb{Z}_m^n)$ is still not well-understood. In this regime, the question is (up to constant factors depending on m) equivalent to Question 5.3. Indeed, using the general upper bound $\mathfrak{s}(\mathbb{Z}_{k\ell}^n) \leq \ell \cdot (\mathfrak{s}(\mathbb{Z}_k^n) - 1) + \mathfrak{s}(\mathbb{Z}_\ell^n)$ (which can be proved with a simple combinatorial argument and was first observed by Harborth [39]), the question can be reduced to the case of $m = p$ being a prime. Furthermore, if we are given a sequence with entries in $\mathbb{F}_p^n$ not containing a subsequence of length p whose entries have sum zero in $\mathbb{F}_p^n$, then every vector can appear at most $p-1$ times in this sequence (otherwise, the p copies of the same vector would form a subsequence of length p summing to zero). Hence the problem of finding the maximum possible length of such a sequence is basically equivalent to Question 5.3 asking about the maximum possible size of a subset of $\mathbb{F}_p^n$ without p distinct vectors summing to zero. In other words, for a fixed prime p, the Erdős–Ginzburg–Ziv constant $\mathfrak{s}(\mathbb{Z}_p^n)$ agrees with the answer to Question 5.3 up to constant factors (depending on p).

In the case of $p=3$, Question 5.3 is precisely the Cap-Set Problem (note that in characteristic 3 the equation $x+y+z=0$ is equivalent to the equation $x-2y+z=0$). The bound of Ellenberg and Gijswijt [22] stated in Theorem 1.1 therefore gives $\mathfrak{s}(\mathbb{Z}_3^n) \leq 2 \cdot 2.756^n + 1$.

As already discussed in Section 5.2, for $p \geq 5$ the proof approach of Ellenberg–Gijswijt or the slice rank polynomial method cannot be easily applied to Question 5.3 (because the corresponding tensor is not diagonal). Various researchers worked on trying to extend the slice rank polynomial method to this setting [31, 42, 55, 70, 72, 80].

For $p \geq 5$, Naslund [55] was the first to exponentially improve upon the trivial upper bound p^n in Question 5.3. He proved that $|X| \leq (2^p - p - 2) \cdot (\Gamma_p)^n$, where $\Gamma_p < p$ is as defined in (1.1). Recalling that $0.84p \leq \Gamma_p \leq 0.92p$, the exponential bound in his base is strictly smaller than p, but still linear in p. Naslund's proof was based on his notion of partition rank from [56], but a simpler proof of the same bound (with a better constant factor) was later found by Fox and the author [31], deducing the bound from the result of Ellenberg and Gijswijt on subsets of $\mathbb{F}_p^n$ without non-trivial three-term arithmetic progressions (see Theorem 1.2).

Naslund's bound was significantly improved by the author [70], who showed that $|X| \leq C_p \cdot (2\sqrt{p})^n$ for some constant $C_p > 0$ depending on p. Here, the exponential base is only on the order of $\sqrt{p}$. The proof combined the slice rank polynomial method (or, more precisely, the p-colored sum-free theorem, i.e. the case $k=p$ of the statement in Exercise 3.4) with additional combinatorial ideas to circumvent the problem that the relevant tensor is not diagonal. The proof also extends to the natural "multi-colored" analogue of Question 5.3 (see Exercise 3.4 for an indication of what is meant by that), and in the multi-colored setting the $\sqrt{p}$ term in the base of the bound is actually best possible.

At this point, the situation was similar to the current state for the Cap-Set Problem and more generally the problem of bounding subsets of $\mathbb{F}_p^n$ for a fixed prime p without a non-trivial three-term arithmetic progression. As discussed in Section 5.1, the exponential bases in the bounds in Theorems 1.1 and 1.2 are not known to be tight, but they are tight in the natural multi-colored generalization of these theorems. This imposes a barrier towards improving the exponential bases in these bounds, since an argument would be needed that does not generalize to the multi-colored setting. After [70], the situation was the same for Question 5.3 (and hence also for the bounding the Erdős–Ginzburg–Ziv constant $\mathfrak{s}(\mathbb{Z}_p^n)$ for fixed p and large n). The $\sqrt{p}$ term in the exponential base is tight for the multi-colored version of Question 5.3, imposing a barrier towards improving the bounds.

Recently, the author and Zakharov [72] overcame this barrier for Question 5.3, proving the following much stronger bound.

Theorem 6.3 *Let $\varepsilon > 0$, and let p be a prime. If $X \subseteq \mathbb{F}_p^n$ does not contain p distinct vectors $x_1, \ldots, x_p \in X$ with $x_1 + \cdots + x_p = 0$, then $|X| \leq D_{\varepsilon,p} \cdot (C_\varepsilon p^\varepsilon)^n$ (where $D_{\varepsilon,p}$ is a constant depending on ε and p, and C_ε is a constant depending only on ε).*

As a consequence, one obtains $\mathfrak{s}(\mathbb{Z}_m^n) \leq D_{\varepsilon,m} \cdot (C_\varepsilon p^\varepsilon)^n$ for every fixed $m \geq 2$ and every fixed $\varepsilon > 0$, where p denotes the largest prime factor of m (and $D_{\varepsilon,m}$ is a constant depending on ε and m, and C_ε is a constant depending only on ε).

The proof of Theorem 6.3 combines the k-colored sum-free theorem stated in Exercise 3.4 (which is a consequence of the slice rank polynomial method) with combinatorial and probabilistic arguments, as well as a higher uniformity version of the Balog–Szemerédi–Gowers Theorem due to Borenstein–Croot [12] (this part of the argument does not carry over to the multi-colored setting, which is crucial for overcoming the barrier discussed above).

In the proof, the k-colored sum-free theorem is used in connection with combinatorial arguments to show the following proposition, which may be of independent interest.

Proposition 6.4 *For every fixed $0 < \lambda \leq 1$, there exists a constant C_λ such that for every prime $p > 1/\lambda$ and every positive integer n the following holds. Every subset $X \subseteq \mathbb{F}_p^n$ of size $|X| \geq p^2 \cdot (C_\lambda)^n$ contains a solution to $x_1 + \cdots + x_p = 0$ with vectors $x_1, \ldots, x_p \in X$, such that every vector in $\mathbb{F}_p^n$ appears among $x_1, \ldots, x_p$ at most λp times.*

While the conclusion here does not guarantee finding a solution to $x_1 + \cdots + x_p = 0$ in X where the vectors $x_1, \ldots, x_p$ are distinct, it does give a solution where not too many of the vectors $x_1, \ldots, x_p$ are equal to each other.

While Theorem 6.3 gives the best known upper bound for Question 5.3 (and thus for the Erdős–Ginzburg–Ziv constant $\mathfrak{s}(\mathbb{Z}_m^n)$ for fixed m and large n), the best known lower bound for Question 5.3 is roughly 2.1398^n and is due to Edel [21]. Thus, there is still a big gap between the upper and lower bound. In particular, the following problem is open.

Question 6.5 *Is there an absolute constant C such that every subset $X \subseteq \mathbb{F}_p^n$ without p distinct elements summing to zero has size at most $C_p \cdot C^n$ (for some constant C_p depending on p)?*

In the opposite parameter range, where n is fixed and p is sufficiently large with respect to n, Zakharov (2020+) proved such a bound, namely $|X| \leq 4^n \cdot p$. However, his methods do not seem to carry over to the range where p is fixed and n is large.

Acknowledgements

The author would like to thank her former PhD advisor Jacob Fox, under whose wonderful guidance she has learnt a lot of the material covered in this survey during her time as a PhD student. She is also grateful to an anonymous reviewer for several helpful comments.

This survey paper was written while the author was at the Massachusetts Institute of Technology (MIT).

References

[1] N. Alon and M. Dubiner, *A lattice point problem and additive number theory*, Combinatorica **15** (1995), 301–309.

[2] N. Alon, A. Shpilka, and C. Umans, *On sunflowers and matrix multiplication*, Comput. Complexity **22** (2013), 219–243.

[3] R. Alweiss, S. Lovett, K. Wu, and J. Zhang, *Improved bounds for the sunflower lemma*, Ann. of Math. **194** (2021), 795–815.

[4] L. Babai and P. Frankl, Linear algebra methods in combinatorics, book draft, 2022.

[5] M. Bateman and N. H. Katz, *New bounds on cap sets*, J. Amer. Math. Soc. **25** (2012), 585–613.

[6] F. A. Behrend, *On sets of integers which contain no three terms in arithmetical progression*, Proc. Nat. Acad. Sci. U.S.A. **32** (1946), 331–332.

[7] A. Bhattacharyya, V. Chen, M. Sudan, and N. Xie, *Testing linear-invariant non-linear properties*, Theory Comput. **7** (2011), 75–99.

[8] A. Bhattacharyya and N. Xie, *Lower Bounds for Testing Triangle-freeness in Boolean Functions*, Comput. Complexity **24** (2015), 65–101.

[9] J. Blasiak, T. Church, H. Cohn, J. A. Grochow, E. Naslund, W. F. Sawin, William, and C. Umans, *On cap sets and the group-theoretic approach to matrix multiplication*, Discrete Anal. (2017), 3.

[10] T. F. Bloom, *A quantitative improvement for Roth's theorem on arithmetic progressions*, J. Lond. Math. Soc. **93** (2016), 643–663.

[11] T. F. Bloom and O. Sisask, *Breaking the logarithmic barrier in Roth's theorem on arithmetic progressions*, arXiv preprint arXiv:2007.03528 (2020).

[12] E. Borenstein and E. Croot, *On a certain generalization of the Balog–Szemerédi–Gowers theorem*, SIAM J. Discrete Math. **25** (2011), 685–694.

[13] J. Bourgain, *On triples in arithmetic progression*, Geom. Funct. Anal. **9** (1999), 968–984.

[14] ———, *Roth's theorem on progressions revisited*, J. Anal. Math. **104** (2008), 155–192.

[15] A. Cohen and G. Moshkovitz, *Partition and analytic rank are equivalent over large fields*, Duke Math. J., in press.

[16] H. Cohn, R. D. Kleinberg, B. Szegedy, and C. Umans, *Group-theoretic algorithms for matrix multiplication*, In: *Proc. 46th Annual IEEE Symposium on Foundations of Computer Science*, 2005, 379–388.

[17] D. Coppersmith and S. Winograd, *Matrix multiplication via arithmetic progressions*, J. Symbolic Comput. **9** (1990), 251–280.

[18] E. Croot, V. Lev, and P. Pach, *Progression-free sets in $\mathbb{Z}_4^n$ are exponentially small*, Ann. of Math. **185** (2017), 331–337.

[19] J. van Dobben de Bruyn and D. Gijswijt, *On the size of subsets of $\mathbb{F}_q^n$ avoiding solutions to linear systems with repeated columns*, arXiv preprint arXiv:2111.09879 (2021).

[20] Y. Edel, *Extensions of generalized product caps*, Des. Codes Cryptogr. **31** (2004), 5–14.

[21] _____, *Sequences in abelian groups G of odd order without zero-sum subsequences of length $\exp(G)$*, Des. Codes Cryptogr. **47** (2008), 125–134.

[22] J. S. Ellenberg and D. Gijswijt, *On large subsets of $\mathbb{F}_q^n$ with no three-term arithmetic progression*, Ann. of Math. **185** (2017), 339–343.

[23] C. Elsholtz and P. P. Pach, *Caps and progression-free sets in $\mathbb{Z}_m^n$*, Des. Codes Cryptogr. **88** (2020), 2133–2170.

[24] P. Erdős, A. Ginzburg, and A. Ziv, *Theorem in the additive number theory*, Bull. Res. Council Israel **10F** (1961), 41–43.

[25] P. Erdős and R. Rado, *Intersection theorems for systems of sets*, J. London Math. Soc. **35** (1960), 85–90.

[26] P. Erdős and E. Szemerédi, *Combinatorial properties of systems of sets*, J. Combinatorial Theory Ser. A **24** (1978), 308–313.

[27] P. Erdős and P. Turán, *On Some Sequences of Integers*, J. London Math. Soc. **11** (1936), 261–264.

[28] J. Fox, *A new proof of the graph removal lemma*, Ann. of Math. **174** (2011), 561–579.

[29] J. Fox and L. M. Lovász, *A tight bound for Green's arithmetic triangle removal lemma in vector spaces*, Adv. Math. **321** (2017), 287–297.

[30] J. Fox, L. M. Lovász, and L. Sauermann, *A polynomial bound for the arithmetic k-cycle removal lemma in vector spaces*, J. Combinatorial Theory Ser. A **160** (2018), 186–201.

[31] J. Fox and L. Sauermann, *Erdős-Ginzburg-Ziv constants by avoiding three-term arithmetic progressions*, Electron. J. Combin. **25** (2018), 2.14.

[32] D. Gijswijt, *Excluding affine configurations over a finite field*, arXiv preprint arXiv:2112.12620 (2021).

[33] W. T. Gowers and J. Wolf, *Linear forms and higher-degree uniformity for functions on $\mathbb{F}_p^n$*, Geom. Funct. Anal. **21** (2011), 36–69.

[34] B. Green, *Finite field models in additive combinatorics*, In: *Surveys in combinatorics 2005*, Cambridge University Press, Cambridge, 2005, pp. 1–27.

[35] _____, *A Szemerédi-type regularity lemma in abelian groups, with applications*, Geom. Funct. Anal. **15** (2005), 340–376.

[36] _____, *Sárközy's theorem in function fields*, Q. J. Math. **68** (2017), 237–242.

[37] J. A. Grochow, *New applications of the polynomial method: the cap set conjecture and beyond*, Bull. Amer. Math. Soc. **56** (2019), 29–64.

[38] L. Guth, *Polynomial methods in combinatorics*, Univ. Lecture Ser. *64*, American Mathematical Society, Providence, RI, 2016.

[39] H. Harborth, *Ein Extremalproblem für Gitterpunkte*, J. Reine Angew. Math. **262** (1973), 356–360.

[40] I. Haviv and N. Xie, *Sunflowers and testing triangle-freeness of functions*, Comput. Complexity **26** (2017), 497–530.

[41] D. R. Heath-Brown, *Integer sets containing no arithmetic progressions*, J. London Math. Soc. **35** (1987), 385–394.

[42] G. Hegedűs, *The Erdős-Ginzburg-Ziv constant and progression-free subsets*, J. Number Theory **186** (2018), 238–247.

[43] O. Janzer, *Polynomial bound for the partition rank vs the analytic rank of tensors*, Discrete Anal. (2020), 7.

[44] R. Kleinberg, W. Sawin, and D. E. Speyer, *The growth rate of tri-colored sum-free sets*, Discrete Anal. (2018), 12.

[45] D. Král', O. Serra, and L. Vena, *A combinatorial proof for the removal lemma for groups*, J. Combin. Theory Ser. A **116** (2009), 971–978.

[46] _____, *A removal lemma for systems of linear equations over finite fields*, Israel J. Math. **187** (2012), 193–207.

[47] A. Li and L. Sauermann, *Sárközy's Theorem in Various Finite Field Settings*, arXiv preprint arXiv:2212.12754 (2022).

[48] L. M. Lovász and L. Sauermann, *A lower bound for the k-multicolored sum-free problem in $\mathbb{Z}_m^n$*, Proceedings of the London Mathematical Society **119** (2019), 55–103.

[49] R. Meshulam, *On subsets of finite abelian groups with no 3-term arithmetic progressions*, J. Combin. Theory Ser. A **71** (1995), 168–172.

[50] L. Milićević, *Polynomial bound for partition rank in terms of analytic rank*, Geom. Funct. Anal. **29** (2019), 1503–1530.

[51] M. Mimura and N. Tokushige, *Avoiding a shape, and the slice rank method for a system of equations*, arXiv preprint arXiv:1909.10509 (2019).

[52] _____, *Solving linear equations in a vector space over a finite field*, Discrete Math. **344** (2021), 112603.

[53] G. Moshkovitz and D. G. Zhu, *Quasi-linear relation between partition and analytic rank*, arXiv preprint arXiv:2211.05780 (2022).

[54] D. Munhá Correia, B. Sudakov, and I. Tomon, *Flattening rank and its combinatorial applications*, Linear Algebra Appl. **625** (2021), 113–125.

[55] E. Naslund, *Exponential bounds for the Erdős-Ginzburg-Ziv constant*, Combin. Theory Ser. A **174** (2020), 105185.

[56] _____, *The partition rank of a tensor and k-right corners in $\mathbb{F}_q^n$*, J. Combin. Theory Ser. A **174** (2020), 105190.

[57] _____, *Monochromatic equilateral triangles in the unit distance graph*, Bull. Lond. Math. Soc. **52** (2020), 687–692.

[58] _____, *The chromatic number of $\mathbb{R}^n$ with multiple forbidden distances*, Mathematika **69** (2023), 692–718.

[59] _____, *Upper Bounds For Families Without Weak Delta-Systems*, Combinatorica, in press.

[60] E. Naslund and W. Sawin, *Upper bounds for sunflower-free sets*, Forum Math. Sigma **5** (2017), e15.

[61] S. Norin, *A distribution on triples with maximum entropy marginal*, Forum Math. Sigma **7** (2019), e46.

[62] M. Omar, *Partition Rank and Partition Lattices*, arXiv preprint arXiv:2208.06932 (2022).

[63] L. Pebody, *Proof of a conjecture of Kleinberg-Sawin-Speyer*, Discrete Anal. (2018), 13.

[64] G. Pellegrino, *Sul massimo ordine delle calotte in $S_{4,3}$*, Matematiche (Catania) **25** (1970), 149–157.

[65] K. F. Roth, *On certain sets of integers*, J. London Math. Soc. **28** (1953), 104–109.

[66] C. Reiher, *On Kemnitz' conjecture concerning lattice-points in the plane*, Ramanujan J. **13** (2007), 333–337.

[67] R. Salem and D. C. Spencer, *On sets of integers which contain no three terms in arithmetical progression*, Proc. Nat. Acad. Sci. U.S.A. **28** (1942), 561–563.

[68] T. Sanders, *On certain other sets of integers*, J. Anal. Math. **116** (2012), 53–82.

[69] _____, *On Roth's theorem on progressions*, Ann. of Math. **174** (2011), 619–636.

[70] L. Sauermann, *On the size of subsets of $\mathbb{F}_p^n$ without p distinct elements summing to zero*, Israel Journal of Mathematics **243** (2021), 63–79.

[71] _____, *Finding solutions with distinct variables to systems of linear equations over $\mathbb{F}_p$*, Mathematische Annalen **386** (2023), 1-33.

[72] L. Sauermann and D. Zakharov, *On the Erdős–Ginzburg–Ziv Problem in large dimension*, arXiv preprint arXiv:2302.14737 (2023).

[73] W. Sawin and T. Tao, *Notes on the "slice rank" of tensors*, https://terrytao.wordpress.com/2016/08/24/n, 2016.

[74] T. Schoen, *Improved bound in Roth's theorem on arithmetic progressions*, Adv. Math. **386** (2021), 107801.

[75] A. Shapira, *A proof of Green's conjecture regarding the removal properties of sets of linear equations*, J. Lond. Math. Soc. **81** (2010), 355–373.

[76] E. Szemerédi, *On sets of integers containing no k elements in arithmetic progression*, Acta Arith. **27** (1975), 199–245.

[77] _____, *Integer sets containing no arithmetic progressions*, Acta Math. Hungar. **56** (1990), 155–158.

[78] T. Tao, *A symmetric formulation of the Croot-Lev-Pach-Ellenberg-Gijswijt capset bound*, http://terrytao.wordpress.com/2016/05/18/a, 2016.

[79] F. Tyrrell, *New lower bounds for cap sets*, Discrete Analysis, in press.

[80] D. Zakharov, *Convex geometry and Erdős–Ginzburg–Ziv problem*, arXiv preprint arXiv:2002.09892 (2020).

Institut für Angewandte Mathematik
Universität Bonn
Endenicher Allee 60
53115 Bonn, Germany
lsauerma@mit.edu

An Introduction to Transshipments Over Time

Martin Skutella

Abstract

Network flows over time are a fascinating generalization of classical (static) network flows, introducing an element of time. They naturally model problems where travel and transmission are not instantaneous and flow may vary over time. Not surprisingly, flow over time problems turn out to be more challenging to solve than their static counterparts. In this survey, we mainly focus on the efficient computation of transshipments over time in networks with several source and sink nodes with given supplies and demands, which is arguably the most difficult flow over time problem that can still be solved in polynomial time.

1 Introduction

The study of flows over time goes back to the work of Ford and Fulkerson [19] who introduced the topic under the name *dynamic flows* and devoted a section of their seminal book [20] to that topic. In contrast to classical static flows, flows over time include an element of time. They model the fluctuation of flow along arcs over time as well as non-instantaneous travel through a network where flow traveling along an arc experiences a certain delay. Such effects play a critical role in many network routing problems such as road, pedestrian, rail, or air traffic control. Further applications include production systems, communication and data networks, and financial flows.; see, e.g., [4, 37, 48].

Flow over time problems turn out to considerably more difficult to solve than their static counterparts. While Ford and Fulkerson [19] show how to reduce the maximum flow over time problem to a static minimum cost flow problem, Klinz and Woeginger [36] prove that finding minimum cost flows over time is already an weakly NP-hard problem. Hall, Hippler, and Skutella [27] show the same for multicommodity flows over time, with some variants even being strongly NP-hard. Hoppe and Tardos [31, 29] study the transshipment over time problem where flow needs to be sent from several source nodes with given supplies to several sink nodes with given demands. For the case of static flows, finding a transshipment can be reduced to finding a maximum flow from a super-source to a super-sink. Unfortunately, this idea no longer works for the transshipment over time problem. Hoppe and Tardos present an efficient algorithm that relies on a sequence of parametric submodular function minimizations and Schlöter and Skutella [52, 50] observe that a single submodular function minimization is sufficient for finding a transshipment over time that sends flow fractionally.

The main purpose of this paper is to provide a gentle introduction to the topic of transshipments over time and related problems, that may also be used for the purpose of teaching in an advanced course on combinatorial optimization. For a broader overview of flows over time we refer to the survey papers by Aronson [4], Powell, Jaillet, and Odoni [48], Kotnyek [39], Lovetskii and Melamed [41], and Skutella [57], as well as the PhD thesis of Schlöter [50] on the topic.

Outline In Section 2 we introduce notation and some basic definitions. This includes a discussion of different time models for flows over time and, in particular, the role of time-expanded networks. In contrast to all previous literature on the considered flow over time problems, we manage to completely avoid the use of time-expanded networks in the remainder of the paper. Section 3 is devoted to the algorithm of Ford and Fulkerson and its analysis for computing maximum flows over time via temporally repeated flows. This is an important building block for the more advanced flow over time problems covered in later sections. In Section 4 we give a novel proof that a natural set function on subsets of terminals given by corresponding maximum flow over time values is submodular. This fact turns out to be important for the efficient solution of the transshipment over time problem. Section 5 is devoted to earliest arrival flows which constitute a fascinating refinement of maximum flows over time. We treat them here since earliest arrival flows require an interesting generalization of temporally repeated flows, so-called chain-decomposable flows, which are needed in more generality for transshipments over time. In Section 6 we present Hoppe and Tardos' algorithm for efficiently computing lexicographically maximum flows over time. This algorithm is the most important subroutine for solving the transshipment over time problem, which is the subject of the final Section 7.

2 Notation and Preliminaries

We consider a network N given by a directed graph (V, A) with node set V and arc set A. Every arc $a \in A$ has a positive *capacity* $u_a \in \mathbf{R}_{>0} \cup \{\infty\}$ and a non-negative *transit time* $\tau_a \in \mathbf{R}_{\geq 0}$. We also refer to transit times as *length* and *cost*. Network N comes with a set of *terminals* $S \subseteq V$ of cardinality $k := |S|$ that is partitioned into two subsets of *sources* S^+ and *sinks* S^-.

We refer to an arc $a \in A$ with start node v and end node w simply as vw. Without loss of generality, there are no parallel or opposite arcs in network N, sources have no incoming, and sinks no outgoing arcs. Having said that, for simplicity of notation, we include for each arc vw its reverse arc wv in A. The reverse arc's capacity is set to $u_{wv} := 0$ and its transit time (or cost or length) to $\tau_{wv} := -\tau_{vw} \leq 0$. One advantage of this convention is that all arcs of the residual network introduced below will be contained in A.

For a path or cycle C, we denote its total transit time (or length or cost) by $\tau(C)$. Whenever we use the term *shortest path* or *shortest cycle*, we mean a path/cycle of minimum transit time. For simplicity only, we assume that the only zero length cycles in A consist of pairs of opposite arcs. This assumption is without loss of generality, because it may be enforced by adding infinitesimal (and thus purely symbolic) values to the arc transit times.

Static Flows A flow (or *static flow*) x in network N assigns a flow value x_{vw} to every arc vw and the negative of this value $x_{wv} = -x_{vw}$ to the opposite arc wv. Intuitively, the flow value x_{vw} is the amount of flow that is sent from node v to node w through arc vw. A flow x must satisfy *flow conservation* at every non-terminal node $v \in V \setminus S$, meaning that the net amount of flow arriving at v is zero. That is, $\sum_w x_{vw} = 0$, where we use the convention that $x_{vw} = 0$ if vw is not an arc. The cost of flow x is $c(x) := \sum_{\{v,w\}} \tau_{vw} x_{vw}$, again setting $\tau_{vw} x_{vw} = 0$ if vw is not an arc. Notice that the sum in the definition of $c(x)$ goes over node pairs $\{v,w\}$ with $v \neq w$, taking only one arc vw of each pair of opposite arcs vw and wv into account; since $\tau_{wv} x_{wv} = (-\tau_{vw})(-x_{vw}) = \tau_{vw} x_{vw}$, the cost $c(x)$ is well-defined.

A *(static) circulation* x is a flow that satisfies flow conservation at every node of the network. A flow or circulation x is *feasible* if it satisfies the capacity constraint $x_{vw} \leq u_{vw}$ for every arc vw. For a feasible flow x in network N, the *residual network* N_x is obtained from network N by replacing the original arc capacity u_{vw} with the *residual capacity* $u_{vw}^x := u_{vw} - x_{vw} \geq 0$, for every arc vw. Due to our convention of reverse arcs wv with $u_{wv} = 0$ and $x_{wv} = -x_{vw}$, this definition of residual capacities is in line with the standard notion of reverse arcs in the residual network, since $u_{wv}^x = u_{wv} - x_{wv} = 0 - (-x_{vw}) = x_{vw}$. Notice that if we take a feasible flow x in N and a feasible flow y in the residual network N_x of x, then their sum $x + y$ is again a feasible flow in network N.

A *chain flow* γ of value $|\gamma|$ sends $|\gamma|$ units of flow along a cycle or a path C_γ. The length (or transit time) of chain flow γ is $\tau(\gamma) := \tau(C_\gamma)$, its cost is $c(\gamma) = |\gamma| \tau(\gamma)$. If C_γ is a cycle, then γ is obviously a circulation. Otherwise, if C_γ is a path, γ only violates flow conservation at its two end nodes, which will always be terminals in this paper.

A multiset of chain flows Γ is a *chain decomposition* of flow x, if the sum of all chain flows in Γ is equal to x, and thus $c(x) = \sum_{\gamma \in \Gamma} c(\gamma) =: c(\Gamma)$. The subset of chain flows $\gamma \in \Gamma$ such that C_γ contains arc vw is denoted by Γ_{vw}. A chain decomposition Γ of x is a *standard chain decomposition* if all cycles or paths C_γ, $\gamma \in \Gamma$, use arcs in the same direction as x does.

Flows Over Time First of all, for an intuitive understanding of flows over time, one can associate arcs of the network with pipes in a pipeline system for transporting some kind of fluid. The length of each pipeline determines the transit time of the corresponding arc (assuming that flow progresses at a constant pace) while the width determines its capacity. For a comprehensive introduction to flows over time we refer to the survey [57].

Formally, a *flow over time* f consists of a function $f_{vw} : \mathbf{R} \to \mathbf{R}$ for every arc vw. We say that $f_{vw}(\theta')$ is the *flow rate* (i.e., amount of flow per time unit) entering arc vw at its tail node v at time θ'. The flow particles entering arc vw at time θ' arrive at the head node w exactly τ_{vw} time units later at time $\theta' + \tau_{vw}$. In particular, the outflow rate at the head node w at time θ' equals $f_{vw}(\theta' - \tau_{vw})$. Accordingly, for pairs of opposite arcs vw and wv we get

$$f_{wv}(\theta') = -f_{vw}(\theta' - \tau_{vw}) \qquad \text{for all } \theta' \in \mathbf{R}.$$

A flow over time f must satisfy *flow conservation* at every non-terminal node, meaning that the net rate of flow arriving at $v \in V \setminus S$ is zero at all times. That is, $\sum_w f_{vw}(\theta') = 0$ for all $\theta' \in \mathbf{R}$. A flow over time f is *feasible* if it satisfies the capacity constraint $f_{vw}(\theta') \leq u_{vw}$ for all $\theta' \in \mathbf{R}$ and for every arc vw. A flow over time f has *time horizon* $\theta \in \mathbf{R}_{\geq 0} \cup \{\infty\}$, if

$$f_{vw}(\theta') = 0 \qquad \text{for all } \theta' \in (-\infty, 0) \cup [\theta, \infty) \text{ and for every arc } vw.$$

All flow over time problems considered in this paper ask for a flow over time with a given time horizon that we always denote by θ.

In order to determine, for example, the amount of flow that a flow over time f sends through arc vw within some finite time interval like $[0, \theta)$, one has to integrate the function f_{vw} over that time interval. In order for this to be well defined, it is usually assumed that all arc functions f_{vw} must be Lebesgue-integrable. All flows over time considered in this paper, however, have piecewise constant flow rates on arcs so that we can safely ignore such technicalities.

Other Flow Over Time Models We mention that several variations of the described flow over time model exist in the literature. For example, we require a very strict notion of flow conservation here, where all flow arriving at a non-terminal node must leave that node again instantaneously. Alternatively, one could allow for some limited or unlimited amount of flow to be temporarily stored at a node. For all flow over time problems considered in this paper, this more relaxed version of flow conservation does not lead to better solutions, such that we stick to the strict and simple flow conservation model. There exist other flow over time problems, however, where the possibility to temporarily store flow at intermediate nodes does make a difference. In this context, multi-commodity flows over time are a prominent example; see [16, 24].

Another, more fundamental distinction of flow over time models is the underlying time model. We work with the so-called *continuous time model* where a flow over time assigns a flow rate $f_{vw}(\theta')$ to every point in time $\theta' \in \mathbf{R}$. The original definition by Ford and Fulkerson [19, 20] is based on a *discrete time model*, where transit times τ_{vw} and time horizon θ are integral, and the flow along an arc vw is described by a function $g_{vw} : \mathbf{Z} \to \mathbf{R}$ where $g_{vw}(\theta')$ denotes the amount of flow sent at time θ' into arc vw and arriving at the head node w at time $\theta' + \tau_{vw}$. Fleischer and Tardos [18] point out a strong connection between the two models. They show that many results and algorithms which have been developed for the discrete time model can be carried over to the continuous time model.

Time-Expanded Networks A distinctive feature of the discrete time model is the existence of *time-expanded networks* with time layers $\theta' = 1, \ldots, \theta$ for time horizon θ, where every layer θ' contains a copy $v_{\theta'}$ of every node v of the underlying flow network. Moreover, for every arc vw there is a copy $v_{\theta'} w_{\theta' + \tau_{vw}}$ in the time-expanded network, for $\theta' = 1, \ldots, \theta - \tau_{vw}$. Every discrete flow over time with time horizon θ thus corresponds to a static flow in the time-expanded network and vice versa. As a consequence, flow over time problems in the discrete time model may be solved as static flow problems in the time-expanded network. The downside of this approach, however, is the huge size of the time-expanded network which is linear in the given time horizon θ and therefore only pseudo-polynomial in the input size. Therefore, time-expanded networks do not lead to efficient algorithms whose running time is polynomially bounded in the input size.

Despite this deficit, time-expanded networks can sometimes be used in the analysis of efficient algorithms; see, e.g., the work of Hoppe and Tardos [31] on transshipments over time. Due to the strong connections between the discrete and the continuous time model mentioned above, time-expanded networks can also play a role in the analysis of algorithms for the latter time model; see, e.g., [18]. This comes at a price, however, since in the first instance the results obtained in this way are usually limited to networks with integral or rational transit times.

One purpose of our paper is to avoid the use of time-expanded networks altogether. On the one hand, this generalizes previously known results to networks with arbitrary (rational or irrational) transit times. On the other hand, it sometimes even leads to arguably simpler proofs.

3 Maximum Flows Over Time

Ford and Fulkerson [19] introduced the *maximum flow over time problem*. Here the task is to send as much flow as possible from the sources S^+ to the sinks S^- within a given time bound θ. This problem can be reduced to a static minimum cost circulation problem in an extended network N^{S^+}. We describe the construction of this network in more generality since it will be used for more complex flow over time problems in later sections as well.

For a subset of terminals $X \subseteq S$, the *extended network* N^X is constructed by adding super-source ψ together with arcs (ψ, s) for all sources $s \in S^+ \cap X$ of infinite capacity $u_{(\psi,s)} = \infty$ and zero transit time $\tau_{(\psi,s)} = 0$, as well as arcs (t, ψ) for all sinks $t \in S^- \setminus X$ of infinite capacity $u_{(t,\psi)} = \infty$ and negative transit time $\tau_{(t,\psi)} = -\theta$; see Figure 1. Network N^X is designed to send flow from ψ via the sources in $S^+ \cap X$ through the original network N into the sinks in $S^- \setminus X$, and then back to ψ. Since a maximum flow over time may use all sources and sinks, in the following we work with network N^{S^+}.

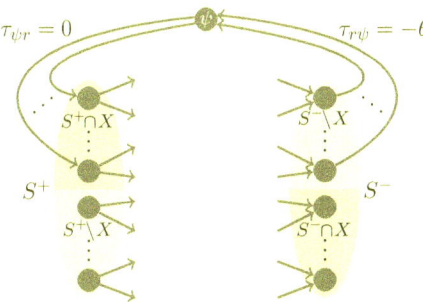

Figure 1: Extended network N^X for a subset of terminals $X \subseteq S$

Remark The problem originally studied by Ford and Fulkerson [19] is to send the maximum possible amount of flow from a single source s to a single sink t within time horizon θ. Notice that this problem is equivalent to our generalized variant of the problem described above: We could simply split up the super-source ψ in network N^{S^+} into two nodes s and t where s is connected to every source in S^+ and every sink in S^- is connected to t. Then, sending as much flow as possible from s to t in this network is equivalent to sending as much flow as possible from S^+ to S^- in N. In the following we thus stick to our variant of the problem.

Ford and Fulkerson's algorithm starts by computing a standard chain decomposition Γ of a minimum cost circulation x in N^{S^+}. Each chain flow $\gamma \in \Gamma$ sends $|\gamma|$ units of flow along a cycle C_γ of nonpositive cost $\tau(C_\gamma) \leq 0$ that contains the super-source ψ. If we delete ψ and its incident arcs from C_γ, what remains is a source sink path P_γ in N with $\tau(P_\gamma) = \tau(C_\gamma) + \theta \leq \theta$.

For each $\gamma \in \Gamma$, Ford and Fulkerson's resulting flow over time $[\Gamma]^\theta$ sends flow at rate $|\gamma|$ into γ's source sink path P_γ during the time interval $[0, \theta - \tau(P_\gamma))$. Notice that $\tau(P_\gamma) \geq 0$ and $\theta - \tau(P_\gamma) = -\tau(C_\gamma) \geq 0$. Then, for each $\gamma \in \Gamma$, flow arrives at the final sink node of P at rate $|\gamma|$ during the time interval $[\tau(P_\gamma), \theta)$ and no flow remains in the network after time θ. Due to its special structure, the computed flow over time $[\Gamma]^\theta$ is called a *temporally repeated flow* or a *standard chain-decomposable flow*.

Definition 3.1 Let Γ be a standard chain decomposition of a static circulation in N^{S^+} such that, for each $\gamma \in \Gamma$, C_γ contains arc ψr_γ for some source $r_\gamma \in S^+$. Let P_γ denote the source sink path starting at r_γ that is obtained from C_γ by deleting ψ and its incident arcs. For each $\gamma \in \Gamma$, the *standard chain-decomposable* (or *temporally repeated*) *flow* $[\Gamma]^\theta$ *with time horizon* θ sends flow at rate $|\gamma|$ into path P_γ during the time interval $[0, \theta - \tau(P_\gamma))$.

Ford and Fulkerson's algorithm is summarized in Algorithm 1.

Algorithm 1 Ford and Fulkerson's maximum flow over time algorithm

Require: network N and time horizon θ
Ensure: maximum flow over time with time horizon θ
 1: compute a static minimum cost circulation x in N^{S^+}
 2: find a standard chain decomposition Γ of x
 3: **return** $[\Gamma]^\theta$

By construction of the computed temporally repeated flow $[\Gamma]^\theta$, the flow rate entering any arc $a \in A$ is always non-negative and bounded by $\sum_{a \in \gamma} |\gamma| = x_a \leq u_a$. Moreover, it satisfies flow conservation constraints at all intermediate nodes of the paths P_γ, $\gamma \in \Gamma$, since all flow arriving at an intermediate node v while traveling along some path P_γ immediately continues its journey on the next arc leaving v. Thus, $[\Gamma]^\theta$ is a feasible flow over time with time horizon θ. The total amount of flow sent from the sources to the sinks is

$$|[\Gamma]^\theta| = \sum_{\gamma \in \Gamma} |\gamma|(\theta - \tau(P_\gamma)) = -\sum_{\gamma \in \Gamma} |\gamma|\tau(C_\gamma) = -c(x). \quad (3.1)$$

Notice that this flow value does not depend on the particular chain decomposition Γ but only on the minimum cost $c(x)$. In view of (3.1), the following theorem states that Algorithm 1 returns a maximum flow over time in N with time horizon θ.

Theorem 3.2 (Ford, Fulkerson [19]; Anderson, Philpott [3]) *The maximum amount of flow that a feasible flow over time with time horizon θ can send from S^+ to S^- is equal to $-c(x)$ where x is a static minimum cost circulation in network N^{S^+}.*

While Ford and Fulkerson [19] proved this result in the discrete time model via minimum cuts in time-expanded networks, the proof of Anderson and Philpott [3] uses a more general concept of *cuts over time* introduced for the special case of zero transit times by Anderson, Nash, and Philpott [2] (see also [1]) and generalized to the setting with non-zero transit times by Philpott [47]; see also Fleischer and Tardos [18].

Definition 3.3 A *cut over time* with time horizon θ in N is given by non-negative values $(\alpha_v)_{v \in V}$, with $\alpha_s \leq 0$ for $s \in S^+$ and $\alpha_t \geq \theta$ for $t \in S^-$. Its capacity is

$$\sum_{vw \in A} \max\{0, \alpha_w - \tau_{vw} - \alpha_v\} u_{vw}.$$

The intuition behind Definition 3.3 is as follows: a node $v \in V$ is on the *source side* of the cut over time from time α_v on. Before time α_v, it is on the *sink side*. In particular, within the time interval $[0, \theta)$, sources in S^+ are always on the source side and sinks in S^- are always on the sink side. Therefore, any flow particle traveling from a source to a sink during that time interval must eventually cross the cut over time from the source to the sink side. The only possibility to do so is to enter an arc vw while v is already on the source side, that is, after time α_v, and to arrive at w while w is still on the sink side, that is, before time α_w. Since it takes τ_{vw} time to traverse arc vw, the critical time interval to enter is $[\alpha_v, \alpha_w - \tau_{vw})$. As the length of the critical time interval is $\max\{0, \alpha_w - \tau_{vw} - \alpha_v\}$ and the flow rate entering arc vw is bounded by its capacity u_{vw}, the total amount of flow that can cross the cut over time from the source to the sink side is bounded by its capacity.

Theorem 3.4 (Anderson et al. [2], Philpott [47]) *For a given network N, the capacity of a cut over time with time horizon θ is an upper bound on the maximum amount of flow that can be sent from S^+ to S^- by a feasible flow over time with time horizon θ.*

With this result at hand, it is not difficult to prove the optimality of the flow over time computed by Algorithm 1.

Proof [of Theorem 3.2] According to (3.1), $-c(x)$ is a lower bound on the maximum flow value. It thus remains to show that it is also an upper bound. This can be done via a suitable cut over time. The value $-c(x)$ of a static minimum cost circulation x in network N^{S^+} is the optimum solution value of the following linear programming (LP) formulation (notice that the LP only uses variables x_a corresponding to the original (i.e., positive capacity) arcs a in N:

$$\max \; -\sum_a \tau_a x_a$$
$$\text{s.t.} \quad \sum_{a \in \delta^+(v)} x_a - \sum_{a \in \delta^-(v)} x_a = 0 \quad \text{for all } v,$$
$$x_a \leq u_a \quad \text{for all } a,$$
$$x_a \geq 0 \quad \text{for all } a.$$

To find a suitable cut over time, we consider the corresponding dual linear program:

$$\min \; \sum_a u_a y_a$$
$$\text{s.t.} \quad y_a + \alpha_v - \alpha_w \geq -\tau_a \quad \text{for all } a = vw, \qquad (3.2)$$
$$y_a \geq 0 \quad \text{for all } a.$$

For an optimum dual solution it holds that $y_a = \max\{0, \alpha_w - \tau_a - \alpha_v\}$ for all $a = vw$. It only remains to show that the optimal dual values (α_v) constitute a cut over time as in Definition 3.3, whose capacity is then equal to the optimum value $-c(x)$ of the primal and dual linear program.

Notice that increasing or decreasing all dual variables α_v by the same amount affects neither feasibility nor the dual solution value. We may therefore assume that $\alpha_\psi = 0$. Since all arcs ψs, $s \in S^+$, and $t\psi$, $t \in S^-$, have infinite capacity, complementary slackness implies that their dual variables $y_{\psi s}$ and $y_{t\psi}$ are zero. Thus, the dual constraints (3.2) imply $\alpha_s = \alpha_s - \tau_{\psi s} \leq \alpha_\psi = 0$ for all $s \in S^+$. Similarly, for $t \in S^-$ the dual constraint (3.2) implies $\alpha_t \geq \alpha_\psi - \tau_{t\psi} = 0$. □

4 Submodularity of the Maximum Flow Over Time Function

In the previous section we have studied the problem of finding in the given network N a flow over time with time horizon θ that maximizes the amount of flow sent from the sources S^+ to the sinks S^-. In the context of transshipments over time, we are also interested in how much flow can be sent from a particular subset of sources to a particular subset of sinks within time θ. More precisely, for some subset of terminals $X \subseteq S$, we denote by $o(X)$ the maximum amount of flow that can be sent from $S^+ \cap X$ to $S^- \setminus X$ within time horizon θ. As a consequence of Theorem 3.2, the value $o(X)$ can be obtained via one static minimum cost circulation computation in network N^X.

Corollary 4.1 *For $X \subseteq S$, the maximum amount of flow $o(X)$ that a feasible flow over time with time horizon θ can send from $S^+ \cap X$ to $S^- \setminus X$ is equal to $-c(x)$ where x is a static minimum cost circulation in network N^X.*

For the discrete time model, the values $o(X)$ are minimum cut capacities in the time-expanded network. In this model, the corresponding *maximum flow over time function* $o : 2^S \to \mathbf{R}$ is thus a cut function, and as such submodular; see Megiddo [44]. For the continuous time model with arbitrary (irrational) transit times and time horizon, however, this line of argument cannot be directly applied and we therefore present a different kind of proof.

Theorem 4.2 *The function $o : 2^S \to \mathbf{R}$ is submodular.*

Proof For any $X \subseteq S$ and terminals $r_1, r_2 \in S$ with $r_1 \neq r_2$ we need to argue that

$$o(X \cup \{r_1, r_2\}) - o(X \cup \{r_1\}) \leq o(X \cup \{r_2\}) - o(X).$$

We distinguish four cases, depending on whether r_1 and r_2 are sources or sinks.

First case: $r_1, r_2 \in S^+$. Let x be a minimum cost circulation in network N^X and y a minimum cost circulation in $N_x^{X \cup \{r_1\}}$. Notice that $N^{X \cup \{r_1\}}$ compared to N^X has the additional arc ψr_1 which might create negative cycles in the residual network $N_x^{X \cup \{r_1\}}$ that must be canceled by y. Then, $x + y$ is a minimum cost circulation in $N^{X \cup \{r_1\}}$ and thus

$$o(X \cup \{r_1\}) = -c(x+y) = -c(x) - c(y)$$

by linearity of the cost function. Moreover, let z be a minimum cost circulation in $N_{x+y}^{X\cup\{r_1,r_2\}}$. Again, $N^{X\cup\{r_1,r_2\}}$ compared to $N^{X\cup\{r_1\}}$ has the additional arc ψr_2 which might create negative cycles in the residual network $N_{x+y}^{X\cup\{r_1,r_2\}}$ that must be canceled by z. Then, $x+y+z$ is a minimum cost circulation in $N^{X\cup\{r_1,r_2\}}$, and therefore

$$o(X\cup\{r_1,r_2\}) = -c(x+y+z) = -c(x) - c(y) - c(z),$$

again by linearity of the cost function. Furthermore, by construction, $y+z$ is a feasible circulation in $N_x^{X\cup\{r_1,r_2\}}$ which can be decomposed into flow along cycles in $N_x^{X\cup\{r_1,r_2\}}$. The circulation given by the flow along cycles containing arc ψr_1 is denoted by y', the circulation given by flow along all remaining cycles is denoted by z'. Since circulation y' does not send any flow along arc ψr_2 in $N_x^{X\cup\{r_1,r_2\}}$, it can be interpreted as a feasible circulation in $N_x^{X\cup\{r_1\}}$. Thus, since y is a minimum cost circulation in $N_x^{X\cup\{r_1\}}$, we get $c(y') \geq c(y)$. This inequality then implies $c(z') \leq c(z)$, since $y+z = y'+z'$, and thus $c(y)+c(z) = c(y')+c(z')$. Moreover, by choice of y', circulation z' does not send any flow along arc ψr_1 in $N_x^{X\cup\{r_1,r_2\}}$ and can thus be interpreted as a feasible circulation in $N_x^{X\cup\{r_2\}}$. In particular, $c(z')$ is an upper bound on the cost of a minimum cost circulation in $N_x^{X\cup\{r_2\}}$, that is,

$$-c(z') \leq o(X\cup\{r_2\}) - o(X).$$

Putting everything together yields

$$o(X\cup\{r_1,r_2\}) - o(X\cup\{r_1\}) = -c(z) \leq -c(z') \leq o(X\cup\{r_2\}) - o(X).$$

Second case: $r_1, r_2 \in S^-$. We may use symmetric arguments as in the first case above.

Third case: $t := r_1 \in S^-, s := r_2 \in S^+$. Let x be a minimum cost circulation in N^X and $x+y$ a minimum cost circulation in $N^{X\cup\{t\}}$. Notice that network $N^{X\cup\{t\}}$ compared to N^X is missing arc $t\psi$. Therefore, circulation y must cancel out the entire flow that x sends through that arc. That is, y is a feasible circulation in the residual network N_x^X with $y_{\psi t} = x_{t\psi}$ of minimum cost. In particular, $(x+y)_{t\psi} = 0$ such that $x+y$ can indeed be interpreted as a feasible circulation in $N^{X\cup\{t\}}$ and therefore also in network $N^{X\cup\{t,s\}}$, which in addition contains arc ψs.

Let z be a minimum cost circulation in $N_{x+y}^{X\cup\{t,s\}}$, such that $x+y+z$ is a minimum cost circulation in $N^{X\cup\{t,s\}}$, and therefore

$$o(X\cup\{t,s\}) = -c(x+y+z) = -c(x) - c(y) - c(z)$$

by linearity of the cost function. By construction, $y+z$ is a feasible circulation in $N_x^{X\cup\{s\}}$ which can be decomposed into flow along cycles in $N_x^{X\cup\{s\}}$. The circulation given by the flow along cycles containing arc ψt is denoted by y', the circulation given by flow along all remaining cycles is denoted by z'. Since circulation y' does not send any flow along arc ψs in $N_x^{X\cup\{s\}}$, it can be interpreted as a feasible circulation in N_x^X. Moreover, $y'_{\psi t} = x_{t\psi}$, and since y is a minimum cost circulation with this property in N_x^X, we get $c(y') \geq c(y)$. This implies $c(z') \leq c(z)$, since $y+z = y'+z'$

and thus $c(y) + c(z) = c(y') + c(z')$. Moreover, $c(z')$ is an upper bound on the cost of a minimum cost circulation in $N_x^{X \cup \{s\}}$, that is,

$$-c(z') \leq o(X \cup \{s\}) - o(X).$$

Putting everything together yields

$$o(X \cup \{t, s\}) - o(X \cup \{t\}) = -c(z) \leq -c(z') \leq o(X \cup \{s\}) - o(X).$$

Fourth case: $r_1 \in S^+, r_2 \in S^-$. We may again use symmetric arguments as in the third case above. □

As discussed in the beginning of this section, for the case of integral (or rational) transit times, the maximum flow over time function $o: 2^S \to \mathbf{R}$ is a cut function in a time-expanded network. Even for the case of arbitrary (irrational) transit times, however, one can argue that $o: 2^S \to \mathbf{R}$ is the cut function of a more elaborate time-expanded network [42].

5 Earliest Arrival Flows

In Section 3 we showed that there is always a maximum flow over time that belongs to the structurally simple class of temporally repeated (or standard chain decomposable) flows. Unfortunately, other flow over time problems, including the transshipment over time problem, generally only allow for solutions that feature a more complex temporal structure and therefore require a richer class of flows over time. An illustrative example are earliest arrival flows, which are a fascinating refinement of maximum flows over time.

Definition 5.1 An *earliest arrival flow* in N is a feasible flow over time maximizing the amount of flow that has arrived at the sinks S^- at each point in time $\theta' \in [0, \theta]$ simultaneously.

The existence of earliest arrival flows has been first established by Gale [21] for the discrete time model and by Philpott [47] for the continuous time model. In general, however, there is no temporally repeated earliest arrival flow; a small counterexample is given in Figure 2.

Wilkinson [60] and Minieka [46] present algorithms for computing earliest arrival flows, which Fleischer and Tardos [18] analyze in the continuous time setting. These algorithms are based on shortest cycle canceling. More precisely, an earliest arrival flow can be obtained by using a nonstandard chain decomposition Γ of a static minimum cost flow x in N^{S^+} where Γ is the collection of chain flows used by the shortest cycle canceling algorithm; see Algorithm 2. The resulting chain-decomposable flow $[\Gamma]^\theta$ is defined analogously to a standard chain-decomposable (or temporally repeated) flow in Definition 3.1, where a standard chain decomposition Γ was used.

Definition 5.2 Let Γ be a nonstandard chain decomposition of some static circulation in N^{S^+} such that, for each $\gamma \in \Gamma$, C_γ contains arc ψr_γ for some $r_\gamma \in S^+$. Let P_γ denote the path starting at r_γ that is obtained from C_γ by deleting ψ and its incident arcs. For each $\gamma \in \Gamma$, the *chain-decomposable flow* $[\Gamma]^\theta$ *with time horizon* θ sends flow at rate $|\gamma|$ into path P_γ during the time interval $[0, \theta - \tau(P_\gamma))$.

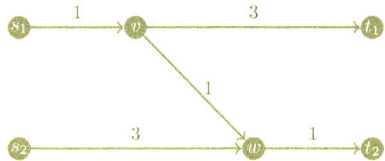

Figure 2: Network with two sources $S^+ = \{s_1, s_2\}$, two sinks $S^- = \{t_1, t_2\}$, and unit arc capacities; numbers at arcs indicate transit times; a maximum flow over time with time horizon $\theta = 4$ has value 1 and must send flow at rate 1 into path s_1, v, w, t_2 during time interval $[0,1)$; a maximum temporally repeated flow with time horizon $\theta = 6$, however, has value 4 and must send flow at rate 1 into both paths s_1, v, t_1 and s_2, w, t_2 during time interval $[0,2)$; notice that there is no temporally repeated flow that is maximum for $\theta = 4$ and $\theta = 6$ simultaneously.

Algorithm 2 Earliest arrival flow algorithm
Require: network N
Ensure: earliest arrival flow
 1: $x^0 \leftarrow$ zero flow in N^{S^+}
 2: $\Gamma^0 \leftarrow \emptyset$
 3: $i \leftarrow 0$
 4: **while** there is a negative cycle in $N^{S^+}_{x^i}$ with positive residual capacity **do**
 5:     find shortest such cycle C^{i+1}
 6:     let γ^{i+1} be the static chain flow in $N^{S^+}_{x^i}$ saturating C^{i+1}
 7:     $x^{i+1} \leftarrow x^i + \gamma^{i+1}$
 8:     $\Gamma^{i+1} \leftarrow \Gamma^i \cup \{\gamma^{i+1}\}$
 9:     $i \leftarrow i + 1$
10: **end while**
11: $\Gamma \leftarrow \Gamma^i$
12: **return** $[\Gamma]^\theta$

This definition, however, raises certain questions which we first illustrate for the example network in Figure 2.

Given the network in Figure 2 with time horizon $\theta = 4$, Algorithm 2 in its first iteration sends a chain flow γ_1 of value $|\gamma_1| = 1$ along path $P_1 := s_1, v, w, t_2$ (or rather along cycle $C_{\gamma_1} := \psi, P_1, \psi$). In its second iteration, it sends a chain flow γ_2 of value $|\gamma_2| = 1$ along path $P_2 = s_2, w, v, t_1$ (i.e., along cycle $C_{\gamma_2} := \psi, P_2, \psi$). Notice that P_2 contains the backward arc wv with negative transit time $\tau_{wv} = -\tau_{vw} = -1$. The resulting chain-decomposable flow $[\{\gamma_1, \gamma_2\}]^4$ is illustrated in Figures 3 and 4. In Figure 3 we illustrate the temporal development of flow along the two paths P_1 and P_2 separately. The interesting snapshot is at time $\theta' = 3$ when three copies of the single flow unit traveling along path P_2 can be seen. This is due to the negative transit time of arc wv which enables the flow unit to travel simultaneously along arcs s_2w and vt_1. But then again, the third copy of the flow unit depicted on arc wv should be rather interpreted as a negative flow unit traveling along the opposite arc vw such that, in total, there is still only one unit of flow.

Obviously, due to the discussed time travel issue, the flow over time along path P_2

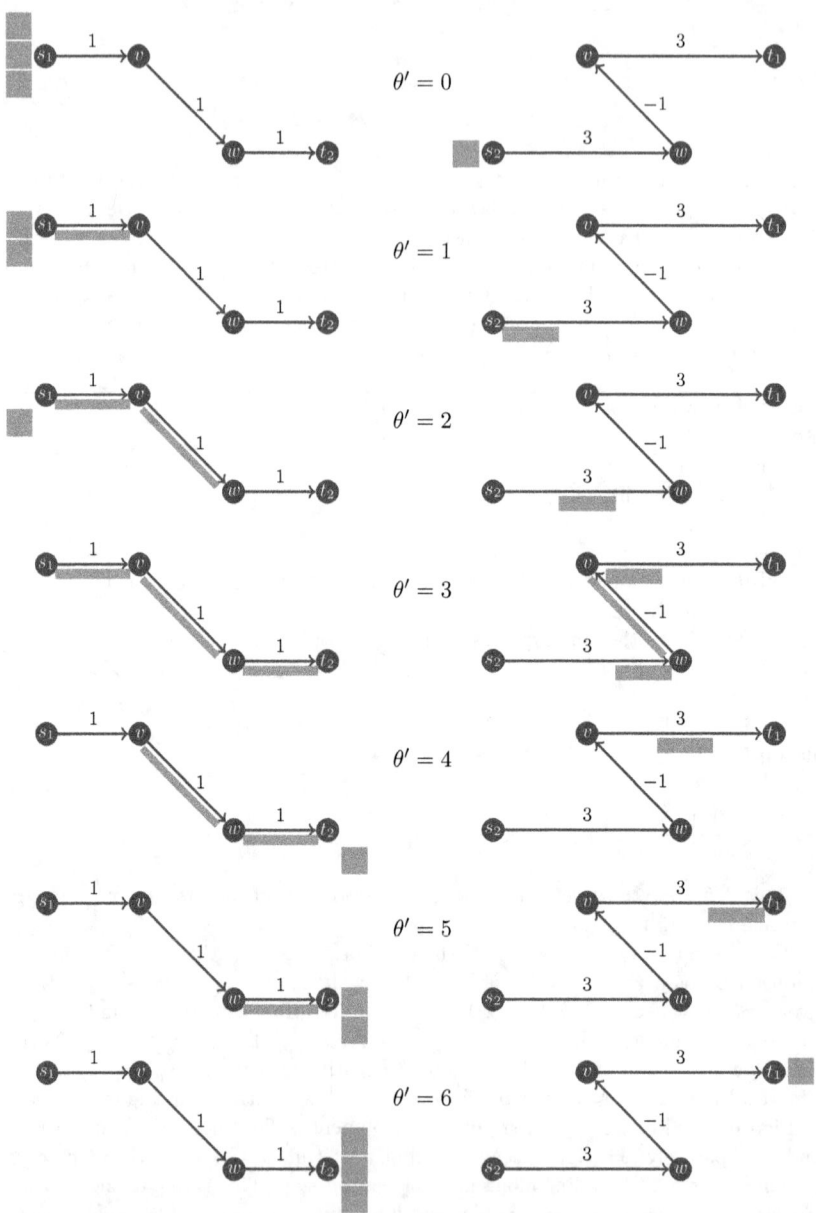

Figure 3: Sending flow at rate 1 into paths $P_1 = s_1, v, w, t_2$ and $P_2 = s_2, w, v, t_1$ during time intervals $[0, 3)$ and $[0, 1)$, respectively

depicted in Figure 3 is not a feasible flow over time on its own. Together with the flow traveling along path P_1, however, we obtain a feasible flow over time since the negative flow unit on arc vw then cancels out a positive flow unit that simultaneously travels along that arc as part of its path P_1; see also Figure 4.

In general, a chain-decomposable flow can only be feasible if flow traveling along a backward arc is always met by flow traveling along the corresponding forward arc at the same (or higher) rate. In Figure 5 we give an example of a chain decomposition that does not meet this condition and therefore yields an infeasible chain-decomposable flow. The condition is, however, always fulfilled for the chain-decomposable flow found by Algorithm 2 as the shortest path distance from ψ to any node v of $N_{x^i}^{S^+}$ is non-decreasing over the iterations i. This follows from the well-known result that augmenting flow along a shortest s-t-path in a residual graph can neither decrease the shortest path distances from s nor the shortest path distances to t; see, e.g., Korte and Vygen [38, Chapter 9.4].

Theorem 5.3 (Wilkinson [60], Minieka [46]) *The chain-decomposable flow* $[\Gamma]^\theta$ *returned by Algorithm 2 is an earliest arrival flow.*

Proof We have already argued that $[\Gamma]^\theta$ is a feasible flow over time. In order to show that the amount of flow arriving at the sinks before some point in time $\theta' \leq \theta$ is optimal, let i be the first iteration such that $N_{x^i}^{S^+}$ does not contain a cycle with positive residual capacity of length at most $\theta' - \theta$. Then x^i is a minimum cost circulation in N^{S^+} for time horizon θ' with chain decomposition Γ^i. Moreover, the amount of flow arriving at the sinks in $[\Gamma]^\theta$ before time θ' is equal to $|[\Gamma^i]^{\theta'}|$. As for the case of standard chain decompositions in (3.1), we again obtain

$$|[\Gamma^i]^{\theta'}| = \sum_{\gamma \in \Gamma^i} |\gamma|(\theta' - \tau(P_\gamma)) = -c(x^i).$$

Theorem 3.2 thus implies that the amount of flow arriving at the sinks in $[\Gamma]^\theta$ before time θ' is maximal. □

Notice that the worst case running time of Algorithm 2 is not polynomial in the input size since shortest cycle canceling performed in its while loop might require exponentially many iterations; see Zadeh [61]. Only recently, Disser and Skutella [12] observed that an earliest arrival flow is indeed NP-hard to find in some sense.

The flow over time found by Algorithm 2 is not only an earliest arrival flow but also a *latest departure flow* with time horizon θ. That is, it maximizes the amount of flow leaving the sources S^+ after time θ' for any θ' with $0 \leq \theta' \leq \theta$.

In the context of earliest arrival flows it is also interesting to consider flows over time with infinite time horizon. Notice that, independently of how large the time horizon θ is chosen, Algorithm 2 terminates after finitely many iterations as soon as there is no more source sink path with positive residual capacity in $N_{x^i}^{S^+}$. The final chain decomposition Γ then yields a flow over time $[\Gamma]$ with infinite time horizon maximizing the amount of flow that has arrived at the sinks for any time $\theta \geq 0$.

Definition 5.4 Let Γ be a nonstandard chain decomposition of some static circulation in N^{S^+} such that, for each $\gamma \in \Gamma$, C_γ contains arc ψr_γ for some terminal $r_\gamma \in S$.

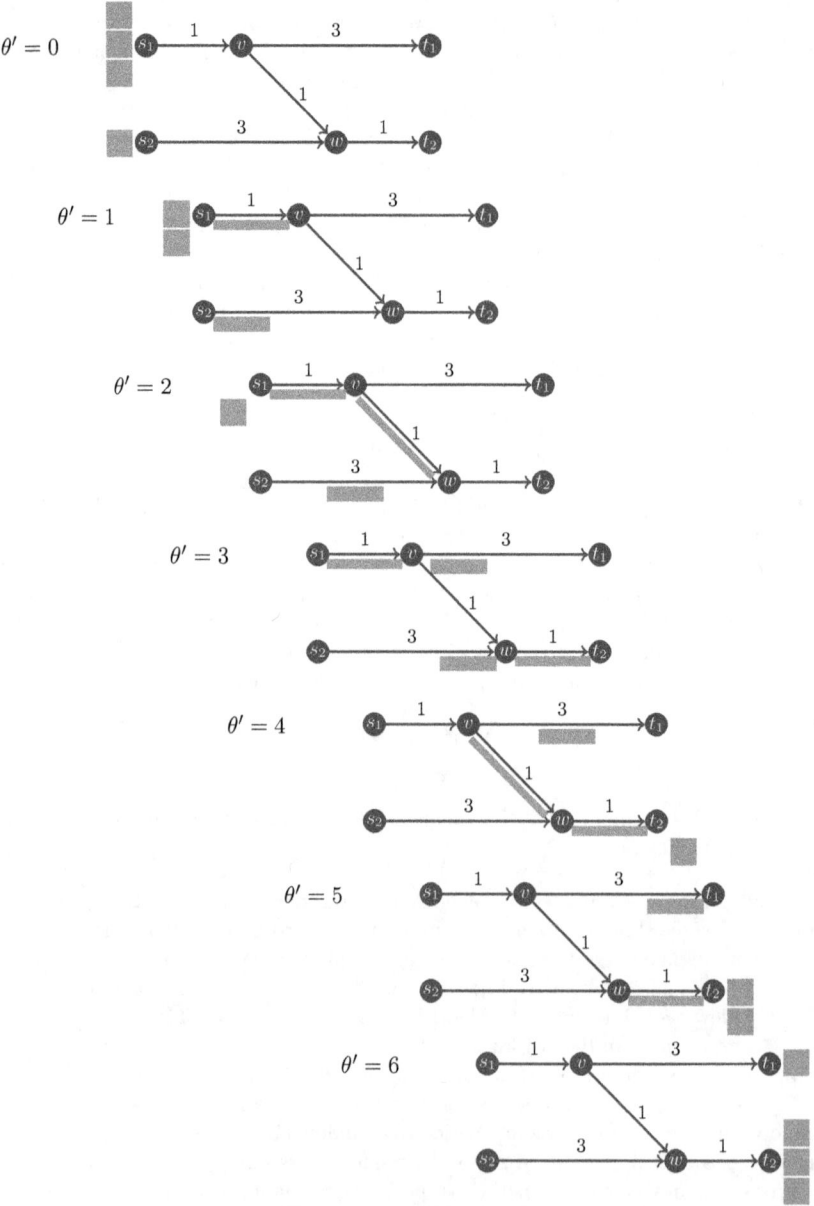

Figure 4: Chain-decomposable earliest arrival flow with time horizon $\theta = 6$, obtained by adding the flow along paths $P_1 = s_1, v, w, t_2$ and $P_2 = s_2, w, v, t_1$ depicted in Figure 3

An Introduction to Transshipments Over Time

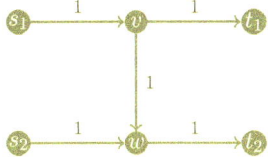

Figure 5: Unit capacity network with sources $S^+ = \{s_1, s_2\}$, sinks $S^- = \{t_1, t_2\}$, and unit transit times; consider the static flow x with $x_{vw} = 0$ and $x_a = 1$ for all other arcs a; a nonstandard chain decomposition Γ may send one unit of flow along each of the paths $P_1 = s_1, v, w, t_2$ and $P_2 = s_2, w, v, t_1$; for $\theta \geq 3$, the corresponding chain-decomposable flow $[\Gamma]^\theta$, however, is not feasible since P_2 uses backward arc wv before P_1 can send flow along the corresponding forward arc vw.

Let P_γ denote the path starting at r_γ that is obtained from C_γ by deleting ψ and its incident arcs. For each $\gamma \in \Gamma$, the *chain-decomposable flow* $[\Gamma]$ *with infinite time horizon* sends flow at rate $|\gamma|$ into path P_γ during the time interval $[\tau_{\psi r_\gamma}, \infty)$.

Further Results on Earliest Arrival Flows From the Literature The following short literature review is an updated version of the corresponding review given in [57]. In view of the exponential worst case running time of Algorithm 2, Hoppe and Tardos [30] present a fully polynomial-time approximation scheme for the earliest arrival flow problem that is based on a clever scaling trick. For a fixed $\varepsilon > 0$, the computed flow over time has the following property. For all $\theta \geq 0$ simultaneously, the flow value at time θ is at least $1 - \varepsilon$ times the maximum possible value for θ.

In a network with given supplies and demands at the terminals that may not be exceeded, flows over time having the earliest arrival property do not necessarily exist; see Fleischer [17]. For the case of several sources with given supplies and a single sink, however, there is always such an earliest arrival flow. In this setting, Hajek and Ogier [26] give the first polynomial time algorithm for computing earliest arrival flows in networks with zero transit times. Fleischer [17] gives an algorithm with improved running time. For the case of multiple sinks with given demands, Schmidt and Skutella [54] characterize which networks with zero transit times always allow for earliest arrival flows. Groß, Kappmeier, Schmidt, and Schmidt [23] present tight approximation results for earliest arrival flows in networks with arbitrary transit times and multiple sinks with given demands. For such networks, Schlöter [51] proves that it is NP-hard to decide whether an earliest arrival flow exists and gives an exact polynomial-space algorithm for a special class of instances.

Baumann and Skutella [5] give an algorithm that computes earliest arrival flows for the case of several sources and arbitrary transit times and whose running time is polynomially bounded in the input plus output size. Schlöter and Skutella [52] present a refined variant of that algorithm that only needs polynomial space. Fleischer and Skutella [16] use so-called condensed time-expanded networks to approximate such earliest arrival flows. They give a fully polynomial-time approximation scheme that approximates the time delay as follows: For every time $\theta \geq 0$ simultaneously, the amount of flow that should have reached the sink in an earliest arrival flow by time θ, reaches the sink at latest at time $(1+\varepsilon)\theta$. Tjandra [59] shows how to compute earliest arrival transshipments in networks with time dependent supplies and capacities in time polynomial in the time horizon and the total supply at sources.

Earliest arrival flows are motivated by applications related to evacuation. In the context of emergency evacuation from buildings, Berlin [6]. Jarvis and Ratliff [33] show that three different objectives of this optimization problem can be achieved simultaneously: (i) Minimizing the total time needed to send the supplies of all sources to the sink, (ii) fulfilling the earliest arrival property, and (iii) minimizing the average time for all flow needed to reach the sink. Hamacher and Tufecki [28] study an evacuation problem and propose solutions which further prevent unnecessary movement within a building. Dressler, Flötteröd, Lämmel, Nagel, and Skutella [13] combine network flow over time with simulation techniques to find good evacuation strategies. Dressler et al. [14] use network flow over time techniques for finding good assignments of evacuees to emergency exits.

6 Lexicographically Maximum Flows Over Time

The efficient computation of lexicographically maximum flows over time is the main building block of all known efficient transshipment over time algorithms.

Definition 6.1 Given an ordering of the k terminals $r_k, r_{k-1}, \ldots, r_2, r_1$, a *lexicographically maximum flow over time* with time horizon θ lexicographically maximizes the flow amounts leaving the terminals in the given order.

As the set of terminals S contains both sources and sinks, notice that maximizing the amount of flow leaving a terminal is equivalent to minimizing the amount of flow entering the terminal. In other words, a lexicographically maximum flow over time lexicographically minimizes the amount of flow entering the terminals in the given order.

Lexicographically maximum static flows have already been studied in the early 1970s by Minieka [46] and Megiddo [44]. They even proved the existence of a feasible flow that simultaneously maximizes the total amount of flow leaving the i highest priority terminals for each $i = 0, 1, \ldots, k-1$. Notice that such a flow is clearly lexicographically maximum. Hoppe and Tardos [31] present an efficient algorithm for computing a lexicographically maximum flow over time that also satisfies this stronger property.

We use the following notation: For a given ordering of terminals $r_k, r_{k-1}, \ldots, r_2, r_1$ and $i \in \{0, 1, \ldots, k-1\}$ let $X_i := \{r_k, r_{k-1}, \ldots, r_{i+1}\}$; moreover, let $X_k := \emptyset$.

Theorem 6.2 (Hoppe, Tardos [31]) *There is an efficient algorithm that, given an ordering of the k terminals $r_k, r_{k-1}, \ldots, r_2, r_1$, finds a feasible flow over time with time horizon θ such that the amount of flow sent from the sources in $S^+ \cap X_i$ to the sinks in $S^- \setminus X_i$ is equal to $o(X_i)$ and thus maximum, for each $i = 0, 1, \ldots, k$.*

Hoppe and Tardos' algorithm consists of a sequence of iterations $i = 1, \ldots, k$, where static minimum cost circulations x^i in networks N^{X_i} are computed. Point of departure is the zero flow x^0 in network $N^{X_0} = N^S$ where all arcs ψs, $s \in S^+$, are present but none of the sinks $t \in S^-$ has an arc $t\psi$. In particular, all arcs in N^{X_0} have non-negative transit time such that the zero flow x^0 is a minimum cost circulation.

An Introduction to Transshipments Over Time

Algorithm 3 Hoppe and Tardos' lex-max flow over time algorithm
Require: ordering of terminals $r_k, r_{k-1}, \ldots, r_2, r_1$
Ensure: lexicographically maximum flow over time
1: $x^0 \leftarrow$ zero flow in N^{X_0}
2: $\Gamma^0 \leftarrow \emptyset$
3: **for** $i \leftarrow 1$ to k **do**
4:     **if** $r_i \in S^-$ **then**
5:         $y^i \leftarrow$ minimum cost circulation in $N^{X_i}_{x^{i-1}}$
6:     **else if** $r_i \in S^+$ **then**
7:         $y^i \leftarrow$ minimum cost circulation in $N^{X_{i-1}}_{x^{i-1}}$ with $y^i_{r_i\psi} = x^{i-1}_{\psi r_i}$
8:     **end if**
9:     $x^i \leftarrow x^{i-1} + y^i$
10:     $\Delta^i \leftarrow$ standard chain decomposition of y^i
11:     $\Gamma^i \leftarrow \Gamma^{i-1} \cup \Delta^i$
12: **end for**
13: $\Gamma \leftarrow \Gamma^k$
14: **return** $[\Gamma]$

In each iteration i, going from network $N^{X_{i-1}}$ to N^{X_i}, either arc $r_i\psi$ is added if r_i is a sink, or arc ψr_i is removed if r_i is a source. In both cases, the previous minimum cost circulation x^{i-1} in $N^{X_{i-1}}$ is augmented into a minimum cost circulation x^i in N^{X_i} by computing an appropriate minimum cost circulation y^i in the residual network of flow x^{i-1}, and setting $x^i := x^{i-1} + y^i$. Moreover, in each iteration the algorithm computes a standard chain decomposition Δ^i of y^i. The flow over time computed by the algorithm is then the chain-decomposable flow $[\Gamma]$ where Γ is the collection of all chain flows in Δ^i for $i = 1, \ldots, k$; see Algorithm 3. We illustrate Algorithm 3 on a small example network in Figures 6 and 7.

Our analysis of Algorithm 3 resembles the original analysis given by Hoppe and Tardos [31]; in particular we use the same or similar notation. There are, however, some crucial differences. While Hoppe and Tardos use cuts in time-expanded networks to prove optimality of the chain-decomposable flow $[\Gamma]$, we provide a shorter and more direct proof based on Corollary 4.1. As a consequence, our analysis works for arbitrary (rational or irrational) time horizon and transit times. Also Fleischer and Tardos [18] observe that the algorithm works for the case of an irrational time horizon, however still assuming integral (or rational) arc transit times.

Before we start to analyze the chain-decomposable flow $[\Gamma]$, we first prove that the circulations x^i in N^{X_i} have indeed minimum cost and Γ^i is a chain decomposition of x^i; notice that this is also stated in Algorithm 3 as comments.

Lemma 6.3 *For $i = 0, \ldots, k$, x^i is a feasible flow and a minimum cost circulation in N^{X_i} with chain decomposition Γ^i; moreover, all chain flows $\gamma \in \Delta^i$ use arc $r_i\psi$.*

Proof We use induction on i. For $i = 0$, the statement is true since all arcs of $N^{X_0} = N^S$ have non-negative transit times, and x^0 is the zero flow with empty chain decomposition Γ^0. In iteration $i \geq 1$, we may assume that x^{i-1} is feasible and a minimum cost circulation in $N^{X_{i-1}}$. In particular, no cycle in the residual graph $N^{X_{i-1}}_{x^{i-1}}$ has negative cost.

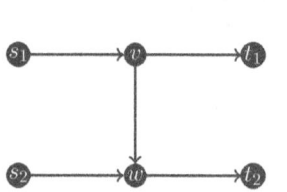
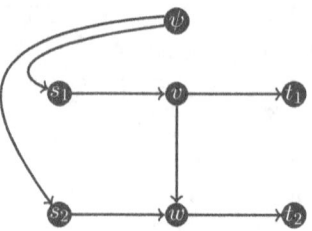

a) Network N with $S^+ = \{s_1, s_2\}$, $S^- = \{t_1, t_2\}$, $u \equiv 1$, and $\tau \equiv 1$

b) Network $N^{X_0} = N^{\{s_1,t_1,s_2,t_2\}}$ with zero flow x^0 before first iteration

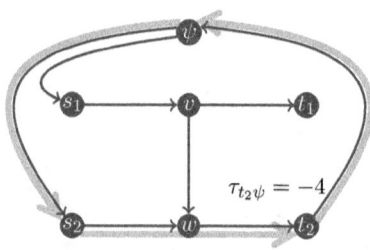
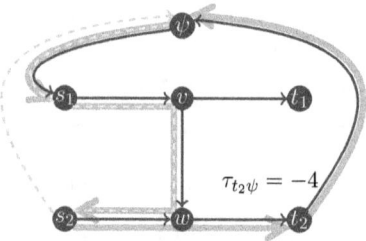

c) Network $N^{X_1} = N^{\{s_1,t_1,s_2\}}$ with flow x^1 and chain decomposition $\Gamma^1 = \Delta^1$ after first iteration;
x^1 is obtained from x^0 by augmenting one unit of flow along cycle $\psi, s_2, w.t_2, \psi$.

d) Network $N^{X_2} = N^{\{s_1,t_1\}}$ with flow x^2 (gray, solid) and chain decomposition $\Gamma^2 = \Delta^1 \cup \Delta^2$ after two iterations;
x^2 is obtained from x^1 by augmenting one unit of flow along cycle $\psi, s_1, v, w.s_2, \psi$ (light gray, dashed).

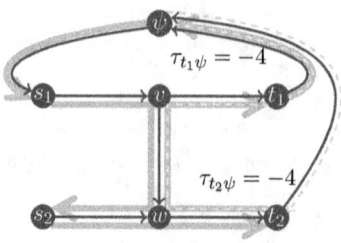
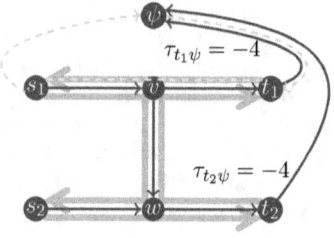

e) Network $N^{X_3} = N^{\{s_1\}}$ with flow x^3 (gray, solid) and the chain decomposition $\Gamma^3 = \Gamma^2 \cup \Delta^3$ after third iteration;
x^3 is obtained from x^2 by augmenting one unit of flow along cycle $\psi, t_2, w, v.t_1, \psi$ (light gray, dashed).

f) Network $N^{X_4} = N^{\emptyset}$ with flow x^4 (gray, solid) and the final chain decomposition $\Gamma^4 = \Gamma^3 \cup \Delta^4$ after fourth iteration;
x^4 is obtained from x^3 by augmenting one unit of flow along cycle $\psi, t_1, v.s_1, \psi$ (light gray, dashed).

Figure 6: Static flows x^i and chain decompositions Γ^i computed by Algorithm 3 for input network N (see Fig. 6a) with unit capacities, unit transit times, time horizon $\theta = 4$, and terminal order s_1, t_1, s_2, t_2.

An Introduction to Transshipments Over Time

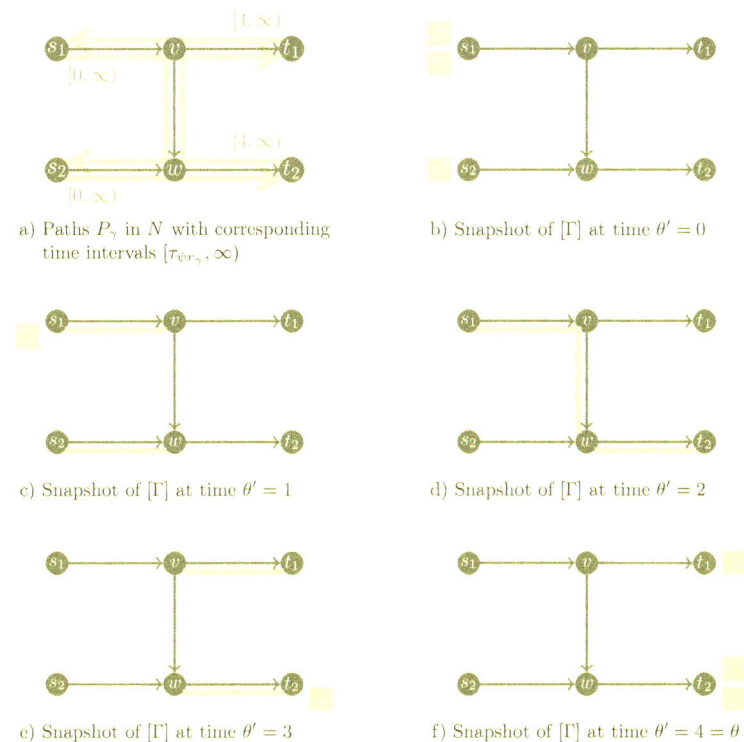

Figure 7: Chain-decomposable flow $[\Gamma]$ computed by Algorithm 3 for input network N and time horizon $\theta = 4$ (cp. Figure 6)

If r_i is a sink, arc $r_i\psi$ is added in iteration i. Then, the minimum cost circulation y^i in $N_{x^{i-1}}^{X_i}$ cancels all negative cycles such that $x^i = x^{i-1} + y^i$ is indeed a minimum cost circulation; moreover, all these negative cycles must contain arc $r_i\psi$.

If r_i is a source, y^i cancels all flow on arc ψr_i such that $x^i = x^{i-1} + y^i$ is indeed a feasible flow in N^{X_i} where arc ψr_i has been removed. Moreover, since y^i is chosen to have minimum cost, no negative cycles can occur in $N_{x^i}^{X_i}$, and x^i is thus a minimum cost circulation. Moreover, since no cycle in the residual graph $N_{x^{i-1}}^{X_{i-1}}$ has negative cost, y^i can only send flow along cycles containing arc $r_i\psi$. $\square$

Notice that the final minimum cost circulation x^k in $N^{X_k} = N^{S^-}$ is the zero flow, since N^{S^-} does not contain cycles of non-positive cost. Therefore $\Gamma = \Gamma^k$ is a chain decomposition of the zero flow. This explains why the resulting chain-decomposable flow $[\Gamma]$ has finite time horizon even though it never stops sending flow into its source sink paths (see Definition 5.4). As we show below, all flow in the network cancels out after time θ such that $[\Gamma]$ actually has time horizon θ.

Moreover, in view of the discussion in Section 5 on canceling flow along backward arcs with negative transit time, we also need to argue that $[\Gamma]$ is indeed a feasible flow over time. To this end, for $i = 0, \ldots, k$ and $v \in V$, let node label $p^i(v)$ denote the minimum cost of a ψ-v-path in $N_{x^i}^{X_i}$.

Lemma 6.4 *The following holds for all $i = 1, \ldots, k$:*

(i) $p^i(v) \geq p^{i-1}(v) \geq 0$ for all $v \in V$.

(ii) For all $\gamma \in \Delta^i$, if γ contains node $v \in V$, then the length of the ψ-v-subpath of γ lies in the interval $[p^{i-1}(v), \min\{p^i(v), \theta\}]$.

In order to prove Lemma 6.4, we will consider the residual network $N_{x^{i-1}}^{X_i}$, but first need to clarify a subtle issue: If $r_i \in S^-$, then network N^{X_i} is obtained from $N^{X_{i-1}}$ by adding arc $r_i\psi$. In particular, x^{i-1} can be naturally interpreted as a circulation in N^{X_i} by simply setting the flow on arc $r_i\psi$ to zero. This is indeed what is meant in Step 5 of Algorithm 3 when the corresponding residual network $N_{x^{i-1}}^{X_i}$ is being considered.

If $r_i \in S^+$, however, network N^{X_i} is obtained from $N^{X_{i-1}}$ by deleting arc ψr_i. Since circulation x^{i-1} may send a positive amount of flow through arc ψr_i, it cannot simply be interpreted as a feasible flow in N^{X_i}. Nevertheless, by slightly overloading notation, we let $N_{x^{i-1}}^{X_i}$ denote the residual network of the flow in N^{X_i} that is obtained from x^{i-1} by ignoring the flow on arc ψr_i, that is, setting it to zero. Notice that the resulting flow in N^{X_i} is no longer a circulation in general but rather an r_i-ψ-flow of value $x_{\psi r_i}^{i-1}$. An alternative way of formulating Step 7 of Algorithm 3 is thus to find a minimum cost ψ-r_i-flow of value $x_{\psi r_i}^{i-1}$ in $N_{x^{i-1}}^{X_i}$ by which the above mentioned r_i-ψ-flow is augmented to yield the minimum cost circulation x^i in N^{X_i}.

Proof [of Lemma 6.4] Let $q^i(v)$ denote the minimum cost of a ψ-v-path in $N_{x^{i-1}}^{X_i}$. Remember that N^{X_i} is obtained from $N^{X_{i-1}}$ by either deleting arc ψr_i (if $r_i \in S^+$) or adding arc $r_i\psi$ (if $r_i \in S^-$). Neither of these changes can create a shorter ψ-v-path for any node v in the residual network $N_{x^{i-1}}^{X_i}$; thus $q^i(v) \geq p^{i-1}(v)$ for all $v \in V$.

As discussed above, circulation x^i can be obtained from x^{i-1} by adding a minimum cost flow in the residual network $N_{x^{i-1}}^{X_i}$. This minimum cost flow can be obtained by repeatedly augmenting flow in the residual network from ψ along a shortest ψ-r_i-path (if $r_i \in S^+$) or cycle (if $r_i \in S^-$), respectively. It is a well-known result in network flow theory that such flow augmentations cannot decrease the shortest path distances from ψ in the residual network; see, for example, [38, Chapter 9.4]. Thus, $p^i(v) \geq q^i(v) \geq p^{i-1}(v)$.

Moreover, $p^0(v) \geq 0$ since network $N_{x^0}^{X_0} = N^S$ only contains arcs of non-negative cost. This implies the non-negativity of the node labels stated in (i) and therefore concludes the proof of this part of the lemma.

In order to prove (ii), we again use the node labels $q^i(v)$. Since the ψ-v-subpath of γ lives in the residual network $N_{x^{i-1}}^{X_i}$, its length is at least $q^i(v) \geq p^{i-1}(v)$. On the other hand, a flow-carrying ψ-v-path of length strictly larger than $p^i(v)$ would induce a negative cycle in $N_{x^i}^{X_i}$, contradicting Lemma 6.3.

It remains to show that the length of the ψ-v-subpath of γ is bounded from above by θ. We distinguish two cases.

First case: x^i sends a positive amount of flow through node v. Since x^i is a minimum cost circulation in N^{X_i}, all flow-carrying cycles in N^{X_i} must have non-positive cost. The only negative cost arcs in N^{X_i} are those entering ψ with transit

An Introduction to Transshipments Over Time   259

time $-\theta$. Therefore, there must exist a flow-carrying v-ψ-path in N^{X_i} whose last arc $r_j\psi$ has transit time $\tau_{r_j\psi} = -\theta$. The corresponding backward ψ-v-path lives in the residual network $N^{X_i}_{x^i}$ and has cost at most θ since ψr_j is its only arc with positive transit time. As a consequence, the length of a shortest ψ-v-path in $N^{X_i}_{x^i}$ is $p^i(v) \leq \theta$. And, as we argued above, $p^i(v)$ is an upper bound on the length of the ψ-v-subpath of γ.

Second case: x^i does not send any flow through node v. Notice that $x^i = x^{i-1} + y^i$ and y^i does send positive flow through v (e.g., along γ). Thus, this flow must cancel out flow that is sent through v by x^{i-1}. Since x^{i-1} is a minimum cost circulation in $N^{X_{i-1}}$, we can use the same arguments as in the first case to show that positive flow through v always implies the existence of a backward path in the residual graph of length at most θ. Since y^i is a minimum cost flow in the residual network $N^{X_i}_{x^{i-1}}$, it (iteratively) augments flow along shortest paths or cycles. In particular, the length of the ψ-v-subpath of γ is bounded from above by θ. □

Lemma 6.4 (ii), in particular, states that flow being sent by the chain decomposable flow $[\Gamma]$ along the source sink path given by chain $\gamma \in \Delta^i \subseteq \Gamma$ arrives at node v from some time θ' on, where $\theta' \in [p^{i-1}(v), \min\{p^i(v), \theta\}]$. With the help of Lemma 6.4, we can now prove feasibility of the chain-decomposable flow $[\Gamma]$.

Theorem 6.5 *The chain-decomposable flow $[\Gamma]$ is a feasible flow over time in N with time horizon θ.*

Proof First of all, $[\Gamma]$ satisfies flow conservation at all non-terminal nodes, since all paths P_γ, $\gamma \in \Gamma$, start and end at a terminal, and all flow arriving at an intermediate node v while traveling along some path P_γ immediately continues its journey on the next arc leaving v.

Next we consider an arc $a = vw$ in N and argue that the flow rate entering arc vw is always within the interval $[0, u_{vw}]$ and equal to 0 before time 0 and from time θ on.

First of all, by Lemma 6.4, the chain-decomposable flow $[\Gamma]$ cannot send any flow into arc vw before time 0. Moreover, after time $\min\{p^k(v), \theta\}$ all flow on arc vw cancels out since from that time on the flow rates on all source sink paths P_γ given by chains $\gamma \in \Gamma$ that contain arc vw sum up to $x^k_{vw} = 0$.

For some point in time $\theta' \in [0, \min\{p^k(v), \theta\}]$, consider i with $\theta' \in [p^i(v), p^{i+1}(v)]$. Then, again by Lemma 6.4, the flow rate entering arc vw at time θ' is within the interval $[\min\{x^i_{vw}, x^{i+1}_{vw}\}, \max\{x^i_{vw}, x^{i+1}_{vw}\}] \subseteq [0, u_{vw}]$. □

We finally analyze the amount of flow that $[\Gamma]$ sends out of a source or into a sink. We use the following notation: For a terminal $r \in S$, let $|[\Gamma]|_r$ denote the net amount of flow leaving r in $[\Gamma]$. Moreover, let $\Gamma_{\psi r}$ and $\Gamma_{r\psi}$ denote the subsets of chain flows in Γ that use arc ψr and arc $r\psi$, respectively.

Lemma 6.6 $|[\Gamma]|_{r_i} = c(\Delta^i)$ *for all* $i = 1, \ldots, k$.

Proof Notice that the only chain flows in Γ that contribute to the net amount of flow leaving terminal r_i in $[\Gamma]$ are those in $\Gamma_{\psi r_i} \cup \Gamma_{r_i \psi}$. Since Γ is a chain decomposition of the zero flow, the chain flows on arcs ψr_i and $r_i \psi$ cancel out, that is,

$$\sum_{\gamma \in \Gamma_{\psi r_i}} |\gamma| = \sum_{\gamma \in \Gamma_{r_i \psi}} |\gamma|. \tag{6.1}$$

Notice that $\Gamma_{r_i \psi} = \Delta^i$ by Lemma 6.3, since every chain flow $\gamma \in \Gamma$ contains exactly one arc entering ψ. We now distinguish two cases.

First case: $r_i \in S^+$. For every chain flow $\gamma \in \Gamma_{\psi r_i}$, the chain-decomposable flow $[\Gamma]$ starts to send flow at rate $|\gamma|$ from source r_i into the network at time 0. Moreover, for every chain flow $\gamma \in \Gamma_{r_i \psi}$, flow at rate $|\gamma|$ arrives at source r_i from time $\tau(\gamma)$ on. Thus, due to equation (6.1), flow rates leaving and entering r_i cancel out from time $\bar{\theta} := \max\{0, \max_{\gamma \in \Gamma_{r_i \psi}} \tau(\gamma)\}$ on. As a consequence, the net amount of flow leaving r_i in $[\Gamma]$ is equal to

$$|[\Gamma]|_{r_i} = \bar{\theta} \sum_{\gamma \in \Gamma_{\psi r_i}} |\gamma| - \sum_{\gamma \in \Gamma_{r_i \psi}} (\bar{\theta} - \tau(\gamma))|\gamma| \stackrel{(6.1)}{=} \sum_{\gamma \in \Gamma_{r_i \psi}} \tau(\gamma)|\gamma| = c(\Gamma_{r_i \psi}) = c(\Delta^i).$$

Second case: $r_i \in S^-$. For every chain flow $\gamma \in \Gamma_{r_i \psi}$, flow at rate $|\gamma|$ arrives at sink r_i from time $\theta + \tau(\gamma)$ on. Moreover, for every chain flow $\gamma \in \Gamma_{\psi r_i}$, flow leaves r_i at rate $|\gamma|$ from time $\theta = \tau_{\psi r_i}$ on. Thus, due to equation (6.1), flow rates entering and leaving r_i cancel out from time $\bar{\theta} := \theta + \max\{0, \max_{\gamma \in \Gamma_{r_i \psi}} \tau(\gamma)\}$ on. As a consequence, the net amount of flow leaving r_i in $[\Gamma]$ is equal to

$$|[\Gamma]|_{r_i} = (\bar{\theta} - \theta) \sum_{\gamma \in \Gamma_{\psi r_i}} |\gamma| - \sum_{\gamma \in \Gamma_{r_i \psi}} (\bar{\theta} - \theta - \tau(\gamma))|\gamma| \stackrel{(6.1)}{=} \sum_{\gamma \in \Gamma_{r_i \psi}} \tau(\gamma)|\gamma| = c(\Delta^i).$$

This concludes the proof. □

Based on Lemma 6.6, it is now easy to prove that $[\Gamma]$ sends the required amount of flow from $S^+ \cap X_i$ to $S^- \setminus X_i$ (cp. Theorem 6.2).

Corollary 6.7 $\sum_{r \in X_i} |[\Gamma]|_r = o(X_i)$ *for all* $i = 0, \ldots, k$.

Proof By flow conservation and Lemma 6.6,

$$\sum_{r \in X_i} |[\Gamma]|_r = \sum_{j=i+1}^{k} |[\Gamma]|_{r_j} = -\sum_{j=1}^{i} |[\Gamma]|_{r_j} = -\sum_{j=1}^{i} c(\Delta^j).$$

Notice that $\bigcup_{j=1}^{i} \Delta^j = \Gamma^i$, which is a chain decomposition of the minimum cost circulation x^i in N^{X_i}. As a consequence

$$\sum_{r \in X_i} |[\Gamma]|_r = -c(x^i) = o(X_i).$$

This concludes the proof. □

We conclude this section with the proof of the main Theorem 6.2 and a remark.

Proof [of Theorem 6.2] By Theorem 6.5, Algorithm 3 computes a feasible flow over time in network N with time horizon θ. Corollary 6.7 states that the computed chain-decomposable flow $[\Gamma]$ is a lexicographically maximum flow over time. Finally notice that the running time of Algorithm 3 is dominated by the k minimum cost circulation computations, one in each iteration of its for-loop, which can be done in strongly polynomial time; see, e.g., [38]. □

Remark Notice that the chain-decomposable flow $[\Gamma]$ computed by Algorithm 3 has the following favorable properties: For integral arc capacities, the values of all chain flows in Γ are integral such that flow rates on arcs are always integral in $[\Gamma]$. Moreover, for the case of integral arc transit times, flow rates on arcs only change at integral points in time in $[\Gamma]$.

7 Transshipments Over Time and Submodular Function Minimization

In this section we turn to the transshipment over time problem that is defined as follows.

Definition 7.1 Given a network N, time horizon θ, supplies $b_r \geq 0$, $r \in S^+$, and demands $b_r \leq 0$, $r \in S^-$, with $\sum_{r \in S} b_r = 0$, a feasible solution to the *transshipment over time problem* is a feasible flow over time with time horizon θ that sends b_r units of flow out of each source $r \in S^+$ and $-b_r$ units into each sink $r \in S^-$.

For the static transshipment problem, already Ford and Fulkerson observed that there is a feasible flow satisfying all supplies and demands if and only if the maximum amount of flow that can be sent from sources in $S^+ \cap X$ to sinks in $S^- \setminus X$ is at least $b(X) := \sum_{r \in X} b_r$. This characterization is a straightforward consequence of the max-flow min-cut theorem in the extended network with a super-source s, connected to all sources $r \in S^+$ via an arc sr of capacity $u_{sr} = b_r$, and a supersink t reachable from all sinks $r \in S^-$ via an arc rt of capacity $u_{rt} = -b_r$.

Unfortunately, the resulting reduction of the static transshipment problem to a static maximum s-t-flow problem cannot be generalized to transshipments over time. Here the problem is that the total amount of flow sent from a super-source through one of the sources $r \in S^+$ into the network cannot be easily bounded by b_r since the capacity of the arc connecting the super-source to r only bounds the flow rate but not the amount of flow sent through this arc.

Nevertheless, Klinz [35] observed that the feasibility criterion mentioned above also holds for the transshipment over time problem.

Theorem 7.2 (Klinz [35]) *A feasible solution to the transshipment over time problem exists if and only if*

$$b(X) \leq o(X) \qquad \text{for each } X \subseteq S. \tag{7.1}$$

While the necessity of Condition (7.1) is straightforward, the sufficiency is less obvious. In the discrete time model considered by Klinz [35], it is merely a consequence of the sufficiency of the cut condition in the time-expanded network. We give a short proof that does not rely on time-expansion and therefore proves the result for networks with arbitrary time horizon and arc transit times.

Proof The two main ingredients of the proof are the submodularity of the function $o : 2^S \to \mathbf{R}$ (Theorem 4.2) and the existence of lexicographically maximum flows over time that simultaneously maximize the total amount of flow leaving the i highest priority terminals for each $i = 0, 1, \ldots, k-1$ (Theorem 6.2). Notice that we gave proofs of both results for networks with arbitrary time horizon and arc transit times.

Consider the *base polytope* $B(o)$ of submodular function $o : 2^S \to \mathbf{R}$ given by

$$B(o) := \{z \in \mathbf{R}^S \mid z(X) \leq o(X) \,\forall\, X \subset S,\ z(S) = o(S)\}$$

Edmonds [15] and Shapley [56] observed that every vertex $\bar{z}$ of the base polytope corresponds to an ordering $r_k, r_{k-1}, \ldots, r_1$ of the terminals from which it can be obtained as follows via a greedy procedure:

$$\bar{z}_i := o(\{r_k, r_{k-1}, \ldots, r_i\}) - o(\{r_k, r_{k-1}, \ldots, r_{i+1}\}) \qquad \text{for } i = k, \ldots, 1.$$

Notice that, by Theorem 6.2, vertex $\bar{z}$ is exactly the vector of supplies and demands satisfied by a lexicographically maximum flow over time.

Moreover, by definition of $B(o)$, a supply and demand vector $b \in \mathbf{R}^S$ with

$$\sum_{r \in S} b_r = 0 = o(S)$$

satisfies Condition (7.1) if and only if $b \in B(o)$. Such a vector b is therefore a convex combination of vertices of $B(o)$. And as such, b is the supply and demand vector satisfied by the corresponding convex combination of lexicographically maximum flows over time. Since a convex combination of feasible flows over time is again feasible, we have thus proved that there exists a feasible solution to the transshipment over time problem, under the assumption that the given supplies and demands satisfy Condition (7.1). $\square$

The proof of Theorem 7.2 raises the question whether and how a feasible solution to the transshipment over time problem can be found, if one exists. First of all, Condition (7.1) can be tested efficiently via submodular function minimization.

Theorem 7.3 (Hoppe, Tardos [31]) *There is a strongly polynomial time algorithm that determines whether a transshipment over time problem has a feasible solution.*

Proof In view of the proof of Theorem 7.2, testing feasibility boils down to testing membership of the supply and demand vector b in the base polytope $B(o)$. It is well known that the separation problem over the base polytope is a submodular function minimization problem; see, e.g., McCormick [43]. More precisely, $b \in B(o)$ if and only if the minimum value of the function

$$X \mapsto o(X) - b(X) \qquad \text{for } X \subseteq S, \tag{7.2}$$

is equal to zero (the minimum value is at most zero since $o(\emptyset) - b(\emptyset) = 0$).

Notice that function (7.2) is submodular, because it is the sum of the submodular function $X \mapsto o(X)$ and the modular function $X \mapsto -b(X)$. Finally, submodular function minimization can be solved in strongly polynomial time. $\square$

The first strongly polynomial algorithm for submodular function minimization is due to Grötschel, Lovász, and Schrijver [25] based on the ellipsoid method. The first combinatorial strongly polynomial algorithms are due to Schrijver [55] and Iwata, Fleischer, and Fujishige [32]; see McCormick [43] for a survey. The combinatorial algorithms of Schrijver and of Iwata, Fleischer, and Fujishige are both based on a maximum flow-style algorithmic framework of Cunningham that earlier led to the first pseudopolynomial algorithm for submodular function minimization [7, 9, 10].

Schlöter and Skutella [52, 50] observe that checking feasibility of a transshipment over time problem, as described in the proof of Theorem 7.3, can directly produce a feasible solution if submodular function minimization is done with an algorithm using Cunningham's framework.

Theorem 7.4 (Schlöter, Skutella [52, 50]) *A feasible solution to the transshipment over time problem can be found via one submodular function minimization with an algorithm using Cunningham's framework, and k calls of Hoppe and Tardos' lexicographically maximum flow over time algorithm (Algorithm 3)*

Proof Cunningham's framework builds on the following strong duality result of Edmonds [15]: For a submodular function $g : 2^S \to \mathbf{R}$

$$\min_{X \subseteq S} g(X) = \max\{z^-(S) \mid z \in B(g)\},$$

where $z^-(S)$ is the sum of all negative entries of vector $z \in B(g) \subset \mathbf{R}^S$. Cunningham's framework finds a subset $X \subseteq S$ minimizing $g(X)$ together with an optimal dual solution $z \in B(g)$ where z is represented by a convex combination of k vertices of $B(g)$; notice that k vertices suffice by Carathéodory's theorem since the dimension of base polytope $B(g)$ is at most $k-1$ due to equality constraint $z(S) = g(S)$.

If, in our case of the submodular function g with $g(X) := o(X) - b(X)$, the minimum function value is 0, an optimal dual solution $z \in B(g)$ has no negative entries. Therefore, since $z(S) = g(S) = o(S) - b(S) = 0$, the only optimal dual solution is the zero vector. In this case, a submodular function minimization algorithm based on Cunningham's framework finds a representation of the zero vector as a convex combination of vertices of $B(g)$. Here, each vertex is given by a linear order on the ground set S.

Moreover, it is not difficult to see that the base polytope $B(g)$ of submodular function g with $g(X) = o(X) - b(X)$ is a translation of base polytope $B(o)$ by vector b, that is, $B(g) = B(o) - b$. Therefore, the representation of the zero vector as a convex combination of k vertices of $B(g)$ yields a representation of b as a convex combination of k vertices of $B(o)$, using the same coefficients.

Thus, what remains to be done algorithmically is to compute lexicographically maximum flows over time for the k linear orders of terminals S corresponding to the k vertices of $B(o)$. Their convex combination, using the same coefficients, is a feasible solution to the transshipment over time problem. □

Remark As we discussed at the very end of Section 6, the lexicographically maximum flows over time found by Algorithm 3 only use integral flow rates if all arc capacities are integral. Unfortunately, this favorable property is not inherited by the transshipment over time in Theorem 7.4 since the coefficients of the convex combination found by the submodular function minimization algorithm will in general be fractional.

The transshipment over time algorithm given by Hoppe and Tardos [31], however, always finds an integral transshipment over time. They employ $2k$ parametric submodular function minimizations to construct a carefully tightened network for which one particular lexicographically maximum flow over time exactly satisfies the supplies and demands and is thus a feasible transshipment over time.

Results on the Quickest Transshipment Problem From the Literature As the title of Hoppe and Tardos' paper [31] suggests, they are not merely interested in the transshipment over time problem but actually solve the more difficult *quickest transshipment problem*. In this problem, the time horizon θ is not given but only the network N, and the task is to determine the minimum time horizon θ together with a feasible transshipment over time with that time horizon.

The *quickest flow problem* is the special case of the quickest transshipment problem with a single source and sink, or alternatively, with several sources and sinks without supplies and demands but only a specified flow value that needs to be sent from the sources to the sinks. Burkard, Dlaska, and Klinz [8] observe that the quickest flow problem can be solved in strongly polynomial time by incorporating Ford and Fulkerson's maximum flow over time algorithm (Algorithm 1) into Megiddo's parametric search framework [45]. Lin and Jaillet [40] present a cost-scaling algorithm that solves the problem in the same running time that Goldberg and Tarjan's cost-scaling algorithm [22] needs to find a static minimum cost flow. Refining Lin and Jaillet's approach, Saho and Shigeno [49] achieve the currently best known algorithm with strongly polynomial running time.

For the quickest transshipment problem, Hoppe and Tardos [31] show how to find the minimum time horizon θ by solving a parametric submodular function minimization problem using Megiddo's parametric search framework [45]. Schlöter, Skutella, and Tran [53] present a sophisticated extension of the discrete Newton method (or Dinkelbach's algorithm [11]) for computing the minimum time horizon which, in terms of its running time, beats the parametric search approach by several orders of magnitude. For the special case of a single source node or a single sink node, an even faster algorithm is given by Schlöter [50] and Kamiyama [34], based on a simpler variant of the discrete Newton method. For the case where each arc also has a cost coefficient (independent of its transit time), it is not difficult to show that finding a quickest transshipment of minimum cost can be reduced to solving a quickest transshipment problem without costs on a carefully chosen subnetwork [58].

Acknowledgements

The author is much obliged to Lizaveta Manzhulina, Tom McCormick, Britta Peis, Miriam Schlöter, and Khai Van Tran for insightful discussions on the topic of this paper. The presentation of the paper has improved thanks to a referee's instructive comments and suggestions.

This work is supported by the Deutsche Forschungsgemeinschaft (DFG, German Research Foundation) under Germany's Excellence Strategy — The Berlin

Mathematics Research Center MATH+ (EXC-2046/1, project ID: 390685689).

References

[1] E. J. Anderson and P. Nash, *Linear programming in infinite-dimensional spaces*, Wiley, New York, 1987.

[2] E. J. Anderson, P. Nash, and A. B. Philpott, *A class of continuous network flow problems*, Mathematics of Operations Research **7** (1982), 501–514.

[3] E. J. Anderson and A. B. Philpott, *Optimisation of flows in networks over time*, Probability, Statistics and Optimisation (F. P. Kelly, ed.), Wiley, New York, 1994, pp. 369–382.

[4] J. E. Aronson, *A survey of dynamic network flows*, Annals of Operations Research **20** (1989), 1–66.

[5] N. Baumann and M. Skutella, *Solving evacuation problems efficiently: Earliest arrival flows with multiple sources*, Mathematics of Operations Research **34** (2009), 499–512.

[6] G. N. Berlin, *The use of directed routes for assessing escape potential*, National Fire Protection Association, Boston, MA, 1979.

[7] R. E. Bixby, W. H. Cunningham, and D. M. Topkis, *The partial order of a polymatroid extreme point*, Mathematics of Operations Research **10** (1985), 367–378.

[8] R. E. Burkard, K. Dlaska, and B. Klinz, *The quickest flow problem*, ZOR — Methods and Models of Operations Research **37** (1993), 31–58.

[9] W. H. Cunningham, *Testing membership in matroid polyhedra*, Journal of Combinatorial Theory, Series B **36** (1984), 161–188.

[10] ———, *On submodular function minimization*, Combinatorica **3** (1985), 185–192.

[11] W. Dinkelbach, *On nonlinear fractional programming*, Management Science **13** (1967), 492–498.

[12] Y. Disser and M. Skutella, *The simplex algorithm is NP-mighty*, ACM Transactions on Algorithms **15** (2019), 5:1–5:19.

[13] D. Dressler, G. Flötteröd, G. Lämmel, K. Nagel, and M. Skutella, *Optimal evacuation solutions for large-scale scenarios*, Operations Research Proceedings 2010 (B. Hu, K. Morasch, S. Pickl, and M. Siegle, eds.), Springer, 2011, pp. 239–244.

[14] D. Dressler, M. Groß, J.-P. Kappmeier, T. Kelter, J. Kulbatzki, D. Plümpe, G. Schlechter, M. Schmidt, M. Skutella, and S. Temme, *On the use of network flow techniques for assigning evacuees to exits*, Proceedings of the International Conference on Evacuation Modeling and Management, Procedia Engineering, vol. 3, Elsevier Ltd, 2010, pp. 205–215.

[15] J. Edmonds, *Submodular functions, matroids and certain polyhedra*, Proceedings of the Calgary International Conference on Combinatorial Structures and Their Applications (R. Guy, H. Hanai, N. Sauer, and J. Schönheim, eds.), 1970, pp. 69–87.

[16] L. Fleischer and M. Skutella, *Quickest flows over time*, SIAM Journal on Computing **36** (2007), 1600–1630.

[17] L. K. Fleischer, *Faster algorithms for the quickest transshipment problem*, SIAM Journal on Optimization **12** (2001), 18–35.

[18] L. K. Fleischer and É. Tardos, *Efficient continuous-time dynamic network flow algorithms*, Operations Research Letters **23** (1998), 71–80.

[19] L. R. Ford and D. R. Fulkerson, *Constructing maximal dynamic flows from static flows*, Operations Research **6** (1958), 419–433.

[20] _____, *Flows in networks*, Princeton University Press, 1962.

[21] D. Gale, *Transient flows in networks*, Michigan Mathematical Journal **6** (1959), 59–63.

[22] A. V. Goldberg and R. E. Tarjan, *Finding minimum-cost circulations by successive approximation*, Mathematics of Operations Research **15** (1990), 430–466.

[23] M. Groß, J.-P. W. Kappmeier, D. R. Schmidt, and M. Schmidt, *Approximating earliest arrival flows in arbitrary networks*, Algorithms - ESA 2012 - 20th Annual European Symposium, Ljubljana, Slovenia, September 10-12, 2012. Proceedings (L. Epstein and P. Ferragina, eds.), Lecture Notes in Computer Science, vol. 7501, Springer, 2012, pp. 551–562.

[24] M. Groß and M. Skutella, *A tight bound on the speed-up through storage for quickest multi-commodity flows*, Operations Research Letters **43** (2015), 93–95.

[25] M. Grötschel, L. Lovász, and A. Schrijver, *Geometric Algorithms and Combinatorial Optimization*, Algorithms and Combinatorics, vol. 2, Springer, 1988.

[26] B. Hajek and R. G. Ogier, *Optimal dynamic routing in communication networks with continuous traffic*, Networks **14** (1984), 457–487.

[27] A. Hall, S. Hippler, and M. Skutella, *Multicommodity flows over time: Efficient algorithms and complexity*, Theoretical Computer Science **379** (2007), 387–404.

[28] H. W. Hamacher and S. Tufecki, *On the use of lexicographic min cost flows in evacuation modeling*, Naval Research Logistics **34** (1987), 487–503.

[29] B. Hoppe, *Efficient dynamic network flow algorithms*, Ph.D. thesis, Cornell University, 1995.

[30] B. Hoppe and É. Tardos, *Polynomial time algorithms for some evacuation problems*, Proceedings of the 5th Annual ACM–SIAM Symposium on Discrete Algorithms (SODA 1994), 1994, pp. 433–441.

[31] B. Hoppe and É. Tardos, *The quickest transshipment problem*, Mathematics of Operations Research **25** (2000), no. 1, 36–62.

[32] S. Iwata, L. Fleischer, and S. Fujishige, *A combinatorial strongly polynomial algorithm for minimizing submodular functions*, Journal of the ACM **48** (2001), 761–777.

[33] J. J. Jarvis and H. D. Ratliff, *Some equivalent objectives for dynamic network flow problems*, Management Science **28** (1982), 106–108.

[34] N. Kamiyama, *Discrete Newton methods for the evacuation problem*, Theoretical Computer Science **795** (2019), 510–519.

[35] B. Klinz, *Personal communication cited in [31]*, 1994.

[36] B. Klinz and G. J. Woeginger, *Minimum-cost dynamic flows: The series-parallel case*, Networks **43** (2004), 153–162.

[37] E. Köhler, R. H. Möhring, and M. Skutella, *Traffic networks and flows over time*, Algorithmics of Large and Complex Networks: Design, Analysis, and Simulation (J. Lerner, D. Wagner, and K. A. Zweig, eds.), Lecture Notes in Computer Science, vol. 5515, Springer, 2009, pp. 166–196.

[38] B. Korte and J. Vygen, *Combinatorial optimization: Theory and algorithms*, 6th ed., Springer, 2018.

[39] B. Kotnyek, *An annotated overview of dynamic network flows*, Rapport de recherche 4936, INRIA Sophia Antipolis, 2003.

[40] M. Lin and P. Jaillet, *On the quickest flow problem in dynamic networks – a parametric min-cost flow approach*, Proceedings of the 26th Annual ACM-SIAM Symposium on Discrete Algorithms (SODA 2015) (P. Indyk, ed.), 2015, pp. 1343–1356.

[41] S. E. Lovetskii and I. I. Melamed, *Dynamic network flows*, Automation and Remote Control **48** (1987), 1417–1434, Translated from Avtomatika i Telemekhanika, 11 (1987), pp. 7–29.

[42] L. Manzhulina, *Personal communication*, August 2023.

[43] S. T. McCormick, *Submodular function minimization*, Discrete Optimization (K. Aardal, G.L. Nemhauser, and R. Weismantel, eds.), Handbooks in Operations Research and Management Science, vol. 12, Elsevier, 2005, pp. 321–391.

[44] N. Megiddo, *Optimal flows in networks with multiple sources and sinks*, Mathematical Programming **7** (1974), 97–107.

[45] _____, *Combinatorial optimization with rational objective functions*, Mathematics of Operations Research **4** (1979), 414–424.

[46] E. Minieka, *Maximal, lexicographic, and dynamic network flows*, Operations Research **21** (1973), 517–527.

[47] A. B. Philpott, *Continuous-time flows in networks*, Mathematics of Operations Research **15** (1990), 640–661.

[48] W. B. Powell, P. Jaillet, and A. Odoni, *Stochastic and dynamic networks and routing*, Network Routing (M. O. Ball, T. L. Magnanti, C. L. Monma, and G. L. Nemhauser, eds.), Handbooks in Operations Research and Management Science, vol. 8, North–Holland, Amsterdam, The Netherlands, 1995, pp. 141–295.

[49] M. Saho and M. Shigeno, *Cancel-and-tighten algorithm for quickest flow problems*, Networks **69** (2017), no. 2, 179–188.

[50] M. Schlöter, *Flows over time and submodular function minimization*, Ph.D. thesis, TU Berlin, Institut für Mathematik, 2018.

[51] _____, *Earliest arrival transshipments in networks with multiple sinks*, Proceedings of the 20th International Conference on Integer Programming and Combinatorial Optimization (IPCO 2019) (A. Lodi and V. Nagarajan, eds.), Lecture Notes in Computer Science, vol. 11480, Springer, 2019, pp. 370–384.

[52] M. Schlöter and M. Skutella, *Fast and memory-efficient algorithms for evacuation problems*, Proceedings of the 28th Annual ACM–SIAM Symposium on Discrete Algorithms (SODA 2017) (P. N. Klein, ed.), 2017, pp. 821–840.

[53] M. Schlöter, M. Skutella, and K. V. Tran, *A faster algorithm for quickest transshipments via an extended discrete Newton method*, Proceedings of the 33rd Annual ACM–SIAM Symposium on Discrete Algorithms (SODA 2022) (S. Naor and N. Buchbinder, eds.), 2022, pp. 90–102.

[54] M. Schmidt and M. Skutella, *Earliest arrival flows in networks with multiple sinks*, Discrete Applied Mathematics **164** (2014), 320–327.

[55] A. Schrijver, *A combinatorial algorithm minimizing submodular functions in strongly polynomial time*, Journal of Combinatorial Theory, Series B **80** (2000), 346–355.

[56] L. S. Shapley, *Cores of convex games*, International Journal of Game Theory **1** (1971), 11–26.

[57] M. Skutella, *An introduction to network flows over time*, Research Trends in Combinatorial Optimization (W. Cook, L. Lovász, and J. Vygen, eds.), Springer, 2009, pp. 451–482.

[58] _____, *A note on the quickest minimum cost transshipment problem*, Operations Research Letters **51** (2023), no. 3, 255–258.

[59] S. Tjandra, *Dynamic network optimization with application to the evacuation problem*, Ph.D. thesis, Universität Kaiserslautern, 2003.

[60] W. L. Wilkinson, *An algorithm for universal maximal dynamic flows in a network*, Operations Research **19** (1971), 1602–1612.

[61] N. Zadeh, *A bad network problem for the simplex method and other minimum cost flow algorithms*, Mathematical Programming **5** (1973), 255–266.

Institut für Mathematik, MA 5-2
Technische Universität Berlin
Straße des 17. Juni 136
10623 Berlin, Germany
martin.skutella@tu-berlin.de

Oriented Trees and Paths in Digraphs

Maya Stein

Abstract

Which conditions ensure that a digraph contains all oriented paths of some given length, or even a all oriented trees of some given size, as a subgraph? One possible condition could be that the host digraph is a tournament of a certain order. In arbitrary digraphs and oriented graphs, conditions on the chromatic number, on the edge density, on the minimum outdegree and on the minimum semidegree have been proposed. In this survey, we review the known results, and highlight some open questions in the area.

1 Introduction

A main focus of extremal graph theory is the question how a bound on some invariant of a graph G can ensure that G contains a certain graph H as a subgraph. One of the easiest cases is when H is a tree, or even a path. For instance, a theorem by Erdős and Gallai [28] states that fro all $n, k \in \mathbb{N}$, every n-vertex graph G with more than $(k-1)n/2$ edges contains a path on k edges, and in the famous Erdős-Sós conjecture [27] it is suggested that with this density condition we do not only find the k-edge path but in fact every k-edge tree in G. Another example is Dirac's theorem [22], which establishes a bound on the minimum degree of a graph G that ensures G contains a spanning cycle (or path), and the Komlós–Sárközy–Szemerédi theorem [48] gives a similar bound that ensures G contains any spanning tree of bounded degree.

This survey is centered on the question how results of this flavour translate to digraphs. We will consider sufficient conditions for a digraph or an oriented graph to contain all oriented paths or oriented trees of some fixed order.

A special type of digraph, which has received much attention, is the tournament. (For all notation see Section 2.) Unlike the complete graph K_n, which contains all graphs of the same order n, a tournament does not contain all oriented graphs of the same or of smaller order. For instance, the transitive tournament fails to contain any directed cycle. On the other hand, Rédei's theorem [58] establishes that every tournament contains a spanning directed path. Moreover, every tournament of order greater than 7 contains every oriented path of the same order [41].

There are many more results and open problems concerning subgraphs of tournaments. For containment of oriented trees in tournaments, there is a famous conjecture by Sumner (see [61]), stating that every $2k$-vertex tournament contains each oriented k-edge tree. This conjecture was solved for large n in [50]. Newer variants take the number of leaves of the tree or its maximum degree into account. We will survey these developments in Section 3.

Another prominent class of digraphs, with respect to containment of oriented trees, are oriented graphs of high chromatic number. The famous Gallai-Hasse-Roy-Vitaver theorem, rediscovered several times, states that there is a directed path of length k in every orientation of a graph G if and only if the chromatic number of G is $k+1$. It was conjectured in [15] that every oriented graph of chromatic number $k+1$ should in fact contain *any* oriented path of length k.

Let us turn from containment of oriented paths to containment of oriented trees. In this context, a well-known conjecture of Burr [15] states that any oriented graph D of chromatic number $2k$ should contain any oriented k-edge tree. This would imply Sumner's conjecture. The impact of the chromatic number of an oriented graph on the appearance of oriented paths and trees in it and similar open questions are discussed in Section 4.

Our next focus are conditions on the edge density of the host digraph. That is, we will look for analogues of the Erdős-Sós conjecture or similar statements for digraphs. However, we immediately encounter a huge restriction. Namely, a condition on the edge density alone can only force *antidirected subdigraphs*, that is, subdigraphs without any vertices of both positive outdegree and positive indegree. To see this, consider a complete bipartite graph on two vertex sets A and B and direct all edges from A to B. The obtained digraph has $|A||B|$ edges, and all its subdigraphs are antidirected.

So, unless we wish to impose extremely high bounds on the number of edges of the host digraph, the only oriented trees that can be forced by a density condition are the antidirected trees. A beautiful conjecture in this direction appeared in [2]. It states that every digraph with more than $(k-1)n$ edges contains each antidirected k-edge tree. In a certain sense, this would unify the Erdős–Sós conjecture and Burr's conjecture for antidirected trees. See Section 5 for more details on this conjecture and other results.

We now turn to conditions on analogues of the minimum degree in digraphs. Conditions on the minimum outdegree that imply the existence of oriented paths have been considered in the literature, but these problems appear to be very difficult. For instance, Thomassé conjectured [35, 67] that every oriented graph of minimum outdegree at least k contains a directed path of length $2k$, and this seems wide open (see e.g. [1] for related results concerning subdivisions).

This changes if we add a condition on the minimum indegree. In other words, we will consider conditions on the *minimum semidegree* of a digraph that can ensure it contains oriented paths or trees of a given order. This will be the topic of Section 6.

In contrast to Thomassé conjecture [35, 67] mentioned above, it is known that for any even $k \in \mathbb{N}$, every oriented graph of minimum semidegree at least $k/2$ contains a directed path of length k, this is a theorem of Jackson [42]. A generalisation of this statement to all oriented paths was recently conjectured in [64]: for every $k \in \mathbb{N}$, every oriented graph of minimum semidegree exceeding $k/2$ should contain any oriented path of length k. The analogous question for hosts that are digraphs instead of oriented graphs is more complicated, because a disjoint union of complete digraphs or order k each has minimum semidegree $k-1$ but contains no oriented subgraph with k edges.

We move on to results and open problems concerning semidegree conditions for finding oriented trees. The central result here is the recent generalisation [43] of the Komlós–Sárközy–Szemerédi theorem [48] to digraphs. This result assures that every large enough digraph obeying a certain minimum degree condition contains each oriented tree of the same order, if the maximum total degree of the underlying tree is bounded. Another recent result concerns balanced antitrees of bounded maximum degree in oriented graphs [65], where the antitree may be of lower order than the host.

Oriented Trees and Paths in Digraphs 273

Finally, in Section 7, we investigate another degree condition, the *minimum pseudo-semidegree*. For a non-empty digraph D, this is the maximum k such that each vertex of D has either outdegree 0 or outdegree at least k, and has either indegree 0 or indegree at least k. The motivation for defining the minimum pseudo-semidegree is that, as we will see in Section 7, any digraph of large edge-density has a subgraph of large minimum pseudo-semidegree, and this is not true if we consider the minimum semidegree instead. Moreover, if we wish to find antidirected subgraphs in a digraph, bounds on the minimum semidegree and bounds on the minimum pseudo-semidegree have almost the same effect (see Lemma 7.2). Consequently, some of the results from Section 6 already hold if we consider the minimum pseudo-semidegree instead of the minimum semidegree.

2 Notation

For easy reference, we gather all definitions here, except for a few that are only used once in specific settings. All of the definitions are standard.

Digraphs and Oriented Graphs A *digraph* D consists of a set $V(D)$ of *vertices*, and a set $E(D)$ of *directed edges* (or *edges* for short) which correspond to pairs of distinct vertices. We write ab for the edge (a,b), that is, the edge from a to b, and note that a digraph can have both edges ab and ba.

A digraph is called *symmetric* if $ab \in E(D)$ implies that $ba \in E(D)$. So, there is a one-to-one correspondence between symmetric digraphs and graphs.

An *oriented graph* is a digraph that has at most one edge between any pair of vertices. Thus, an oriented graph D corresponds to an *orientation* of an undirected graph G. We also call G the *underlying graph* of D.

Degrees In a digraph D, the *outdegree* $d^+(v)$ of a vertex v is the number of edges of D that are directed from v to another vertex. The *indegree* $d^-(v)$ of a vertex is defined analogously.

An oriented graph is called ℓ-*regular* if each of its vertices has outdegree ℓ and indegree ℓ. A *regular* tournament is an n-vertex tournament which is $(n-1)/2$-regular, for any odd $n \in \mathbb{N}$. Using induction, it is easy to see that regular tournaments on n vertices exist for each odd $n \in \mathbb{N}$.

The *minimum outdegree* $\delta^+(D)$ of a digraph D is defined as the minimum over all the outdegrees of the vertices, and the *minimum indegree* $d^-(D)$ is defined analogously. The *total degree* of a vertex is the sum of its in- and its outdegree. The *maximum total degree* $\Delta(D)$ of an oriented graph D is defined as the maximum over the total degrees of the vertices, that is, the maximum degree of the underlying undirected graph. Another widely used degree notion is the *minimum semidegree* $\delta^0(D)$ of a digraph D, which is defined as the minimum over all the in- and all the out-degrees of the vertices.

Orientations In graphs, if a path has k edges, we say it is a k-*edge path* or that it has *length* k, and use the same terminology for oriented paths. For $k \in \mathbb{N}$, the *directed* k-*edge path* (or *directed path on k edges*) is the k-edge path having all its edges oriented in the same direction. A *directed cycle* is a cycle with all edges oriented in the same direction.

An *out-arborescence* is an oriented tree having all its edges directed away from a specific vertex. More precisely, an arborescence is an orientation of a tree T with a root r such that an edge xy is directed from x to y if and only if x lies on the unique path from y to r. An *in-arborescence* is defined analogously. Note that a directed path qualifies as both an out- and an in-arborescence

Another example of an out-arborescence is the k-edge star with all its edges directed outwards, that is, away from its centre. This oriented star is denoted by $K_{1\to k}$. Similarly, $K_{1\leftarrow k}$ is the k-edge star with all its edges directed towards its centre.

An oriented graph D is called *antidirected* if no vertex has both positive outdegree and positive indegree. So, any antidirected graph has a bipartition (A, B) which induces the directions on the edges: all edges are directed from A to B. Note that if T is an oriented tree, this bipartition coincides with the natural bipartition of the underlying tree. Finally, a tree, or an oriented tree, is called *balanced* if its bipartiton classes have the same size.

Complete Digraphs and Tournaments A digraph is called *complete* if it contains all possible edges, that is, between each pair of vertices, both edges are present.

An oriented graph is called a *tournament* if it contains exactly one edge between each pair of vertices. The *transitive tournament* is the tournament on vertices $v_1, \ldots, v_n$, with all edges directed towards the endpoint with the larger index. Alternatively, one can define a transitive tournament as a tournament that contains no directed cycles.

Other Notation The *chromatic number* of an oriented graph is the chromatic number of the underlying graph. The chromatic number of an arbitrary digraph D is defined analogously, considering the graph obtained by identifying multiple edges of the underlying multigraph of D.

A vertex or a set of a digraph D is said to *dominate* another set or vertex if all edges from the former to the latter are present. A digraph is *strong* if for each pair x, y of distinct vertices there is a directed path starting in x and ending in y.

A *blow-up* of a digraph D is obtained by replacing each vertex with an independent set of vertices, and adding all edges from such a set X to a set Y, if X and Y originated from vertices x and y belonging to an edge xy of D. We call a blow-up D' of D an ℓ-*blow-up of* D if each original vertex from D was replaced with exactly ℓ vertices in D'.

3 Tournaments

3.1 Paths in Tournaments

The first results on oriented trees in tournaments concern oriented paths, and more specifically, directed paths. One of the earliest results in this respect is Rédei's theorem [58] from 1934, on spanning directed paths in tournaments.

Theorem 3.1 (Rédei 1934 [58]) *For each $k \in \mathbb{N}$, every tournament on $k+1$ vertices contains the directed k-edge path.*

The proof of this theorem is simple: Take a longest directed path $P = v_1 v_2 \ldots v_\ell$, and assume there is a vertex $x \notin V(P)$. Let i be the first index such that xv_i is an edge, and if no such index exists, set $i := \ell + 1$. Then we can insert x between v_{i-1} and v_i (or before v_i, or after v_{i-1} if $i=1$ or $i = \ell+1$) to obtain a longer directed path than P, a contradiction.

Subsequently, containment of other spanning oriented paths was studied. Grünbaum [32] showed in 1971 that each antidirected k-edge path is contained in each tournament on $k+1$ vertices, with three exceptions: the regular tournaments on 3 and 5 vertices, and the Paley tournament on 7 vertices. Let us call the set of these three tournaments $\mathcal{T}^*$. In 1972, Rosenfeld [62] gave an easier proof of Grünbaum's result and showed that moreover, any vertex of the tournament can be chosen as the starting point of the path.

Further, Rosenfeld [62] conjectured that there is a $n > 7$ such that every tournament on at least $k+1 \geq n$ vertices contains all oriented paths on k edges. Progress towards this conjecture was achieved in [5, 29, 61, 66, 75], and finally, in 1986, the conjecture was settled by Thomason [69]. He showed that $n = 2^{128}$ suffices, while expressing the belief that $n=8$ should be the correct number. In 2000, Havet and Thomassé [41] confirmed this, and showed that the only exceptional tournaments not containing all oriented paths of the corresponding length are the ones found by Grünbaum in 1971:

Theorem 3.2 (Havet and Thomassé 2000 [41]) *For each $k \in \mathbb{N}$, every tournament T on $k+1$ vertices contains each oriented k-edge path P, unless $T \in \mathcal{T}^*$ and P is antidirected.*

A shorter proof of Theorem 3.2 was recently given by Hanna [34]. In [41], Havet and Thomassé also obtain the additional result that for each vertex v of the tournament T and each oriented k-edge path P, there is a copy of P in T either starting or ending at v, as long as $|T| \geq 9$.

There are some related questions which have been studied in the literature. To mention a few, the first result on the number of directed Hamilton paths in a tournament is from 1943 and due to Szele [68]. Szele's lower bound is complemented by an upper bound found by Alon [4] in 1990. See e.g. [26] for some more recent results. In a slightly different direction, Thomassen [71] characterised in 1982 the tournaments that do not have two directed edge-disjoint spanning paths. Thomassen [70, 71, 72] also showed several results on spanning directed paths starting or ending at specified vertices, or through specified edges. See the survey [7] for more directions.

3.2 Trees in Tournaments

If instead of any oriented path, we wish to guarantee any oriented tree of a certain size in a tournament, it turns out that we need the tournament to be substantially larger than the tree. In order to see this, let us consider $K_{1 \to k}$, the k-edge star with all its edges directed outwards. For $n < 2k$, there are n-vertex tournaments of maximum outdegree at most $k-1$. Clearly, these do not contain $K_{1 \to k}$ as a subgraph.

So, if we aim for the statement that all tournaments on $f(k)$ vertices contain all oriented k-edge trees, we need to choose $f(k) \geq 2k$.

It is easy to see that $f(k) = 2k$ is sufficient for finding $K_{1 \to k}$, for any $k \in \mathbb{N}^+$, as every tournament on at least $2k$ vertices necessarily has a vertex of outdegree at least k. Clearly, any tournament T on $2k$ vertices also contains the other antidirected star, $K_{1 \leftarrow k}$. In fact, it contains any oriented k-edge star. Indeed, say S is a k-edge oriented star with ℓ 'out-leaves' (vertices of in-degree 1). If T fails to contain S, then each vertex of T either has outdegree less than ℓ or in-degree less than $k - \ell$ (in cannot fulfil both these conditions, as its total degree is $2k - 1$). So $V(T)$ splits into the two sets $V_1 = \{v \in V(T) : d^+(v) < \ell\}$ and $V_2 = \{v \in V(T) : d^-(v) < k - \ell\}$. As V_1 has $|V_1|(|V_1| - 1)/2$ edges, and therefore some $v \in V_1$ sends edges to at least $\lceil (|V_1| - 1)/2 \rceil$ vertices of V_1, we know that $\lceil (|V_1| - 1)/2 \rceil < \ell$ or $|V_1| = 0 = \ell$. It follows that $|V_1| \leq 2\ell - 1$ or $V_1 = \emptyset$. Similarly, we see that $|V_2| \leq 2k - 2\ell - 1$ or $V_2 = \emptyset$. So $2k = |V_1| + |V_2| < 2k$, a contradiction. Note that if $\ell \neq 0 \neq k - \ell$, then this argument actually shows that any tournament on $2k - 1$ vertices contains the k-edge oriented star with ℓ out-leaves.

Sumner conjectured in 1971 that the same bound on the order of the host tournament would be sufficient to find any oriented tree having k edges:

Conjecture 3.3 (Sumner 1971, see [61]) *Every tournament on $2k$ vertices contains every oriented k-edge tree.*

Conjecture 3.3 has also been called 'Sumner's universal tournament conjecture' in the literature. Much of the early progress towards Conjecture 3.3 consists of proofs for specific classes of oriented trees, or by replacing the bound $2k$ from Conjecture 3.3 with a larger number (see for instance [17, 23, 25, 33, 37, 40, 74]).

The first result showing that any tournament on $O(k)$ vertices contains every oriented k-edge tree is due to Häggkvist and Thomason [33]. The currently best bound on the order of the host tournament was established in 2021 by Dross and Havet [23]: They proved that for each $k \in \mathbb{N}$, every tournament on $\lceil 2.625k - 2.9375 \rceil$ vertices contains each oriented k-edge tree. Furthermore, in 2000, Havet and Thomassé [40] showed that Conjecture 3.3 holds for all trees that are arborescences, which could be viewed as an analogue of Rédei's theorem for trees.

In 2011, Kühn, Mycroft and Osthus proved an approximate version [49] of Sumner's conjecture (Conjecture 3.3), and shortly afterwards, the same authors confirmed the conjecture for large k:

Theorem 3.4 (Kühn, Mycroft and Osthus 2011 [50]) *There is a $k_0 \in \mathbb{N}$ such that for every $k \geq k_0$, every tournament on $2k$ vertices contains every oriented k-edge tree.*

Interestingly, for most trees, the host tournament can be much smaller. If we wish to find a random oriented tree in a tournament, we have the following result, which was conjectured in 1988 by Bender and Wormald [9] and was proved thirty years later by Mycroft and Naia [55].

Theorem 3.5 (Mycroft and Naia 2018 [55]) *Let T be chosen uniformly at random from the set of all labelled oriented trees with k edges. Then asymptotically almost surely every tournament on $k + 1$ vertices contains T.*

So there is a large class of trees which allow for a better bound on the size of the host tournament in Conjecture 3.3. Since the known extremal examples for Conjecture 3.3 are the antidirected stars, and in view of Theorems 3.1 and 3.2, which show that Conjecture 3.3 is far from tight for oriented paths, it seems natural to suspect that trees which are very different from stars may be good candidates for such trees. One natural possibility would be to consider a restriction on the maximum degree of the underlying tree T. And indeed, it turns out that for oriented trees of bounded maximum degree, the size of the host tournament can be lowered almost to k.

The first results in this direction are for a type of trees that has been called 'claws' in the literature. A *claw* is a collection of directed paths whose first vertices have been identified. Improving previous results by Saks and Sós [63] and by Lu [51, 52], it was shown by Lu, Wang and Wong [53] in 1998 that every tournament on $k+1$ vertices contains each k-edge claw of maximum degree at most $19(k+1)/50$, and that, on the other hand, there is a family of k-edge claws with maximum degree converging to $11(k+1)/23$ which are not contained in all tournaments on $k+1$ vertices.

Arbitrary trees of bounded maximum degree were first studied in [49]. Kühn, Mycroft and Osthus [49] proved in 2011 that if k is sufficiently large, then every tournament on $(1+o(1))k$ vertices contains every orientation of every k-edge tree T with $\Delta(T) \leq \Delta$, where Δ is some constant. Mycroft and Naia [55] improved this bound to $\Delta(T) \leq (\log k)^{\Delta'}$, where Δ' is a constant. Very recently, Benford and Montgomery gave a linear bound on the maximum degree of the tree.

Theorem 3.6 (Benford and Montgomery [11]) *For each ε, there is a $k_0 \in \mathbb{N}$ and a constant c such that for every $k \geq k_0$, every tournament on $(1+\varepsilon)k$ vertices contains every orientation of every k-edge tree T with $\Delta(T) \leq ck$.*

The following related question first appears explicitly in [55]:

Question 3.7 (Mycroft and Naia [55]) *Is there a function g such that every tournament on $k + g(\Delta)$ vertices contains every orientation of every k-edge tree T with $\Delta(T) \leq \Delta$?*

Mycroft and Naia [55] also ask whether one can choose $g(\Delta) = 2\Delta - 4$ if k is much larger than Δ. Earlier examples by Allen and Cooley (see [49]) show that this would be best possible: Let $h_1, h_2 \in \mathbb{N}$. For this, consider the oriented tree T obtained by taking a directed path $P = v_0 v_1 \ldots v_{h_2}$, and adding $2h_1$ new vertices, of which h_1 vertices send an edge to v_0, and h_1 vertices receive an edge from v_{h_1}. Note that T has $k := 2h_1 + h_2$ edges and $\Delta(T) = h_1 + 1$. Now, take the disjoint union of three tournaments H_1, H_2 and H_3, where H_1 and H_3 are (h_1-1)-regular tournaments on $2h_1 - 1$ vertices, and H_2 is any tournament on h_2 vertices. For $1 \leq i < j \leq 3$, add all edges between H_i and H_j, orienting them from H_i to H_j. It is not hard to see that the resulting tournament H does not contain T, and H has $4h_1 + h_2 - 3 = k + 2\Delta(T) - 5$ vertices.

Looking for conditions that make a tree resemble more a path than a star, one could, instead of restricting the maximum degree of T, impose a condition on the number of leaves of T. In this very natural direction, Häggkvist and Thomason [33] showed in 1991 that there is a function f (which is exponential in ℓ^3) such that every tournament on $k + f(\ell)$ contains each oriented tree with k edges and at most ℓ leaves. Havet and Thomassé proposed in 1996 that it should be possible to drop the size of the tournament from Sumner's conjecture to $k + \ell$ if we are looking for an oriented tree having at most ℓ leaves. In other words, they proposed the following generalisation of Conjecture 3.3.

Conjecture 3.8 (Havet and Thomassé 1996, see [38]) *Let T be an oriented k-edge tree with ℓ leaves. Then every tournament on $k + \ell$ vertices contains a copy of T.*

In 2021, Dross and Havet [23] showed that there is a function f which is quadratic in ℓ such that every tournament on $k + f(\ell)$ vertices contains each oriented tree with k edges and at most ℓ leaves. Recently, Benford and Montgomery [10] proved a linear bound on the number of 'extra' vertices in the tournament.

Theorem 3.9 (Benford and Montgomery 2022 [10]) *There is some C such that every $(k + C\ell)$-vertex tournament contains a copy of any k-edge oriented tree with ℓ leaves.*

Let us turn back to Conjecture 3.8. Clearly, for $K_{1 \to k}$ and $K_{1 \leftarrow k}$, the bound from Conjecture 3.8 coincides with the bound from Conjecture 3.3, and for all other trees, it is lower. Moreover, the bound from Conjecture 3.8 is not tight for small ℓ: As we saw above in Theorem 3.2, if T is a path, i.e. a tree with $\ell = 2$ leaves, and if $k \geq 7$, then already any tournament on at least $k + 1 = k + \ell - 1$ vertices contains T, which is one smaller than required by Conjecture 3.8. Ceroi and Havet [16] showed that for any tree T with $\ell = 3$ leaves, if $k \geq 4$, then every tournament on $k + 2 = k + \ell - 1$ vertices contains T.

Dross and Havet [23] conjectured that the same holds for all trees with few leaves, i.e. they conjectured that the bound of $k + \ell - 1$ is sufficient whenever k, the number of edges of the tree is of sufficiently larger order than ℓ, its number of leaves. This conjecture was confirmed recently by Benford and Montgomery [10]:

Theorem 3.10 (Benford and Montgomery 2022 [10]) *For every ℓ there is a k_0 such that for every $k \geq k_0$, every tournament on $k + \ell - 1$ vertices contains each oriented k-edge tree with ℓ leaves.*

One can understand the number k_0 in Theorem 3.10 being a function of ℓ. The authors of [10] state that their k_0 can be chosen as $\ell^{O(\ell)}$, but also express their belief that the theorem may stay correct with k_0 being of order $O(\ell)$.

4 Chromatic Number

4.1 Chromatic Number and Oriented Paths

Recall that the chromatic number of an oriented graph is the chromatic number of the underlying undirected graph. So the chromatic number of a tournament equals its number of vertices. This motivates the question whether the results from Section 3 do not only apply to tournaments, but more generally to digraphs or oriented graphs of large chromatic number.

An early and quite famous result in this direction is the Gallai-Hasse-Roy-Vitaver theorem (see, e.g. [8]), discovered several times during the 1960's, which we state next. The theorem actually is a duality result.

Theorem 4.1 (Gallai-Hasse-Roy-Vitaver theorem, 1960's) *Let G be a graph. Then G has chromatic number $k+1$ if and only if in every orientation of G there is a directed k-edge path.*

We are interested in the following immediate corollary of Theorem 4.1, which can be seen as a generalisation of Rédei's theorem.

Corollary 4.2 *Every $(k+1)$-chromatic oriented graph contains the directed path with k edges.*

For large k, a version of Corollary 4.2, replacing directed paths with oriented paths hase been proposed by Burr in [15] (see also [17]):

Conjecture 4.3 (Burr 1980) *Every $(k+1)$-chromatic oriented graph contains each oriented path with k edges.*

There is some support for this conjecture, apart from it being true for tournaments. El Sahili [24] proved in 2004 that Conjecture 4.3 holds for $k=3$ and, if the path is antidirected, for $k=4$. Addario-Berry, Havet, and Thomassé [3] proved in 2007 that the conjecture holds for every oriented path with $k \geq 3$ edges that changes direction at most once.

4.2 Chromatic Number and Oriented Trees

Let us now turn to oriented trees in digraphs of large chromatic number. A variant of Conjecture 4.3 replacing oriented paths with oriented stars is not true (we already saw this for tournaments in Section 3.2). However, as in Sumner's conjecture, it seems natural to consider larger host tournaments. In this spirit, Burr [15] suggested the following conjecture, which, if true, would imply Sumner's conjecture.

Conjecture 4.4 (Burr 1980 [15]) *Every $2k$-chromatic digraph contains each oriented k-edge tree.*

Conjecture 4.4 is only known for some specific classes of oriented paths (see above) and for all oriented stars [56]. Burr [15] showed that Conjecture 4.4 is true if we replace $2k$ with k^2. Addario-Berry, Havet, Linhares Sales, Thomassé and Reed [2] improved this bound, roughly by a factor of $1/2$. Naia [57] gives a bound which is better for all oriented graphs with large chromatic number, by showing that every $((k+2)\log_2 n)$-chromatic oriented graph contains each oriented k-edge tree.

A generalisation of Conjecture 3.8 in the spirit of Conjecture 4.4 has been proposed by Havet [36] in 2013, and appears in [23]:

Conjecture 4.5 (Havet 2013 [23, 36]) *Every $(k+\ell)$-chromatic digraph contains each oriented k-edge tree having ℓ leaves.*

In view of the results for tournaments exhibited in the previous section, it seems natural to ask whether the maximum degree of the oriented tree could have a similar influence on the bound on the chromatic number. In analogy to Theorem 3.6 and Question 3.7, it seems natural to ask the following.

Question 4.6 *For each ε, are there is a k_0 and c such that for every $k \geq k_0$, every $(1 + \varepsilon)k$-chromatic digraph contains every orientation of every k-edge tree T with $\Delta(T) \leq ck$?*

Question 4.7 *Is there a function g such that every $(k + g(\Delta))$-chromatic digraph contains every orientation of every k-edge tree T with $\Delta(T) \leq \Delta$?*

We add that Naia [57] explicitly conjectures that for every oriented k-edge tree T, the minimum n such that every tournament of order n contains T coincides with the minimum n such that every n-chromatic oriented graph contains T. We close the section by referring the reader to [6] for a survey of more open problems related to tournaments and containment of trees and other subdigraphs.

5 Edge Density

For graphs, the influence of the edge density of a graph G on the appearance of paths and other trees in G has been much studied. The first result in this direction is a theorem by Erdős and Gallai from 1959 [28], which states that for each $k \in \mathbb{N}$, each graph of average degree greater than $k - 1$ contains the path on k edges. Of course, such a graph also contains the k-edge star.

In 1963, Erdős and Sós (see [27]) suggested that the same should hold for each tree: They conjectured that for each $k \in \mathbb{N}$, each graph of average degree greater than $k-1$ contains each tree with k edges. The Erdős–Sós conjecture has been shown to be true for large dense graphs: in an sharp version for large trees of bounded maximum degree [14], and very recently, in an approximate version for all large trees [19]. See [64] for an overview of this conjecture.

Note that the condition on the average degree in the Erdős–Gallai theorem and in the Erdős–Sós conjecture is best possible as a complete graph on k vertices fails to contain a k-edge path or tree. Also, it is clear that the condition on the average degree can be reformulated as a condition on the edge density of the host graph G, namely the condition that $|E(G)|/|V(G)| > (k - 1)/2$.

If one wishes to extend these statements to digraphs, the first obstacle one encounters is the fact that for all $n \in \mathbb{N}$ there are oriented graphs D on $2n$ vertices and of edge density $|E(D)|/|V(D)| = n/2$ that are antidirected, and thus only have antidirected subgraphs. These are the oriented graphs obtained from the complete bipartite graph $K_{n,n}$, for any $n \in \mathbb{N}$, with all edges directed from the first partition class to the second. So the only oriented k-edge trees that can be guaranteed by imposing a lower bound $f(k)$ on the edge density of the host digraph are the antidirected trees. (This observation is attributed in [15] to de Bruijn, and the explicit example above appears in [2].)

Thus motivated, it becomes natural to ask which edge density is sufficient to guarantee that a digraph contains all antidirected trees of a certain size. In this direction, Graham [31] confirmed in 1970 a conjecture he attributes to Erdős: for every antidirected tree T there is a constant c_T such that every sufficiently large directed graph D on n vertices and with at least $c_T n$ edges contains T. In 1982, Burr [15] gave an improvement of Graham's result:

Theorem 5.1 (Burr [15]) *Every n-vertex digraph D with more than $4kn$ edges contains each antidirected tree T on k edges.*

Burr obtains Theorem 5.1 by taking a maximal cut, which contains at least half of the edges of D, and then omitting all edges from one of the possible directions in this cut. This gives an oriented bipartite subgraph $D' = (A, B)$ of D which has more than kn edges, which are all directed in the same way, say from A to B. Then the average total degree of the vertices of D' exceeds $2k$. A standard argument yields a subgraph D'' of D' such that D'' has minimum total degree at least k. So in D'', each vertex from A has outdegree at least k, and each vertex from B has indegree at least k. Now, one can embed the k-edge tree T greedily into D'', thus completing the proof of Theorem 5.1.

Burr writes in [15] that the bound $4kn$ on the number of edges can 'almost certainly be made rather smaller'. He also provides an example of an oriented graph on n vertices wit $(k-1)n$ edges that fails to contain $K_{1 \to k}$, the k-edge star with all edges directed outwards. Namely, $K_{1 \to k}$ is not contained in the complete bipartite graph $K_{2k-2,2k-2}$ with half of the edges oriented in either direction in an appropriate way. Note that the graph $K_{2k-2,2k-2}$ has $n = 4k-4$ vertices and $(2k-2)^2 = (k-1)n$ edges.

In the previous example, instead of $K_{2k-2,2k-2}$, one could also consider any $(k-1)$-regular tournament on n vertices or the $(k-1)$-blow-up of a directed cycle on n vertices which each have $(k-1)n$ edges but do not contain $K_{1 \to k}$. Moreover, the complete digraph on $n = k$ vertices has $(k-1)n$ edges and does not contain *any* oriented k-edge tree.

In 2013, Addario-Berry, Havet, Linhares Sales, Reed and Thomassé [2] conjectured that the bound $(k-1)n$, which was shown to be necessary by Burr, is indeed the correct bound.

Conjecture 5.2 (Addario-Berry et al. 2013 [2]) *Every digraph D with more than $(k-1)|V(D)|$ edges contains each antidirected k-edge tree.*

It is observed in [2] is that Conjecture 5.2 implies Conjecture 4.4 for antidirected trees, because every $2k$-chromatic graph has a subgraph H of minimum degree at least $2k-1$, and thus H has more than $(k-1)|V(H)|$ edges. Moreover, if we restrict Conjecture 5.2 to symmetric digraphs, then it becomes equivalent to the Erdős–Sós conjecture mentioned at the beginning of this section.

Evidence for Conjecture 5.2 was given in [2], where it is verified for all antidirected trees of diameter at most 3. In [2], it is also noted that in Theorem 5.1 one can replace the antidirected k-edge tree with any antidirected tree whose largest partition class has at most k vertices. In particular, every n-vertex digraph D with more than $2kn$ edges contains each balanced antidirected k-edge tree. For the antidirected path, Klimošová and the author recently improved the bound on the number of edges of D to roughly $3kn/2$.

Theorem 5.3 (Klimošová and Stein 2023 [47]) *For each $k \geq 3$, every oriented n-vertex graph having at least $(3k-4)n/2$ edges contains each antidirected path of length k.*

Also, an approximate version of Conjecture 5.2 for large dense oriented host graphs and balanced antidirected trees of bounded maximum degree holds.

Theorem 5.4 (Stein and Zárate-Guerén [65]) *For all $\eta \in (0,1)$ and $c \in \mathbb{N}$, there is a number $n_0 \in \mathbb{N}$ such that for every $n \geq n_0$ and for every $k \geq \eta n$, every oriented n-vertex graph with more than $(1+\eta)(k-1)n$ edges contains each balanced antidirected tree T with k edges and $\Delta(T) \leq (\log(n))^c$.*

6 Minimum Semidegree

6.1 Minimum Semidegree and Oriented Paths in Oriented Graphs

Many results in extremal graph theory connect bounds on the minimum degree with the existence of certain subgraphs. One prominent example is Dirac's theorem from 1952 [22], which states that any graph G on $n \geq 3$ vertices and of minimum degree at least $n/2$ contains a Hamilton cycle. If we replace the Hamilton cycle with a Hamilton path, i.e. a path with $n-1$ edges, then the condition on the minimum degree of G can be lowered to $(n-1)/2$. For shorter paths, Dirac, and independently Erdős and Gallai (see [28]) observed the following.

Proposition 6.1 *Let $k \in \mathbb{N}$. If a graph G has minimum degree at least $k/2$ and a connected component on at least $k+1$ vertices then G contains the path with k edges.*

Proposition 6.1 can be proved with an argument very similar to one of the standard proofs of Dirac's theorem, taking a longest path, and noting that the vertices of this path actually span a cycle, which can then be used to make the path longer. Observe that the bound on the minimum degree in Proposition 6.1 is a factor of $1/2$ below the bound needed for a greedy embedding argument.

If we wish to consider the same problem for oriented graphs D, we can replace the minimum degree with the minimum semidegree $\delta^0(D)$. Note that if D is any oriented graph of minimum semidegree $\delta^0(D) \geq \frac{k}{2}$, then the vertices of D have total degree at least k, and therefore, each of the components of the underlying undirected graph of D has more than k vertices. This phenomenon might be interpreted as an indication that the extra condition in Lemma 6.1 on the graph having a large component is not necessary for oriented graphs (although for arbitrary digraphs it may still be needed).

Jackson [42] showed a variant of Proposition 6.1 for directed paths in oriented graphs:

Theorem 6.2 (Jackson 1981, Corollary 3 in [42]) *Let $k \in \mathbb{N}$. Every oriented graph D with $\delta^0(D) \geq k$ contains the directed $2k$-edge path.*

Jackson [42] remarks that this bound on the minimum semidegree is best possible, because of the existence of regular tournaments on $2k+1$ vertices. He also obtains slightly better bounds for strongly connected tournaments.

Note that if we wish to find an odd directed path, say of length $2k-1$, Theorem 6.2 ensures that a minimum semidegree of at least k is sufficient. In other words, for a directed path of length k (of either parity), a minimum degree exceeding $k/2$ will always be sufficient.

In this spirit, the author of this survey conjectured in [64] that the following variant of Jackson's result holds for any orientation of the path.

Conjecture 6.3 (Stein [64]) *Let $k \in \mathbb{N}$. Every oriented graph with $\delta^0(D) > k/2$ contains every oriented path with k edges.*

The bound on the minimum semidegree in Conjecture 6.3 is best possible. To see this, we can consider, for odd k, a $(k-1)/2$-regular tournament on k vertices which does not contain any oriented path with k edges. For even k, there are k-vertex tournaments of minimum semidegree $k/2 - 1$.

A different type of example works for even k and antidirected paths. Consider the $k/2$-blow-up of a directed cycle of length $\ell \geq 3$, where each vertex v of C_ℓ is replaced by an independent set S_v of size $k/2$, and S_v, S_w span a complete bipartite graph with all edges directed from S_u to S_v whenever $vw \in E(C_\ell)$. Any largest antidirected path in this graph has k vertices, and thus, length $k-1$, while the minimum semidegree of the graph is $k/2$. Note that for this example it is necessary that $n = \ell k/2$ for some $\ell \geq 3$, in particular, we need $k \leq 2n/3$.

Note that Conjecture 6.3 becomes very easy if the bound on the minimum semidegree is replaced with $\delta^0(D) \geq k$, as then we can embed any oriented path using a greedy strategy. For antidirected paths, Klimošová and the author showed in 2023 [47] that for each $k \in \mathbb{N}$ with $k \neq 2$, every oriented graph D with $\delta^0(D) \geq \frac{3k-2}{4}$ contains each antidirected path with k edges. We remark that the case $k = 2$ is excluded from Theorem 7.4, because the bound $\delta^0(D) \geq (6-2)/4 = 1$ is below the bound from Conjecture 6.3 and is not sufficient to guarantee an antipath of length two in D, since D could be a directed cycle.

Moreover, an asymptotic version of Conjecture 6.3 is true for antidirected paths whose length is linear in the order of the host digraph. Namely, for all $\eta \in (0,1)$ there is n_0 such that for all $n \geq n_0$ and $k \geq \eta n$ every oriented graph D on n vertices with $\delta^0(D) > (1+\eta)\frac{k}{2}$ contains every antidirected path with k edges [65]. This is a direct consequence of Theorem 6.9 below.

Some more evidence for Conjecture 6.3 can be deduced from results on semidegree conditions for oriented cycles in oriented graphs. It was shown by Keevash, Kühn and Osthus [44] that every sufficiently large n-vertex oriented graph G of minimum semidegree $\delta^0(G) \geq \frac{3n}{8} - 4$ contains a directed Hamilton cycle. Moreover, Kelly [45] proved that every oriented graph D of minimum semidegree $\delta^0(D) \geq \frac{3n}{8} + o(n)$ contains every orientation of a Hamilton cycle (which is tight by examples of Häggkvist, see [45], and has been extended to a pancyclicity result in [46]). With every oriented Hamilton cycle, D also contains every oriented spanning path, implying that if k is relatively close to the order of the host, then Conjecture 6.3 is true, although probably not best possible.

Question 6.4 *Let $k, n \in \mathbb{N}$ with $2n/3 < k < n$. What is the smallest function $f(k)$ such that every n-vertex oriented graph with $\delta^0(D) > f(k)$ contains every oriented path with k edges?*

We only ask Question 6.4 for $k > 2n/3$ because for smaller values of k, if $k/2$ divides n, there is the example of the blow-up of the directed $2n/k$-cycle, which fails to contain any antidirected path with k edges.

6.2 Minimum Semidegree and Oriented Paths in Digraphs

Let us start this section with the known results for semidegrees and Hamilton cycles in digraphs. In 1960, Ghoulia-Houri [30] proved that a minimum semidegree of at least $n/2$ suffices to guarantee a directed Hamilton cycle in any n-vertex digraph D. DeBiasio, Kühn, Molla, Osthus and Taylor [20] showed in 2015 that that same is true for other orientations of the Hamilton cycle, except for antidirected Hamilton cycles. The threshold for antidirected Hamilton cycles is $n/2 + 1$, as was shown shortly before by DeBiasio and Molla [21]. All of these results are best possible.

The results from the previous paragraph imply that every n-vertex digraph of minimum semidegree at least $n/2$ contains every oriented Hamilton path, since every oriented path can be completed to an oriented cycle which is not antidirected. On the other hand, an n-vertex digraph of minimum semidegree $n/2 - 1$ could be the disjoint union of two complete digraphs of order $n/2$, and therefore not contain any oriented paths with more than $n/2 - 1$ edges. Similarly, we can consider the disjoint union of complete digraphs of order $k + 1$ to see that for all k and n satisfying the obvious divisibility conditions, a minimum semidegree of k in an n-vertex digraph is not enough to ensure any oriented path with more than k edges.

So, for possible extensions of Conjecture 6.3 to digraphs, it will be necessary to require some other condition in addition to the minimum semidegree condition. For instance, one could ask for a lower bound on the order of the largest component of the underlying graph, or equivalently, require that the underlying graph is connected. The next result shows that for directed paths, this approach works. Actually, the lower bound on the minimum semidegree can be replaced by a weaker condition: the average of the minimum outdegree and the minimum indegree.

Theorem 6.5 (Bermond, Germa, Heydemann and Sotteau 1981 [12]) *Let D be a digraph whose underlying graph is connected. Then D has a directed path of length $\min\{n - 1, \delta^+(D) + \delta^-(D)\}$.*

Unfortunately, an analogue of Theorem 6.5 does not hold for antipaths, not even if we require the underlying graph to be strong and of sufficiently large order. To see this, consider two copies of the complete digraph on $k - 2$ vertices, K_1 and K_2. Add two new vertices v_1 and v_2, and let v_i dominate $V(K_i)$ and be dominated by $V(K_{3-i})$, for each $i = 1, 2$. The obtained digraph is strong, has order $2k - 2$ and minimum semidegree $k - 2$, while its longest antidirected path has only $k - 1$ edges. (Note that for $\ell < k/2$, this example can easily be modified to an ℓ-strong digraph on $2k - 2\ell$ vertices and of minimum semidegree $k - \ell$ whose longest antidirected path has $k - 1$ edges.)

So perhaps a different condition on the host digraph has to be added, if we are looking for oriented paths other than directed paths. For antidirected paths, we propose the following.

Question 6.6 *Let D be a digraph with $\delta^0(D) > k/2$ such that each pair of vertices is connected by an antidirected path. Does D have an antidirected path of length k?*

This would even be interesting if we replace $k/2$ with $f(k)$ for some function f with $k/2 < f(k) < k$. Note that for $\delta^0(D) \geq k$ the greedy embedding strategy yields an embedding of any oriented path with k edges.

In a slightly different direction, there have been attempts to use bounds on the minimum semidegree to find connectivity-preserving directed paths in digraphs. The next conjecture would be an analogue of results of Mader for graphs. A digraph D is called *k-connected* if $|D| \geq k+1$ and for every pair a, b of distinct vertices, there are at least k internally disjoint directed paths starting at a and ending in b.

Conjecture 6.7 (Mader 2012 [54]) *For each $m \in \mathbb{N}^+$, every k-connected digraph D with $\delta^0(D) \geq 2k + m - 1$ contains a k-edge directed path P such that deleting $V(P)$ from D leaves a k-connected digraph.*

For progress on this conjecture, and an overview of corresponding results in graphs, see [73].

6.3 Minimum Semidegree and Oriented Trees

In this section, we will try to find oriented trees, other than paths, in oriented graphs and in arbitrary digraphs, employing bounds on the minimum semidegree, and when necessary, additional conditions.

First of all, we observe that the greedy embedding strategy works for finding any k-edge oriented tree, showing that every digraph of minimum semidegree at least k contains every oriented tree with k edges. This is no longer true if we change the bound on the minimum semidegree to $k - 1$, because of the following example: The $(k-1)$-blow-up of the directed triangle has minimum semidegree $k - 1$, but no antidirected star with k edges is present. However, the antidirected star is very unbalanced, and the example depends on this feature. It seems natural to suspect that balanced oriented trees with k edges already appear in digraph of minimum semidegree somewhere below k. Moreover, for balanced antidirected trees T and oriented host graphs we suspect this bound should be close to the one we conjecture for antidirected paths.

Conjecture 6.8 *Every oriented n-vertex graph D with $\delta^0(D) > k/2$ contains each balanced antidirected k-edge tree of total maximum degree at most $o(n)$.*

A weaker version would be the conjecture that every oriented n-vertex graph D with $\delta^0(D) > k/2 + o(k)$ contains each balanced antidirected k-edge tree of total maximum degree at most a power of $\log n$. This is true if D is large, and k is large compared to the order of D, as we will see in the next result.

Theorem 6.9 (Stein and Zárate-Guerén [65]) *For all $\eta \in (0, 1)$ and $c \in \mathbb{N}$ there is n_0 such that for all $n \geq n_0$ and $k \geq \eta n$, every oriented graph D on n vertices with $\delta^0(D) > (1+\eta)k/2$ contains every balanced antidirected tree T with k edges and with maximum total degree at most $(\log(n))^c$.*

Let us remark that we have no reason to believe that Theorem 6.9's bound on the maximum total degree of T is best possible. We also note that a more general version of Theorem 6.9, substituting 'oriented graph' with 'digraph', is not true, even if we consider a higher bound on the semidegree, as D may decompose into components of order one larger than the semidegree.

Moreover, in Theorem 6.9 we cannot omit the condition that T is balanced. The ℓ-blow-up of the directed triangle, which has semidegree ℓ, does not contain any antidirected tree where one bipartition class has more than ℓ vertices. Such antidirected trees exist, with any maximum total degree $\Delta \geq 3$ and $\ell = (\Delta - 1)k/\Delta$ (for instance, consider adding $\Delta - 2$ leaves to every second vertex of an odd-length path and giving the resulting caterpillar an antidirected orientation).

For $k = n - 1$, Theorem 6.9 follows from a recent result by Kathapurkar and Montgomery [43]. Extending a previous result of Mycroft and Naia [55] where the tree had constant degree, the authors of [43] show the following.

Theorem 6.10 (Kathapurkar and Montgomery 2022 [43]) *For each $\eta > 0$, there are $c > 0$ and $n_0 \in \mathbb{N}$ such that every n-vertex directed graph on $n \geq n_0$ vertices and with minimum semidegree at least $(1/2 + \eta)n$ contains a copy of every n-vertex oriented tree of maximum total degree at most $cn/\log n$.*

Theorem 6.10 generalises a well-known theorem of Komlós, Sárközy and Szemerédi [48] for graphs. Indeed, the statement of the result from [48] is obtained from the statement of Theorem 6.10 by omitting the words 'directed', 'oriented' and 'total', and substituting 'semidegree' with 'degree'.

The Komlós–Sárközy–Szemerédi theorem [48] is tight in the sense that the maximum degree of the tree could not be of lower order (even if the bound on the minimum degree of the host graph is relaxed). This can be seen by considering the random graph with edge probability 0.9 and a tree obtained from taking a set of $\log n/c$ stars $K_{1,cn/\log n}$ and joining their centres by new edges in any way to form an n-vertex tree, where c is some constant. (See [48] for details.) A direct translation of this construction to digraph shows that also in Theorem 6.10, the bound on the maximum degree of the tree is tight.

Csaba, Levitt, Nagy-György, and Szemerédi [18] showed in 2010 a variant of the Komlós–Sárközy–Szemerédi theorem [48], with a slightly relaxed bound on the minimum degree of the host graph and a stricter bound on the maximum degree of the tree. Namely, they showed that for each $\Delta \in \mathbb{N}$, there are $c > 0$ and $n_0 \in \mathbb{N}$ such that every n-vertex graph on $n \geq n_0$ vertices and with minimum degree at least $\frac{n}{2} + c \log n$ contains every n-vertex tree of maximum degree at most Δ. They also showed that this bound on the minimum degree of the host graph is best possible (even for $\Delta = 4$).

A variant of the result from [18] for digraphs seems to be missing yet. We propose the following.

Conjecture 6.11 *For every Δ there is a c such that every large enough n-vertex digraph of minimum semidegree at least $n/2 + c \log n$ contains every n-vertex oriented tree of maximum total degree at most Δ.*

So far, all the results and open problems in this section only applied to oriented trees of bounded degree. However, in graphs, there are also results for containment of trees of unbounded maximum degree. In particular, let us focus on n-vertex trees of unbounded maximum degree in n-vertex host graphs of large minimum degree. Of course, if we wish to find n-vertex stars, then in the host graph, a vertex of degree $n-1$ is needed. It turns out that the presence of such a universal vertex, together with a corresponding condition on the minimum degree, suffices to guarantee all n-vertex trees. Indeed, Reed and the author proved in 2023 [59, 60] that every n-vertex graph of minimum degree exceeding $2n/3$ and maximum degree $n-1$ contains each n-vertex tree, if n is large enough.

A result of this type might also hold for digraphs. Call a vertex v of an n-vertex digraph *universal* if $d^+(v) = d^-(v) = n-1$.

Question 6.12 *Is it true that every large enough n-vertex digraph of minimum semidegree exceeding $2n/3$ that has a universal vertex contains every oriented tree on n vertices?*

The result from [59, 60] is a special case of an earlier conjecture regarding k-edge trees in n-vertex host graphs obeying certain minimum/maximum degree conditions. In 2020, Havet, Reed, Wood and the author [39] conjectured that for every $k \in \mathbb{N}$, every graph of minimum degree exceeding $2(k+1)/3$ and maximum degree k contains each k-edge tree. This bound on the minimum degree is best possible [39] and an asymptotic version for trees of bounded degree has been established in [13].

Thus motivated, we pose the same problem for digraphs. (One could also see Problem 6.13 as a version of Theorem 6.9 for oriented trees that not necessarily balanced or antidirected.)

Problem 6.13 *Determine the smallest $f(k)$ such that for every $k \in \mathbb{N}$, every oriented graph (or digraph) of minimum semidegree exceeding $f(k)$ that has a vertex v with $\min\{d^+(v), d^-(v)\} \geq k$ contains each oriented k-edge tree.*

Clearly, $f(k) \leq k$, and easy modifications of the extremal examples from [39] show that $f(k) \geq 2(k+1)/3$.

7 Minimum Pseudo-Semidegree

It is well known that in graphs, a lower bound on the edge density can be used to deduce a lower bound on the minimum degree of a subgraph. As we will see now, this is no longer true if we use the semidegree notion in digraphs. However, Lemma 7.1 below guarantees that a lower bound on the edge density implies a lower bound on the *minimum pseudo-semidegree* of a subgraph, where the minimum pseudo-semidegree is a natural variant of the minimum pseudo-semidegree and will be defined below. Later, in Lemma 7.2, we see that for embedding antidirected subgraphs, lower bounds on the minimum semidegree have a very similar effect to lower bounds on the minimum pseudo-semidegree.

We define the *minimum pseudo-semidegree* $\bar\delta^0(D)$ of a digraph D with at least one edge as the maximum d such that for each vertex $v \in V(D)$ we have $d^+(v) \in$

$\{0\} \cup [d, \infty)$ and $d^-(v) \in \{0\} \cup [d, \infty)$. In other words, the minimum pseudo-semidegree of a non-empty digraph is the maximum k such that each vertex has either outdegree 0 or outdegree at least k, and has either indegree 0 or indegree at least k. The minimum pseudo-semidegree of an empty (i.e. edgeless) digraph is 0. Note that a digraph of positive minimum pseudo-semidegree can have isolated vertices, i.e. vertices of in- and outdegree 0, but not all its vertices can be like this.

7.1 Edge Density and Minimum Pseudo-Semidegree

We start by recalling the relation of edge density and minimum degree of a subgraph in the case of undirected graphs. A folklore lemma states that any graph of high edge density has a subgraph of large minimum degree. More precisely, for each $\ell \in \mathbb{N}$, every n-vertex graph with more than ℓn edges has a subgraph of minimum degree at least $(\ell+1)/2$. This subgraph can be found by successively deleting vertices of low degree.

Because of the example of the complete bipartite graph on sets A and B, with all edges directed from A to B, such a lemma cannot exist for digraphs if we replace the subgraph of large minimum degree with a subgraph of large minimum semidegree (or of large minimum outdegree). However, a similar result is true by Lemma 7.1. This lemma also appears in [47, 65], and is one of the ingredients for the proof of Theorem 5.4 in [65]. It provides a subdigraph of large minimum pseudo-semidegree in a dense digraph.

Lemma 7.1 *Let $\ell \in \mathbb{N}$. If a digraph D has more than $\ell|V(D)|$ edges, then it contains a digraph D' with $\bar{\delta}^0(D') \geq (\ell+1)/2$.*

The proof of this lemma relies on the observation that on average, the vertices of D have in-degree greater than ℓ. Also, on average they have out-degree greater than ℓ. We construct an auxiliary bipartite graph B by dividing each vertex $v \in V(D)$ into two vertices v_{in} and v_{out}, and letting v_{in} (v_{out}) be adjacent to all edges ending (starting) at v. We then omit all directions on edges. As the average degree of B is greater than ℓ, the folklore fact mentioned above shows that B has a non-empty subgraph B' of minimum degree at least $(\ell+1)/2$. Translating B' back to the digraph setting, it follows that D has a subdigraph D' with minimum pseudo-semidegree at least $(\ell+1)/2$. This proves Lemma 7.1.

7.2 Minimum Semidegree and Minimum Pseudo-Semidegree

The next result, Lemma 7.2 below, states that the minimum semidegree and the minimum pseudo-semidegree are practically equivalent for the purpose of finding an antidirected subgraph A in an oriented graph D, if there is some control over the placement of A in D. For digraphs A and D, we write $A \subseteq_\gamma D$ if for each set $V^* \subseteq V(D)$ of size at least $\gamma|V(D)|$ and for each $x \in V(A)$, there is an embedding of A in D with x mapped to V^*.

Part (i) of the following lemma appears as Lemma 7.2 in [65], and part (ii) can be proved in the same way.

Lemma 7.2 *For any antidirected graph A whose underlying graph is connected and for any $\ell, n_0 \in \mathbb{N}$ the following holds.*

(i) If for each oriented graph D on at least n_0 vertices with $\delta^0(D) \geq \ell$ we have $A \subseteq_{1/8} D$, then each oriented graph D' on at least n_0 vertices with $\bar{\delta}^0(D') \geq \ell$ contains A.

(ii) If for each digraph D on at least n_0 vertices with $\delta^0(D) \geq \ell$ we have $A \subseteq_{1/8} D$, then each digraph D' on at least n_0 vertices with $\bar{\delta}^0(D') \geq \ell$ contains A.

The idea of the proof of Lemma 7.2 is as follows. Given D' and A, we can assume that D' has no isolated vertices, and we will construct a graph D by gluing together four copies of D', two of them with reversed edge directions.

More precisely, we take a copy of D' as it is and another copy of D' with all edge directions reversed. We identify the vertices of first copy that have outdegree 0 with the corresponding vertices in the second digraph (in that graph, these vertices have indegree 0). This gives a digraph D'' that still has minimum pseudo-semidegree ℓ. We now repeat the procedure from above, taking a second copy of D'' with all edge direction reversed, and identifying the vertices of outdegree 0 from the first copy with their copies in the second digraph. We also identify the vertices of indegree 0 from the first copy with their copies in the second digraph. The obtained digraph D has minimum semidegree ℓ. As $A \subseteq_{1/8} D$, it is not hard to see that we can find A in D, guaranteeing that it is fully contained in one of the 'properly oriented' copies of D'. For details, see [65].

7.3 Minimum Pseudo-Semidegree and Oriented Paths and Trees

The results of the previous subsection, in particular Lemma 7.2, hint at the possibility that if we are looking for antidirected paths or trees, then instead of a lower bound on the minimum semidegree of the host digraph, as in Section 6, we could work with a lower bound on the minimum pseudo-semidegree of the host digraph, which would be a weaker condition.

So, recalling Conjecture 6.3 from Section 6, the following conjecture seems plausible.

Conjecture 7.3 *For each $k \in \mathbb{N}$, every oriented graph with $\bar{\delta}^0(D) > k/2$ contains every antidirected path with k edges.*

Observe that Conjecture 7.3 is not true for other orientations of the path, because of the example of the complete bipartite graph (A, B) with all edges directed from A to B. Also note that replacing the bound of $k/2$ in Conjecture 7.3 with k makes the conjecture trivial, because of the greedy embedding argument.

As evidence for Conjecture 7.3, we now present analogues of some of the results from Section 6. The first one of these uses a bound on the minimum pseudo-semidegree that is close to $3k/4$, which lies halfway between the trivial bound and the bound from the conjecture.

Theorem 7.4 (Klimošová and Stein 2023 [47]) *Let $k \in \mathbb{N}$ with $k \neq 2$. Every oriented graph D with $\bar{\delta}^0(D) > \frac{3k-2}{4}$ contains each antidirected path with k edges.*

We remark that one can use Lemma 7.1 to see that Theorem 7.4 implies Theorem 5.3.

Moreover, for all $\eta \in (0,1)$ there is n_0 such that for all $n \geq n_0$ and $k \geq \eta n$ every oriented graph D on n vertices with $\bar\delta^0(D) > (1+\eta)\frac{k}{2}$ contains every antidirected path with k edges. This follows directly from Theorem 7.5, which we state next.

Theorem 7.5 *For all $\eta \in (0,1)$ and $c \in \mathbb{N}$ there is n_0 such that for all $n \geq n_0$ and $k \geq \eta n$, every oriented graph D on n vertices with $\bar\delta^0(D) > (1+\eta)k/2$ contains every balanced antidirected tree T with k edges and with maximum total degree at most $(\log(n))^c$.*

Theorem 7.5 follows immediately from Theorem 5.2 in [65] and from Lemma 7.2 above.

We believe that the approximation and the condition that k and n are large can be omitted from Theorem 7.5:

Conjecture 7.6 *For each $k \in \mathbb{N}$, every oriented graph on n vertices with $\bar\delta^0(D) > k/2$ contains every antidirected balanced tree having k edges and maximum total degree $o(n)$.*

Acknowledgements

The author acknowledges support by FONDECYT Regular Grant 1221905 and by ANID Basal Grant CMM FB210005.

References

[1] P. Aboulker, N. Cohen, F. Havet, W. Lochet, P. F. S. Moura, and S. Thomassé, *Subdivisions in digraphs of large out-degree or large dichromatic number*, Electronic Journal of Combinatorics **26** (2019), P3.19.

[2] L. Addario-Berry, F. Havet, C. Linhares Sales, B. Reed, and S. Thomassé, *Oriented trees in digraphs*, Discrete Mathematics **313** (2013), 967–974.

[3] L. Addario-Berry, F. Havet, and S. Thomassé, *Paths with two blocks in k-chromatic digraphs*, Journal of Combinatorial Theory, Series B **97** (2007), 620–626.

[4] N. Alon, *The maximum number of hamiltonian paths in tournaments*, Combinatorica **10** (1990), 319–324.

[5] B. Alspach and M. Rosenfeld, *Realisation of certain generalized paths in tournaments*, Discrete Mathematics **34** (1981), 199–202.

[6] J. Bang-Jensen, *Problems and conjectures concerning connectivity, paths, trees and cycles in tournament-like digraphs*, Discrete Mathematics **309** (2009), 5655–5667.

[7] J. Bang-Jensen and G. Gutin, *Paths, trees and cycles in tournaments*, Congressus Numerantium (1996), 131–170.

[8] J. Bang-Jensen and F. Havet, *Tournaments and semicomplete digraphs*, In: *Classes of Directed Graphs*, Springer International Publishing, 2018, pp. 35–124.

[9] E.A. Bender and N.C. Wormald, *Random trees in random graphs*, Proceedings of the American Mathematical Society **103** (1988), no. 1, 314–320.

[10] A. Benford and R. Montgomery, *Trees with few leaves in tournaments*, Journal of Combinatorial Theory, Series B **155** (2022), 141–170.

[11] _____, *Trees with many leaves in tournaments*, Preprint arXiv:2207.06384, 2022.

[12] J. C. Bermond, A. Germa, M. C. Heydemann, and D. Sotteau, *Longest paths in digraphs*, Combinatorica **1** (1981), 337–341.

[13] G. Besomi, M. Pavez-Signé, and M. Stein, *Degree conditions for embedding trees*, SIAM Journal on Discrete Mathematics **33** (2019), 1521–1555.

[14] G. Besomi, M. Pavez-Signé, and M. Stein, *On the Erdős–Sós conjecture for trees with bounded degree*, Combinatorics, Probability and Computing **30** (2021), no. 5, 741–761.

[15] S. A. Burr, *Subtrees of directed graphs and hypergraphs*, Proceedings of the Eleventh Southeastern Conference on Combinatorics, Graph Theory and Computing (Boca Raton, Fla., 1982), vol. 28, 1980, pp. 227–239.

[16] S. Ceroi and F. Havet, *Trees with three leaves are (n+1)-unavoidable*, Discrete Applied Mathematics **141** (2004), no. 1, 19 – 39, Brazilian Symposium on Graphs, Algorithms and Combinatorics.

[17] F.R.K. Chung, *A note on subtrees in tournaments*, Internal Memorandum, Bell Laboratories, 1981.

[18] B. Csaba, I. Levitt, J. Nagy-György, and E. Szemerédi, *Tight bounds for embedding bounded degree trees*, Katona G.O.H., Schrijver A., Szenyi T., Sági G. (eds) Fête of Combinatorics and Computer Science, vol. 20, 2010, pp. 95–137.

[19] A. Davoodi, D. Piguet, H. Řada, and N. Sanhueza-Matamala, *Beyond the Erdős–Sós conjecture*, Proceedings of the 12th European Conference on Combinatorics, Graph Theory and Applications EUROCOMB'23.

[20] L. DeBiasio, D. Kühn, T. Molla, D. Osthus, and A. Taylor, *Arbitrary orientations of hamilton cycles in digraphs*, SIAM Journal on Discrete Mathematics **29** (2015), no. 3, 1553–1584.

[21] L. DeBiasio and T. Molla, *Semi-degree threshold for anti-directed hamiltonian cycles*, Electronic Journal of Combinatorics **22** (2015), no. 4, P4.34.

[22] G. A. Dirac, *Some theorems on abstract graphs*, Proceedings of the London Mathematical Society **2** (1952), 69–81.

[23] F. Dross and F. Havet, *On the unavoidability of oriented trees*, Journal of Combinatorial Theory, Series B **151** (2021), 83–110.

[24] A. El Sahili, *Paths with two blocks in k-chromatic digraphs*, Discrete Mathematics **287** (2004), 151–153.

[25] _____, *Trees in tournaments*, Journal of Combinatorial Theory, Series B **92** (2004), 183–187.

[26] A. El Sahili and Z. Ghazo Hanna, *About the number of oriented hamiltonian paths and cycles in tournaments*, Journal of Graph Theory **102** (2023), no. 4, 684–701.

[27] P. Erdős, *Extremal problems in graph theory*, Theory of graphs and its applications, Proc. Sympos. Smolenice, 1964, pp. 29–36.

[28] P. Erdős and T. Gallai, *On maximal paths and circuits of graphs*, Acta Math. Acad. Sci. Hungar **10** (1959), 337–356.

[29] R. Forcade, *Parity of paths and circuits in tournaments*, Discrete Mathematics **6** (1973), 115–118.

[30] A. Ghouila-Houri, *Une condition suffisante d'existence d'un circuit Hamiltonien*, Comptes rendus de l'Académie des Sciences **251** (1960), 495–497.

[31] R. L. Graham, *On subtrees of directed graphs with no path of length exceeding one*, Canadian Mathematical Bulletin **13** (1970), 329–332.

[32] B. Grünbaum, *Antidirected Hamiltonian paths in tournaments*, Journal of Combinatorial Theory, Series B **11** (1971), 249–257.

[33] R. Häggkvist and A. Thomason, *Trees in tournaments*, Combinatorica **11** (1991), 123–130.

[34] C. B. Hanna, *Paths in tournaments: a simple proof of Rosenfeld's conjecture*, Preprint arXiv:2011.14394, 2020.

[35] F. Havet, *Directed path of length twice the minimum outdegree*, Open Problem Garden, http://www.openproblemgarden.org/op/directed_cycle_of_length_twice_the_minimum_outdegree, retrieved 20-06-2023.

[36] _____, *Oriented trees in n-chromatic digraphs*, Open Problem Garden, http://www.openproblemgarden.org/op/oriented_trees_in_n_chromatic_digraphs, retrieved 20-06-2023.

[37] _____, *Trees in tournaments*, Discrete Mathematics **243** (2002), 121–134.

[38] _____, *On unavoidability of trees with k leaves*, Graphs and Combinatorics **19** (2003), 101–110.

[39] F. Havet, B. Reed, M. Stein, and D. R. Wood, *A variant of the Erdős–Sós conjecture*, Journal of Graph Theory **94** (2020), 131–158.

[40] F. Havet and S. Thomassé, *Median orders of tournaments: a tool for the second neighbourhood problem and Sumner's conjecture*, Journal of Graph Theory **35** (2000), 244–256.

[41] _____, *Oriented Hamiltonian paths in tournaments: a proof of Rosenfeld's conjecture*, Journal of Combinatorial Theory, Series B **78** (2000), 243–273.

[42] B. Jackson, *Long paths and cycles in oriented graphs*, Journal of Graph Theory **5** (1981), no. 2, 145–157.

[43] A. Kathapurkar and R. Montgomery, *Spanning trees in dense directed graphs*, Journal of Combinatorial Theory, Series B **156** (2022), 223–249.

[44] P. Keevash, D. Kühn, and O. Osthus, *An exact minimum degree condition for hamilton cycles in oriented graphs*, Journal of the London Mathematical Society **79** (2009), 144–166.

[45] L. Kelly, *Arbitrary orientations of hamilton cycles in oriented graphs*, Electronic Journal of Combinatorics **18** (2011), P.44.

[46] L. Kelly, D. Kühn, and D. Osthus, *Cycles of given length in oriented graphs*, Journal of Combinatorial Theory Series B **100** (2010), 251–264.

[47] T. Klimošová and M. Stein, *Antipaths in oriented graphs*, Discrete Mathematics **346** (2023), 113515.

[48] J. Komlós, G. N. Sárközy, and E. Szemerédi, *Spanning trees in dense graphs*, Combinatorics, Probability and Computing **10** (2001), no. 5, 397–416.

[49] D. Kühn, R. Mycroft, and D. Osthus, *An approximate version of Sumner's universal tournament conjecture*, J. Combin. Theory Ser. B **101** (2011), no. 6, 415–447.

[50] _____, *A proof of Sumner's universal tournament conjecture for large tournaments*, Proc. Lond. Math. Soc. (3) **102** (2011), no. 4, 731–766.

[51] X. Lu, *On claws belonging to every tournament*, Combinatorica **11** (1991), 173–179.

[52] _____, *Claws contained in all n-tournaments*, Discrete Mathematics **119** (1993), 107–111.

[53] X. Lu, D.-W. Wang, and C.-K. Wong, *On avoidable and unavoidable claws*, Discrete Mathematics **184** (1998), 259–265.

[54] W. Mader, *Connectivity keeping trees in k-connected graphs*, Journal of Graph Theory **69** (2012), no. 3, 324–329.

[55] R. Mycroft and T. Naia, *Unavoidable trees in tournaments*, Random Structures & Algorithms **53** (2018), no. 2, 352–385.

[56] T. Naia, *Large structures in dense directed graphs*, PhD thesis, University of Birmingham, 2018.

[57] _____, *Trees contained in every orientation of a graph*, Electronic Journal of Combinatorics **29** (2022), no. 2, P2.26.

[58] L. Rédei, *Ein kombinatorischer Satz*, Acta Universitatis Szegediensis **7** (1934), 39–43.

[59] B. Reed and M. Stein, *Spanning trees in graphs of high minimum degree with a universal vertex I: An asymptotic result*, Journal of Graph Theory **102** (2023), no. 4, 737–783.

[60] ———, *Spanning trees in graphs of high minimum degree with a universal vertex II: A tight result*, Journal of Graph Theory **102** (2023), no. 4, 797–821.

[61] K. Reid and N. Wormald, *Embedding oriented n-trees in tournaments*, Studis Scientiarum Mathematicarum Hungarica **18** (1983), 377–387.

[62] M. Rosenfeld, *Antidirected Hamiltonian paths in tournaments*, Journal of Combinatorial Theory, Series B **12** (1971), 93–99.

[63] M.E. Saks and V. T. Sós, *On avoidable subgraphs of tournaments*, Colloquia Mathematica Societatis János Bolyai (L. Lovász A. Hajnal and V.T.Sós, eds.), vol. 2, North Holland, 1981, pp. 663–674.

[64] M. Stein, *Tree containment and degree conditions*, Discrete Mathematics and Applications (A. M. Raigorodskii and M. Th. Rassias, eds.), Springer International Publishing, Cham, 2020, pp. 459–486.

[65] M. Stein and C. Zárate-Guerén, *Antidirected subgraphs of oriented graphs*, Preprint 2022, arXiv 2212.00769.

[66] H. J. Straight, *The existence of certain type of semi-walks in tournaments*, Proceedings, XI Southeastern Conference on Combinatorics, Graph Theory and Computing, Cong. Numer. **29** (1980), 901–908.

[67] B. D. Sullivan, *A summary of results and problems related to the Caccetta-Häggkvist conjecture*, Technical Report 2006-13. American Institute of Mathematics, Palo Alto, CA, 2006.

[68] T. Szele, *Kombinatorikai vizsgalatok az iranyitott teljes graffal kapcsolatban*, Matematikai és fizikai lapok **50** (1943), no. 1943, 223–256.

[69] A. Thomason, *Antidirected Hamiltonian paths in tournaments*, Transactions of the American Mathematical Society **296** (1986), 167–180.

[70] C. Thomassen, *Hamiltonian connected tournaments*, Journal of Combinatorial Theory, Series B **28** (1980), 142–163.

[71] ———, *Edge-disjoint hamiltonian paths and cycles in tournaments*, Proceedings of the London Mathematical Society **45** (1982), 151–168.

[72] ———, *Connectivity in tournaments*, Graph Theory and Combinatorics (Bela Bollobás, ed.), Acad. Press, 1984, pp. 305–313.

[73] Y. Tian, H.-J. Lai, L. Xu, and J. Meng, *Nonseparating trees in 2-connected graphs and oriented trees in strongly connected digraphs*, Discrete Mathematics **342** (2019), no. 2, 344–351.

[74] N. C. Wormald, *Subtrees of large tournaments*, Combinatorial Mathematics X (Louis Reynolds Antoine Casse, ed.), Springer Berlin Heidelberg, 1983, pp. 417–419.

[75] C.-Q. Zhang, *Some results on tournaments*, Journal of Qufu Teachers College **1** (1985), 51–53.

Department for Mathematical Engineering and
Center for Mathematical Modeling
University of Chile
Beauchef 851, Piso 7
Santiago, Chile
mstein@dim.uchile.cl

For EU product safety concerns, contact us at Calle de José Abascal, 56–1°, 28003 Madrid, Spain or eugpsr@cambridge.org.

www.ingramcontent.com/pod-product-compliance
Lightning Source LLC
LaVergne TN
LVHW012059070526
838200LV00074BA/3664